# Ex Situ Biological Treatment Technologies

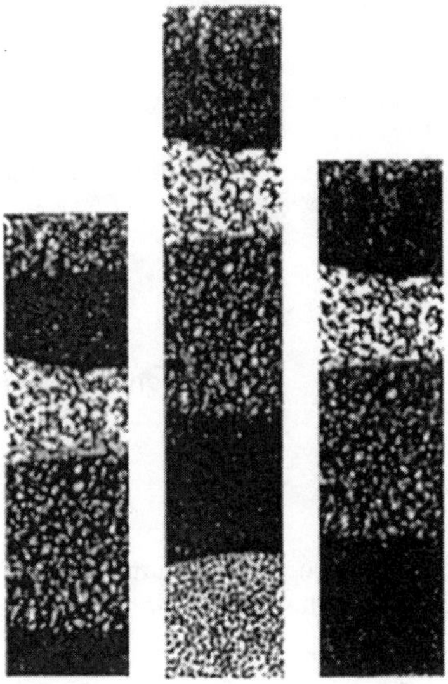

*Editors*

Victor S. Magar,
F. Michael von Fahnestock,
and Andrea Leeson

The Sixth International In Situ and On-Site Bioremediation Symposium

San Diego, California, June 4–7, 2001

BATTELLE PRESS
Columbus • Richland

**Library of Congress Cataloging-in-Publication Data**

International In Situ and On-Site Bioremediation Symposium (6th : 2001 : San Diego, Calif.)
Ex situ biological treatment technologies : the Sixth International In Situ and On-Site Bioremediation Symposium : San Diego, California, June 4-7, 2001 / editors, V. Magar … [et al.].
p. cm. -- (The Sixth International In Situ and On-Site Bioremediation Symposium ; 6)
Includes bibliographical references and index.
ISBN 1-57477-116-7 (hc. : alk. paper)
1. Soil remediation--Congresses. 2. Bioremediation--Congresses. I. Magar, V. (Victor), 1964- . II. Title. III. Series: International In Situ and On-Site Bioremediation Symposium (6th : 2001 : San Diego, Calif.). Sixth International In Situ and On-Site Bioremediation Symposium ; 6.
TD192.5.I56 2001 vol. 6
[TD878]
628.5 s--dc21
[628.5'5]

2001041199

Printed in the United States of America



Battelle Press
505 King Avenue
Columbus, Ohio 43201, USA
614-424-6393 or 1-800-451-3543
Fax: 1-614-424-3819
Internet: press@battelle.org
Website: www.battelle.org/bookstore

For information on future environmental conferences, write to:
Battelle
Environmental Restoration Department, Room 10-123B
505 King Avenue
Columbus, Ohio 43201-2693
Phone: 614-424-7604
Fax: 614-424-3667
Website: www.battelle.org/conferences

# CONTENTS

## *Wood-Preserving and Phenolic Wastes*

## *Ex Situ Bioreactors, Kinetics, and Modeling*

# FOREWORD

The papers in this volume correspond to presentations made at the Sixth International In Situ and On-Site Bioremediation Symposium (San Diego, California, June 4-7 2001). The program included approximately 600 presentations in 50 sessions on a variety of bioremediation and supporting technologies used for a wide range of contaminants.

This volume focuses on *Ex Situ Biological Treatment Technologies*. Papers discuss several active land treatment technologies that rely on biological processes, including landfarming, biopiles, and composting. Techniques that involve the stimulation and augmentation of bacteria and fungi to achieve the destruction of a wide variety of contaminants are presented. Bioreactors continue to find a niche at hazardous waste sites, albeit limited, undoubtedly due to the availability of other less costly and less complex treatment alternatives.

The author of each presentation accepted for the symposium program was invited to prepare an eight-page paper. According to its topic, each paper received was tentatively assigned to one of ten volumes and subsequently was reviewed by the editors of that volume and by the Symposium chairs. We appreciate the significant commitment of time by each of the volume editors, each of whom reviewed as many as 40 papers. The result of the review was that 352 papers were accepted for publication and assembled into the following ten volumes:

*Bioremediation of MTBE, Alcohols, and Ethers* — 6(1). Eds: Victor S. Magar, James T. Gibbs, Kirk T. O'Reilly, Michael R. Hyman, and Andrea Leeson.

*Natural Attenuation of Environmental Contaminants* — 6(2). Eds: Andrea Leeson, Mark E. Kelley, Hanadi S. Rifai, and Victor S. Magar.

*Bioremediation of Energetics, Phenolics, and Polycyclic Aromatic Hydrocarbons* — 6(3). Eds: Victor S. Magar, Glenn Johnson, Say Kee Ong, and Andrea Leeson.

*Innovative Methods in Support of Bioremediation* — 6(4). Eds: Victor S. Magar, Timothy M. Vogel, C. Marjorie Aelion, and Andrea Leeson.

*Phytoremediation, Wetlands, and Sediments* — 6(5). Eds: Andrea Leeson, Eric A. Foote, M. Katherine Banks, and Victor S. Magar.

*Ex Situ Biological Treatment Technologies* — 6(6). Eds: Victor S. Magar, F. Michael von Fahnestock, and Andrea Leeson.

*Anaerobic Degradation of Chlorinated Solvents*— 6(7). Eds: Victor S. Magar, Donna E. Fennell, Jeffrey J. Morse, Bruce C. Alleman, and Andrea Leeson.

*Bioaugmentation, Biobarriers, and Biogeochemistry* — 6(8). Eds: Andrea Leeson, Bruce C. Alleman, Pedro J. Alvarez, and Victor S. Magar.

*Bioremediation of Inorganic Compounds* — 6(9). Eds: Andrea Leeson, Brent M. Peyton, Jeffrey L. Means, and Victor S. Magar.
*In Situ Aeration and Aerobic Remediation* — 6(10). Eds: Andrea Leeson, Paul C. Johnson, Robert E. Hinchee, Lewis Semprini, and Victor S. Magar.

In addition to the volume editors, we would like to thank the Battelle staff who assembled the ten volumes and prepared them for printing: Lori Helsel, Carol Young, Loretta Bahn, Regina Lynch, and Gina Melaragno. Joseph Sheldrick, manager of Battelle Press, provided valuable production-planning advice and coordinated with the printer; he and Gar Dingess designed the covers.

The Bioremediation Symposium is sponsored and organized by Battelle Memorial Institute, with the assistance of a number of environmental remediation organizations. In 2001, the following co-sponsors made financial contributions toward the Symposium:

Geomatrix Consultants, Inc.
The IT Group, Inc.
Parsons
Regenesis
U.S. Air Force Center for Environmental Excellence (AFCEE)
U.S. Naval Facilities Engineering Command (NAVFAC)

Additional participating organizations assisted with distribution of information about the Symposium:

Ajou University, College of Engineering
American Petroleum Institute
Asian Institute of Technology
National Center for Integrated Bioremediation Research & Development (University of Michigan)
U.S. Air Force Research Laboratory, Air Expeditionary Forces Technologies Division
U.S. Environmental Protection Agency
Western Region Hazardous Substance Research Center (Stanford University and Oregon State University)

Although the technical review provided guidance to the authors to help clarify their presentations, the materials in these volumes ultimately represent the authors' results and interpretations. The support provided to the Symposium by Battelle, the co-sponsors, and the participating organizations should not be construed as their endorsement of the content of these volumes.

Andrea Leeson & Victor Magar, Battelle
2001 Bioremediation Symposium Co-Chairs

# CONCEPTUAL DESCRIPTION OF LANDFARMING FOR SUSTAINABLE RESTORATION OF SOILS WORLDWIDE

Ronald C. Sims (Utah Water Research Laboratory, Utah State University, USA)
Joop Harmsen (Alterra, Wageningen-UR, Wageningen, The Netherlands)

## ABSTRACT:

Landfarming technology utilizes principles of biological engineering and economics for the management of soil science to accomplish the destruction and detoxification of chemicals that present unacceptable risks to human health and the environment. In addition, landfarming reestablishes the soil as a natural resource that supports nutrient cycling and vegetative growth that contributes to a sustainable biosphere, and therefore is a sustainable technology. Sustainable refers to the use of the soil resource so that the resource is not depleted or permanently damaged. Thus sustainable landfarming emphasizes treatment rather than disposal of added chemicals. Sustainable landfarming is especially applicable as a low cost alternative for restoration of large areas of contaminated soils in developing nations and in demilitarized areas containing residual soil contamination. Sustainable restoration of soil may also be utilized for stimulation of economic development in those impacted areas. There is a need for the development of international manuals of practice to address planning, design, operation and management, and monitoring of these systems. In addition, improved understanding of chemical-soil-microbe interactions through developing assays and methods to measure soil solid-phase toxicity changes, bioavailability, and biological formation of bound residues will result in improved techniques and methods for accomplishing the sustainable restoration of soils worldwide with beneficial and sustainable economic development of restored land.

## DEFINITION AND DESIGN

Landfarming for accomplishing sustainable restoration of soil is a technology based on the utilization of soil microorganisms and sustainable agricultural methods in an aerobic environment to treat soil contamination through transformation, immobilization, and detoxification processes in order to protect public health and the environment. Sustainable landfarming accomplishes recycling of organic carbon and nutrients within the biosphere and maintains basic soil characteristics necessary to support plant growth (forestation, vegetation, etc.), and thus contributes to environmental natural resource sustainability and economic utilization of restored land. The success of landfarming technology with regard to remediation of contaminated soil depends upon efficient utilization of microorganisms in a soil farm environment to treat soil contamination. The development and availability of tools and methods for effective utilization of microorganisms within a landfarm environment is critical.

## TECHNICAL AND REGULATORY DEVELOPMENT

During the 1970s, oil-containing waste from the oil industry in the United States was disposed by mixing it with soil (Dibble and Bartha,1979). It is estimated that in 1973 less than 10% of facilities with oil refinery sludges utilized land treatment, and by 1979 from 34% to 50% of refinery facilities used land treatment (Arora et al., 1982). However, in 1985 it was reported that there were few definitive data in the literature quantifying waste biodegradation rates in full-scale land treatment systems (Huddleston et al., 1986). This technique, referred to as sludgefarming, was considered effective for the disposal of the refinery waste. During this period the environmental impact of this type of waste management on the long-term behavior of soil and groundwater resources was not a primary design, management, or performance criterion. In 1979 Dr. Michael Overcash made the statement that "... land application is not viewed as a technology by industrial waste engineers, but rather as a nondescript system of irrigation or spreading. The lack of firm design criteria has made land treatment the equivalent of shoveling smoke." (Overcash and Pal, 1979).

In the decade from 1980 to 1990 design of land treatment systems in the U.S. for industrial wastes classified as hazardous waste was based on regulatory mandates of the Resource Conservation and Recovery Act (RCRA), 1976, described in the Code of Federal Regulations (CFR) part 40, Section 264 that required a permitting process. Requirements of the regulations specified that to obtain a permit it must be demonstrated that any hazardous constituents contained in the waste to be land applied must be degraded, transformed, and/or immobilized in the soil treatment zone, generally considered to be the upper layer (five feet) of soil (U.S. EPA, 1986).

An important design criterion that recognized the sustainability issue for soil restoration was identified in 1979, and was based on the constraint of nondegradation of the soil through determining the soil-site assimilative capacity (SSAC) for each target chemical in a waste. That is, the waste when considered on a constituent-by-constituent basis is applied to the plant-soil system at such loading rates and frequencies, and over such limited time spans, that no land is irreversibly removed from some other potential societal usage such as forestation, agriculture, and development (Overcash and Pal, 1979). These design issues were developed within the context of ensuring protection of public health and the environment. The nondegradation constraint and the determination of SSACs for target chemicals were used as the basis for development of the U.S. Environmental Protection Agency Permit Guidance Manual on Hazardous Waste Land Treatment, published in1986 (U.S. EPA, 1986). Land treatment of hazardous waste was banned in the late 1980s because it was defined as a form of land disposal without pretreatment, rather than recognized for the treatment and sustainability bases that guided the design, management, and performance of land treatment units in the early 1980s (U.S. EPA, 1987).

In the decade from 1990 to 2000, the emphasis in the application of hazardous waste landfarming technology evolved from treatment of industrial liquid and sludge wastes to treatment of contaminated soil, not only in the U.S., but worldwide. In the U.S., contaminated soil identified under the U.S. EPA Superfund program, titled the Comprehensive Environmental Response, Compensation, and Liability Act of 1980 (CERCLA) was frequently remediated to target cleanup levels using landfarming technology (Sims, 1990, Sims et al., 1990). A more detailed summary of the practices and effectiveness of landfarming technology in the treatment of petroleum contaminated soils were presented by Sims and Sims (1999). Often, however, the target levels for cleanup identified in the Records of Decision under CERCLA did not consider the design and performance constraints necessary for sustainable restoration of soil (Sims, et. al. 1996).

In 2000, a re-focus on the sustainable restoration of contaminated soil occurred at the NATO Advanced Research Workshop (NATO, 2000). The topic of the conference was the utilization of bioremediation to reduce soil contamination. The topic of using the landfarming framework for soil bioremediation emphasized the importance of designing and operating soil bioremediation systems with the goal of achieving sustainable restoration of soil based on the nondegradation constraint and SSAC described previously (Sims and Sims, 2001).

## SUSTAINABLE LANDFARMING ISSUES

**Optimization.** The landfarming process has been optimized through using tillage to enhance aeration, adding nutrients, and enhancing temperature (Martin et al., 1986, Nieman et al., 1999), and also by modeling the processes in the landfarm in order to predict effects of optimization (Freijer, 1994, Kivanç, 1999). Optimization techniques are applied in order to decrease treatment time required to achieve target soil concentrations. The most successful optimization will lead to a bioreactor in which the desired biodegradation will occur within days or a few weeks. Types of optimization bioreactors may include fixed bed, biopiles, compost beds, and greenhouse farming systems. In the USA all of these landfarming technology-based systems are considered as different technologies. In the Netherlands these are considered as different ways of applying landfarming technology, and distinction is made with regard to intensive versus extensive methods of landfarming. Intensive (proactive and aggressive) landfarming is used to address high chemical concentrations and high bioavailability regardless of contaminant type for treatment in a shorter period of time. Extensive (passive and non-aggressive) landfarming is used to address low bioavailability and often low contaminant concentrations and usually requires longer treatment times. For example, companies in The Netherlands are exploring greenhouses or halls in which mineral oil polluted soils are cleaned within several months utilizing amendments and active landfarming.

**Extension of landfarming.** In addition to mineral oil, in principle all contaminants that need aerobic circumstances for biodegradation can be removed on a landfarm. There is only one restriction. Volatile compounds will partly be removed by biodegradation, but volatilization will be responsible for a significant fraction of the total removal. In practice landfarming is useful for all relatively non volatile contaminants like the non volatile types of mineral oil, polycyclic aromatic hydrocarbons (PAH), and pentachlorophenol. After dewatering, dredged sediments become a soil that can be treated on a landfarm.

Sediments are often polluted with PAH and mineral oil. More time is necessary, because dewatering also takes time and the degradability of mineral oil in sediment is often low (Harmsen et al., 1997).

**Limits of landfarming.** A residual concentration of target chemical(s) is consistently observed following biodegradation that can be higher than the target value. Much time and effort is usually required in order to remove the residual concentration. This concentration can be explained by the limited bioavailability of chemicals following active biodegradation. The contaminant is not present on a site with active microorganisms and therefore must diffuse through the soil matrix to these bioavailable sites. The limited bioavailability is identified as a problem to be solved and as a soil mechanism that limits the application of landfarming because of the long times required to achieve target cleanup levels for sustainable restoration of soil and risk reduction to humans and to the environment (Harmsen, 1991, 1993 and 1999). For this situation, i.e., low bioavailability and low chemical concentrations, extensive landfarming is appropriate.

## TARGET VALUES AND LANDFARMING

Landfarming treatment through bioremediation can be measured and reported in several ways, including percent of contaminant reduction, comparison with target cleanup concentrations, and/or residual risk to human health or the environment. Different standards exist in different countries and different geographical areas. For example, 10 years ago landfarming of soil polluted with initial pentachlorophenol concentration of 100 mg/kg (dry weight) to a final concentration of 10 mg/kg could be considered a success in Canada because a reduction of 90% was achieved. However this would be considered a failure in the Netherlands because the target value was not achieved. Risk-based criteria and the use of the soil following landfarming treatment will become increasingly important in soil remediation, similar to the goal of treating industrial wastes with landfarming using the nondegradation design constraint and the SSAC concepts.

### Dutch example

Ten years ago contaminated soil had to be cleaned to address two criteria for multifunctional use, which required treatment to 50 mg/kg mineral oil and 1 mg/kg PAH. Conventional landfarming could not achieve this level of cleanup, only if relatively volatile hydrocarbons were present. However degradation does not stop after removal of the bioavailable fraction, but continues at a greatly reduced rate and will lead eventually to lower concentrations. The PAH class of compounds presented a problem. In the late 1990s, criteria for reuse as building materials were developed that allowed 500 mg/kg mineral oil and 40 mg/kg PAH. Using these criteria, the restricted use of remediated soil became possible. Cleaning by landfarming using intensive techniques became effective, and as a result bioremediation became more popular. For diffuse polluted sites, for example, the flood planes of the river Rhine, this would not be an acceptable solution, because on these sites a more multifunctional use (nature development) was the wish of the community and the goal of treatment.

## DEVELOPMENTS

With the context of different goals (target values) for sustainable landfarming, including residential and industrial reuse of site, school properties, forestation, agriculture, and building materials, this technology will continue to be used in several ways.

**Landfarming fast-track to produce an industrial product.** Several soil cleaning companies use landfarming as a production process; that is, they produce clean soil for multiple purposes. Because of the extended time requirements to reach criteria for multifunctional use, the focus of industrial treatment companies is on the target cleanup value for limited reuse of soil. Based on industrial process criteria of consistency of treatment, rapid treatment, and predictability of product quality, such companies utilize soil treatment optimization techniques to achieve the most rapid treatment. Reduction of the time required for cleaning the soil to target levels is the most important tool for reducing costs within the industry. Nonetheless, for some soils, even optimized sustainable restoration may require several months to one year.

**Landfarming with flexible time and cleanup constraints** . Companies or communities with available space and without time or development pressure may not need industrial assistance to remediate contaminated soil using landfarming. In this case an efficient or optimized process is not a criterion. The assurance that a community is seeking is that target levels for intended use, or "clean" soil, be in a timeframe that is generally less than 5 years. Due to this still limited time, they may reuse the soil as building material in future projects. The available time for treatment is generally not sufficient to reach criteria for more multifunctional uses of restored soil.

**Landfarming as a natural process that supports vegetation growth.** In the extensive mode of landfarming, optimization of the process is not the priority, and bioavailability is the limiting factor. The soil is required to be on the landfarm for treatment for an extended period of time due to the low target levels for sustainable restoration and the limited bioavailability of the associated target chemicals. Intended uses include development of natural vegetation or special vegetation. Examples include the production of biomass for energy supply, or using the soil to support vegetation in a noise wall, or to cover a landfill. The treatment time may require several years to decades and the soil may even reach the target values for multifunctional land use.

**Landfarming for diffuse polluted sites.** Many contaminated soils have contaminant concentrations slightly above levels for de minimus risks. People often avoid areas containing these soils, and these soils are often a problem in new developments. They are "too clean to be cleaned, but too dirty to be used for everything." The methods of extensive landfarming can be used to improve the quality of these soils and to make sustainable use possible. A possible management approach is to mix or blend this soil with treated soil that contains adapted microorganisms and nutrients in order to inoculate the diffuse polluted soil and also to aerate the diffuse contaminated soil.

**Landfarming for quality improvement during use**. Depending on the initial concentrations of contaminants, soil reuse may be limited. However, as slow but continuous treatment occurs (possibly for decades), a clean soil may be produced for multifunctional societal purposes.

**Sustainable reuse of restored land for sustainable economic development.** In the United States the development of the Environmental Protection Agency Brownfields Program under Superfund has targeted contaminated sites for economic development. These sites contain residual contamination that is considered to present an acceptable risk for limited functions. Utilizing a parallel approach with landfarming for sustainable restoration of soil, it may be possible to utilize the restored land for economic development that would provide sustainable resources to repay the costs incurred by landfarming and to develop additional economic resources for the associated community or country. Examples include the sale of vegetation for energy, production and sale of compost as a soil conditioner, silviculture applications for furniture and electric power poles, and structures including domestic houses and businesses.

## SUMMARY

Landfarming technology is very appropriate for the sustainable restoration of soils in many worldwide situations where it is not currently being "exploited" for such use. A philosophy of nondegradation of soil through identifying the site soil assimilative capacity provides the basis for extending the technology of landfarming beyond remediation to sustainable restoration. In addition, sustainable restoration of soil takes into account risk assessment, reduction, and management issues and criteria. While it may not be possible to bring all contaminated soil into the sustainable restoration category, this alternative should be considered with regard to feasibility in both the short-term and long-term, cost-effectiveness, and value added to a community or industry in a similar manner to a cost-benefit analysis for any project that affects the environment and public health. The economic benefits that may accrue due to sustainable economic development of the restored soil/land provide a worldwide technology where manuals of practice are needed appropriate to international conditions and sites. Drs. Sims and Harmsen are working together in the development of institutional frameworks to facilitate the development of such manuals of practice, and if you are interested you are encouraged to contact Dr. Sims at email address: ronsims@cc.usu.edu.

## REFERENCES

Arora, H.S., R.C. Cantor, and J.C. Nemeth. 1982. land Treatment: A Viable and Successful Method of Treating Petroleum Industry Wastes. Environment International, vol. 7, pp. 285-291.

Dibble, J.T. and R. Bartha,1979. Effects of Environmental Parameters on the Biodegradation of Oil Sludge. Applied and Environmental Microbiology 37: 729-739.

Freijer, J.E., 1994. Mineralization of Hydrocarbons and Gas Dynamics in Oil-Contaminated Soil. Ph.D. Thesis, Univ. of Amsterdam, ISBN 90-6787-038-2.

Harmsen, J. 1991. Possibilities and Limitations of Landfarming for Cleaning Contaminated Soils. In: Hinchee, R.E. and Olfenbuttel (Eds.), On-Site Bioreclamation, processes for Xenobiotic and Hydrocarbon Treatment, pp. 255-272. Butterworth-Heinemann, Boston.

Harmsen, J. 1993. Managing bio-availability: An effective element in the improvement of biological soil cleaning. In: Eijsackers H.J.E. and T. Hamers (Eds.), Integrated Soil and Sediment Research: A Basic for Proper Soil Protection, pp 235-240.Kluwer academic Publishers, Dordrecht.

Harmsen, J., H.J.J. Wieggers, J.J.H. van de Akker, O.M. van Dijk-Hooyer, A. van den Toorn and A.J. Zweers, 1997. Intensive and extensive treatment of dredged sediments on landfarms. Proceedings In Situ and On-Site Bioremediation: Volume 2, 153-158. Battelle Press.

Harmsen, J. and M. Ferdinandy, 1999. Measured Bioavailability as a Tool for Managing Clean-up and Risks on Landfarms. A.Leeson and B.C. Alleman (eds), *Bioremediation Technologies for Polycyclic Aromatic Hydrocarbon Compounds*, 57-62. Battelle Press.

Huddleston, R.L., C.A. Bleckmann, and J.R. Wolfe. 1986. Land Treatment Biological Degradation Processes. Pp. 41-61. In: Land Treatment – A Hazardous Waste Management Alternative. [eds. R.C. Loehn and J.F. Malina, Jr.] Center for Water Resources Research, University of Texas at Austin, TX.

Kivanç, S., and J. Kahraman, 1999. Optimal Design and Operation of Land-treatment systems for Petroleum Hydrocarbons."In B.C. Alleman and A. Leeson (eds). Bioreactor and Ex Situ Biological treatment Technologies. Battelle Press. 81-86.

Martin, J.P., R.C. Sims and J Mattews, 1986. Revieuw and Evaluation of Current Design and Management Practices for Land treatment Units Receiving Petroleum Wastes. Hazardous Waste & Hazarous materials 3: 261-180

NATO. 2000. Utilization of Bioremediation to Reduce Soil Contamination: Problems and Solutions. North Atlantic Treaty Organization. Kluwer Academic Publishers, Dordrecht, The Netherlands (In Press).

Nieman, J.K.C., R.C. Sims and D.M. Cosgriff. 1999. Regulatory and Management Issues in Prepared Bed Land Treatment: Libby Groundwater Site. In B.C. Alleman and A. Leeson (eds). Bioreactor and Ex Situ Biological treatment Technologies. Battelle Press. 97-102.

Overcash, M.R. and D. Pal. 1979. Design of Land Treatment Systems for Industrial Wastes – Theory and Practice. Ann Arbor Science Publishers, Inc., Ann Arbor, Michigan. 684p.

Sims, J.L., R.C. Sims, and J.E. Matthews. 1990. Bioremediation of Contaminated Surface Soils. Hazardous Waste & Hazardous Materials 7 (3):117-149.

Sims, R.C. 1990. Soil Remediation Techniques at Uncontrolled Hazardous Waste Sites. Journal Air & Waste Management Association 40 (5):703-732.

Sims, R.C., D.L. Sorensen, J.L. Sims, and J.E. McLean. 1996. Champion International Superfund Site, Libby, Montana: Bioremediation Field Performance Evaluation of the Prepared Bed Land Treatment System. Volumes I and II EPA-600/R-95/156, U.S Environmental Protection Agency, Ada, OK,

Sims, R.C. and J.L. Sims. 1999. Landfarming of Petroleum Contaminated Soils. Chapter 27, (pp. 767-782) In: Bioremediation of Contaminated Soils, Monograph 37, ASA, CSSA, SSSA, Madison, WI.

Sims, R.C. and J.L. Sims. 2001. Landfarming Framework for Soil Bioremediation. In: Utilization of Bioremediation to Reduce Soil Contamination: Problems and Solutions. North Atlantic Treaty Organization. June. Kluwer Academic Publishers, Dordrecht, The Netherlands (in Press).

*U.S. EPA*. 1986. Permit Guidance Manual for Hazardous Waste Land Treatment Demonstrations. EPA-530/SW-86-032, Office of Solid Waste and Emergency Response, U.S. Environmental Protection Agency, Washington, DC.

*U.S. EPA*. 1987. Permit Guidance Manual on Hazardous Waste Land Treatment Closure and Post-Closure. U.S. EPA Directive No. 9476.00-9, Office of Solid Waste and Emergency Response, U.S. Environmental Protection Agency, Washington, DC.

# BIOREMEDIATION STRATEGIES FOR PESTICIDE-CONTAMINATED SOIL: I. MICROBIAL CONSORTIUM ISOLATION

***Anastasia E. M. Chirnside***, Diane S. Herson, Chin-Pao Huang, Mark Radosevich, and William F. Ritter (University of Delaware, Newark, DE 19717)

**ABSTRACT:** Development of feasible bioremediation strategies for a pesticide-contaminated soil from an agricultural mix-load site was initiated utilizing both fungal and bacterial technologies. One strategy was to isolate a bacterial culture from the mix-load site soil to use as a tool for bioaugmentation of the contaminated site. Three different locations from the site were chosen for experimentation based on herbicide handling activities. A selected microbial consortium (SMC) that was able to degrade two specific herbicides, alachlor (2-chloro-2', 6'-diethyl-N-[methoxymethyl]-acetanilide; AL) and atrazine (2-chloro-4-ethylamino-6-isopropylamino-S-triazine; AT) was isolated from one of the sample locations. Considerable differences in soil parameters of the three sample locations were found that might have had an effect on the ability of the indigenous microbial populations within the soils to degrade AT and AL.

## INTRODUCTION

Typically non-point sources of pesticide use have been identified as the major contributor of pesticide contamination of groundwater . However, "point" uses such as agrochemical dealerships, soil disposal pits, and pesticide mixing and loading sites may actually have contributed to groundwater pollution to a greater extent than non-point processes such as drift, runoff, or soil erosion (Bollag and Liu, 1990; Cheng, 1990; Miles et al., 1990). Pesticide mix-load sites have contained extremely high concentrations of many different pesticides within the soil. These complex mixtures of pesticides and other miscible and immiscible organic solvents used in pesticide formulations were usually found at elevated levels often greater than their solubility limit. The heterogeneous matrix of the contaminated soil has become quite disadvantageous to the remediation efforts. The costs of chemical treatment procedures have been exorbitant. Bioremediation, as a low cost alternative, has often been unsuccessful due to complexity of the interactions between the microbial degraders, the soil, the high concentration of pesticides and the wide array of compounds present.

The two most common herbicides used in the United States are AT and AL. Because of their long term and widespread use these compounds are frequently found in high concentration at mixed-load sites and have been known to migrate from these sites to groundwater wells several kilometers both up and down gradient from the facility (Miles et al., 1990). The overall objective of the proposed research was to assess the biological degradation and mineralization of AT and AL contained within a mix-load site soil.

Bioaugentation remediation techniques have often involved the addition of acclimated indigenous microbes that can degrade the target compounds at accelerated rates. The microbes are isolated from the contaminated soil in the laboratory through a selection and enrichment process. Utilizing microbes isolated from mix-load site soils may overcome many of the difficulties associated with these sites. Another bioaugmentation technique that has been investigated involves an unique group of fungi, the white rot fungi. The ligninolytic system of white rot fungi has produced nonspecific extra-cellular enzymes and hydrogen peroxide ($H_2O_2$) that interact to produce cation radicals capable of oxidizing a variety of xenobiotic compounds. Studies have shown that *Phanerochaete chrysosporium* has the ability to mineralize soil contaminates such as chlorinated biphenyls, polycyclic aromatic hydrocarbons and some insecticides (Hammel, 1989). Because of the extra cellular nature of the ligninolytic system of *P.chrysosporium*, the fungus may be ideal for the bioremediation of contaminated soil from mix-load sites.

**Objective.** The objective of this research was to develop feasible bioremediation strategies that can overcome the difficulties of remediation of contaminated mix-load site soils. The specific objective was to isolate a SMC from the pesticide-contaminated soil capable of degrading AT and AL. The SMC developed was also evaluated as a remediation strategy for the degradation of AT and AL in the contaminated mix-load site soil using laboratory biometer flasks.

## MATERIALS AND METHODS

Soil was acquired from a contaminated mix-load site that is characterized by high concentrations of a wide mix of a variety of pesticides including s-triazine and acetanilide herbicides. Due to the gravelly matrix, soil samples were taken to a depth of 46 cm utilizing both a pick and a shovel that were scrubbed clean between samples. The locations of the three sample sites were selected based on previous data indicating high concentrations of AT and AL. The sample sites (S-5A, S-5B, and S-6) were located in close proximity to the bulk pesticide storage building and loading dock. Soils were air dried, ground and passed through a 4-mm sieve. Twenty grams of soil were extracted with 2 N potassium chloride (KCl) and analyzed for ammonia-nitrogen ($NH_4$-N) and nitrate/nitrite-nitrogen ($NO_3$-N) utilizing steam distillation/acid titration (Standard Methods for the Examination of Water and Wastewater 17th Edition, Method 350.1/417D [Clesceri et al., 1989]) and Devardo's Alloy distillation/acid titration (Method 350.2/417D), respectively. Table 1 lists the standard physical and chemical properties of the soil that were determined by the University of Delaware Soil Testing Laboratory (Sims and Heckendorn, 1991). Fifty-gram aliquots of the dried, ground soil were extracted for AT and AL using a solid phase extraction method previously described (Chirnside and Ritter, 1992) followed by analysis by HPLC. Results are presented in Table 2.

**Development of the SMC.** Standard enrichment techniques were used to obtain indigenous microbial cultures capable of degrading AT and AL following the methods of Radosevich et al. (1993). The soil samples collected from three separate locations on the mix-load site were assessed individually for the presence of microbial degradation of AT and AL. The following three herbicide treatment schemes were utilized for the development of the SMC: 1) AT and AL as the only source of carbon (C) and nitrogen (N), 2) AL as the sole source of C and N and 3) AT as the sole source of C and N. The herbicide treatments were made by dissolving 27.0 mg AL and/or 21.6 mg of AT in one liter of deionized distilled water containing 10 mL of concentrated phosphate buffer. The buffered solutions were stirred overnight to ensure complete dissolution of the herbicides and then autoclaved at 120°C for 20 minutes. A concentrated salt solution and a concentrated trace element solution were made individually and filter sterilized using a 0.20 μm filter system.

**TABLE 1. Soil characterization.**

| Parameter | S-5A | S-5B | S-6 |
|---|---|---|---|
| Location | East side of loading dock | South side of loading dock | West side between RR spurs |
| Textural Class | Loamy sand | Clay loam | Sandy loam |
| SOM % | 1.1 (0)[1] | 1.4 (0) | 2.75 (0.07) |
| pH | 7.33 | 7.60 | 6.13 |
| $NH_4$-N (mg/kg) | 10.49 (0.31) | 0.272 (0.36) | 1.02 (0.41) |
| $NO_3$-N (mg/kg) | 50.24 (3.12) | 22.75 (0.67) | 26.69 (4.08) |
| P (mg/kg) | 454.9 (94) | 138.7 (6.4) | 1320 (5.2) |
| K (mg/kg) | 254.3 (3.9) | 194.8 (2.6) | 1054.7 (16.4) |
| Ca (mg/kg) | 4907.1 (78.9) | 4957.5 (27.6) | 4046.2 (69.0) |
| Mg (mg/kg) | 304.5 (68.3) | 145.6 (2.4) | 111.5 (2.6) |
| Mn (mg/kg) | 65.6 (4.4) | 7.6 (0.31) | 76.8 (2.1) |
| Fe (mg/kg) | 11.5 (6.3) | 0 (0) | 66.1 (4.4) |

**TABLE 2. Parent pesticide concentrations of mix-load site soil**

| Site Location | Atrazine (mg/kg) | Alachlor (mg/kg) |
|---|---|---|
| S-5A | 391 ± 0.88 | 3896 ±19.1 |
| S-5B | 0.389± 0.29 | 5.8 ± 0.46 |
| S- 6 | 1.70 ± 0.14 | 0.602 ± 0.07 |

[1]Number in parentheses represents standard deviation of 3 values.

In a laminar flow hood under sterile conditions, 39 mL of the appropriate buffered herbicide solution and 0.5 mL of each the concentrated filter sterilized solutions were added to 6 autoclaved 250 mL Erlenmeyer flasks stoppered with foam plugs. Two flasks of each treatment solution were inoculated with 10 g of the 3 soil types. A control flask for each soil type contained the combined herbicide treatment (AT & AL) and 10 g of autoclaved soil. The resulting 21 flasks were incubated at 25°C on an orbital shaker at 150 rpm. The flasks were sampled periodically for loss of parent compound. Prior to sampling, the flasks were weighed and corrected for evaporative water loss by addition of sterile water. A 0.5 ml aliquot was removed and mix with 0.5 ml of acetonitrile in a microcentrifuge tube and centrifuged at approximately 12,000 x g for 10 minutes to remove any suspended soil particles. The supernatant was transferred to HPLC vials for analysis. The concentrations of AT and AL were quantified utilizing an HPLC system (Thermal Separation Products, West Palm Beach, FL) under the following conditions: a) mobile phase- 65% acetonitrile in deionized distilled water; b) flow rate- 1.3 ml $min^{-1}$; c) column- RP-18 reverse phase, 25 mm x 4.6 cm, 5µm (E Merk Separations Technology, Darmstadt, Germany); detector- UV dioarray set at 220 nm.

After six weeks, all flasks were subcultured by transferring 10 mL of the original culture into new flasks with fresh media containing the identical herbicide treatments. Flasks were subcultured again after 22 weeks. Selected treatment flasks from these initial cultures that exhibited herbicide degradation were subcultured into fresh media and monitored for 22 weeks as outlined above. At this point cultures exhibiting herbicide degradation were maintained by transferring to flasks containing media consisting of the 3 original herbicide treatment schemes.

Based on the lack of AL degradation in the treatment flasks containing only AL, further subcultures were made into media containing the original treatment 1, i.e. AT and AL as the only source of carbon and energy.

## RESULTS AND DISCUSSION

In the initial SMC cultures, degradation occurred in the AT and AL treatment only. There were considerable differences in the results for the three sample locations. Site 5A soil exhibited no degradation in all treatments (Figure 1). Site 5B soil exhibited degradation of both compounds in the combined treatment demonstrating a time lag in the degradation of the AL (Figure 2). Finally, Site 6 soil exhibited AT degradation only in the combined treatment (Figure 3). Preliminary conclusions can be drawn from these results based on the soil characterization data. Site 5A soil has very high concentrations of $NH_4$-N and $NO_3$-N and of the two herbicides whereas Site 5B soil has low concentrations of both the nitrogen and the herbicides. The mix-load site had been monitored for pesticide contaminants three other times dating back to 1988. Previous analysis of Site 5B revealed that the herbicides were present at high concentrations indicating that degradation did occur at the site over time, thus indicating the possibility that microbial degraders with the ability to degrade AT and AL are present in the soil at Site 5B. Therefore, the introduction of the herbicide treatment in the treatment

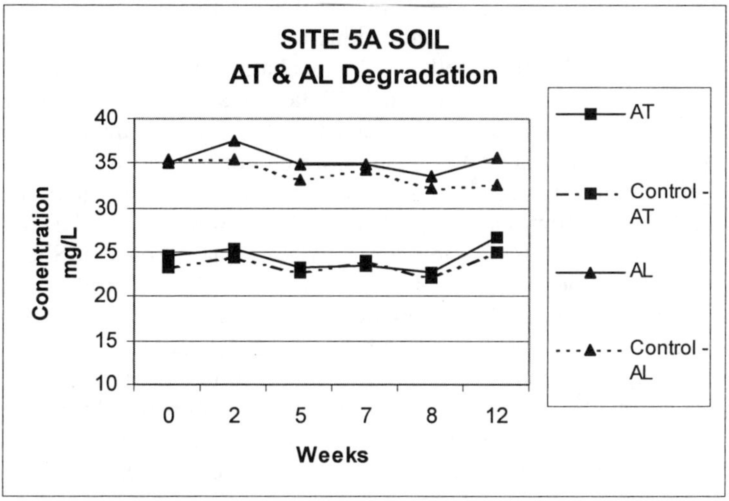


**FIGURE 1. Disappearance of AT and AL from Site 5A SMC flasks**

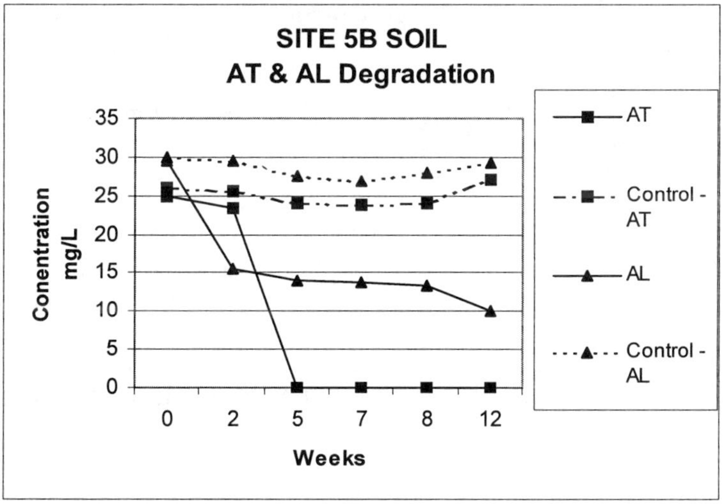


**FIGURE 2. Disappearance of AT and AL from Site 5B SMC flasks**

flask could have triggered the induction of the enzyme(s) necessary for degradation of AT and AL.

Stolpe and Shea (1995) examined the degradation of atrazine at various depths in soil with a history of application versus a soil without recent application. Degradation was slowest in the subsoils (t ½ = 107-248 d). Very rapid mineralization occurred in surface soils from the "history" field (60% in 28 d). Only 11% mineralization occurred in 28 days in the non-history surface soils.

Dissipation of atrazine in the "history" surface soils was primarily due to mineralization whereas it was primarily due to bound residue formation in the non-history surface soils. Bound residue formation was the major transformation process for all subsurface soils. As incubation time increased, the amount of atrazine bound to the soil increased. Polar metabolite concentrations were low throughout and remained relatively constant over time. Atrazine degradation rate was positively correlated to organic carbon, ortho phosphorus, total Kjeldahl nitrogen and phospholipids. Negative correlations were found with inorganic nitrogen concentrations suggesting AT nitrogen utilization by the microorganisms. Thus, the low levels of inorganic nitrogen at Site 5B may have contributed to the degradation of AT and AL by the indigenous microbial population over time. The formation of bound residues may limit total AT mineralization. However, the authors conclude that some of the soil bound residues may be mineralized due to the smaller amounts of both polar metabolites and bound residue fractions where AT mineralization was high. The enhanced AT degradation in the soils with a history of application was attributed to adaptation of the microbes to the substrate. The number of AT degraders was significantly higher in the "history" soil.

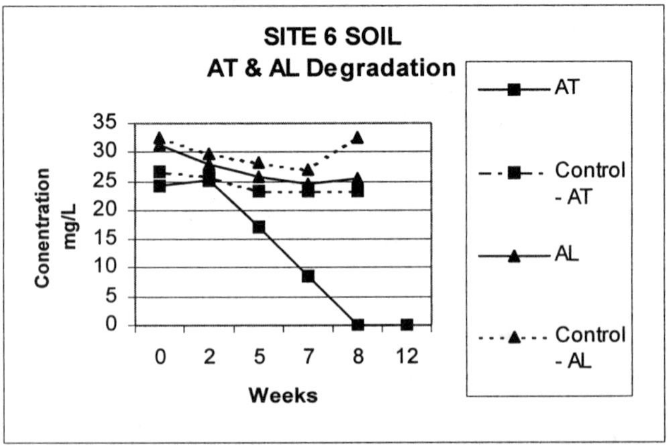


**FIGURE 3. Disappearance of AT and AL from Site 6 SMC flasks**

Site 6 soil had a high organic matter content and contained about 5 cm of gravel fill that could cause bioavailability interferences in herbicide degradation due to adsorption processes which resulted in the slower rate of AT degradation. Gravel fill has been shown to act as a low-organic carbon adsorbent capable of adsorbing large quantities of pesticides. Relatively large amounts of AT (3.60 to 13.4 mg $kg^{-1}$) were adsorbed to the < 2 mm fraction of a mix-load site gravel fill showing hysteric adsorption patterns (Roy et al., 1995). This enhanced adsorption capacity of the gravel fill may have contributed to bound residue formation of AT causing slower rates of biodegradation. This low rate of AT degradation could be responsible for a greater lag in AL degradation. Subcultures of the flasks with evidence of degradation have been successful with degradation occurring in all three treatments (data not shown). At present we are in the process of isolating the degradative microbes.

**CONCLUSIONS**

Conclusions based on the results of the initial SMC experiment are outlined as follows:

1. Degradation in the AT & AL treatments only.
    a. Alachlor alone was not a sufficient N source for microbes.
2. Site 5A soil exhibited no degradation.
    a. High $NH_4$-H and $NO_3$-N – microbes don't need AT/AL for N.
    b. High concentrations- possible toxicity effects.
    c. High concentrations- no enzyme induction, no previous activity.
3. Site 5B soil exhibited AT & AL degradation.
    a. Low $NH_4$-H and $NO_3$-N- use as N source.
    b. Alachlor provides C for complete AT degradation.
    c. Lag in Al degradation- after N-dealkylation of AT.
    d. Low concentrations- degraders present at site, addition of AT and AL triggers induction.
4. Site 6 soil exhibited AT degradation only.
    a. Highest SOM- possible adsorption processes, interferences- slower rate.
    b. Low concentrations- degraders present at site.
    c. Low rate of AT degradation led to lag in AL degradation- (N-dealkylation).

The SMC isolated from Site 5B soil was assessed as an inoculum for bioaugmentation of Site 5A soil that contained high concentrations of AT and AL. Preliminary laboratory results suggested that the SMC was capable of degrading the herbicides and that evaluation of the SMC for use as a field bioremediation tool of mix-load site soils was warranted.

**REFERENCES**

Bollag, J. M., and S.Y. Liu. 1990. "Biological transformation processes of pesticides". In H.H.Cheng (Ed), *Pesticides in the soil environment,* SSSA Book Series, No.2, pp. 169-211. SSSA, Inc., WI.

Cheng, H.H. 1990. "Pesticides in the soil environment - an overview". In H.H. Cheng (Ed), *Pesticides in the soil environment,* SSSA Book Series, No.2, pp. 1-5. SSSA, Inc., WI.

Chirnside, A. E. M., and W. F. Ritter. 1993. "Development of a solid-phase extraction method for herbicide residue analysis of soil samples." In K.B. Hoddinott and T.A. O'Shay, (Eds), *Application of agricultural analysis in environmental studies,* ASTM STP 1162, pp. 92-97. American Society for Testing and Materials, Philadelphia, PA.

Clesceri, L. S., A. E. Greenberg and R. R. Trussell (Eds). 1989. *Standard Methods for the Examination of Water and Wastewater.* $17^{th}$ ed., American Public Health Association, American Water Works Association and Water Pollution Control Federation, Washington, D.C.

Hammel, K.E. 1989, "Organopollutant degradation by ligninolytic fungi." *Enzyme Microbiology Technology*, 11: 776-777.

Miles, C.J., K. Yanagihara, S. Ogata, G. Van de Verg, and R. Boesch. 1990. "Soil and water contamination at pesticide mixing and loading sites on Oahu, Hawaii". *Bull. Environ. Contam. Toxicol.,* 44:955-962.

Radosevich, M., Crawford, J.J., Traina, S.J., Hao, Y-L. and Tuovinen, O.H. 1993. "Biodegradation of atrazine and alachlor in subsurface sediments." In D. M. Linn (Ed), *Sorption and degradation of pesticides and organic chemicals in soil.* Special Publication No. 32, pp. 33-41. SSSA, Inc., ASA, Inc., WI.

Roy, W.R., and I.G.Krapac. 1994. "Adsorption and desorption of atrazine and deethylatrazine by low organic carbon geologic materials". *Journal of Environmental Quality,* 23:549-556.

Sims, J. T. and S. E. Heckendorn. 1991. *Methods of Soil Analysis,* University of Delaware Cooperative Extension Bulletin #10, Newark, DE

Stolpe, N.B., and P.J.Shea. 1995. "Alachlor and atrazine degradation in the Nebraska soil and underlying sediments". *Soil Science,* 160:359-370.

# BIOCATALYTIC OXIDATION OF ORGANOPHOSPHOROUS PESTICIDES BY WHITE ROT FUNGI

***Juan Jauregui*, Brenda Valderrama and Rafael Vazquez-Duhalt**
(Instituto de Biotecnología, Universidad Nacional Autónoma de México.
AP 510-3 Cuernavaca, Mor. 62250 México, jjaureg@ibt.unam.mx)

**ABSTRACT:** The enzymatic mechanism for the degradation of organophosphorus (OPs) pesticides by different white-rot fungi strains was studied. A collection of 17 different fungi metabolized 10% to 97% of added parathion in 96 hrs. Three strains showing the highest activities were selected for the degradation of other OPs pesticides: *Bjerkandera adusta* 8258, *Pleurotus ostreatus* 7989 and *Phanerocheate chrysosporium* 3641. All strains degraded terbufos, azinphos-methyl, phosmet and DEF, from 50% to 96% of these compounds in 96 hrs. In order to identify the origin of the degradation activity, extracellular fungal extracts induced for ligninolytic enzymes were evaluated. While the activity of lignin peroxidase, manganese peroxidase, and laccase were detected in the extracts, no enzymatic modification of five OPs pesticides tested could be found suggesting the intracellular origin of the degradation activity. Microsomal preparations from *B. adusta* 8258 were obtained and tested, the degregadation rates for the microsomal fraction were: 10 μmol mg-prot$^{-1}$ h$^{-1}$ for phosmet, 5.7 μmol mg-prot$^{-1}$ h$^{-1}$ for terbufos, and 2.2 μmol mg-prot$^{-1}$ h$^{-1}$ for azinphos-methyl. These results indicate that in ligninolytic fungi the OP pesticides degradation activity is intracellular. The stringent requirement for NADPH during the degradation suggest cytochrome P450 might be involved.

## INTRODUCTION

Organophosphorus (OPs) and carbamate pesticides are widely used in agricultural and residential applications as insecticides, herbicides, fungicides and rodenticides. This family of chemicals replaced the organochlorine pesticides banned for use in the United States since the 1970's. Unlike organochlorine pesticides, which are persistent in the environment and cause biological damage when they accumulate in an organism over time, OPs and carbamate pesticides are short-lived (several months to years) in the environment and fast-acting on their "target pest". Direct mortality of wildlife from organochlorine pesticides was uncommon (Hayes and Wayland, 1975); however, mortality is the primary documented effect on wildlife from OPs and carbamate pesticides (Grue et al., 1983). Organophosphorus and carbamate pesticide toxicity is not specific to a "target pest" and lethal effects are seen in non-target organisms. In addition to birds, which appear to be the most sensitive class of animals affected by these compounds, OPs are the most likely pesticides to be involved in acute human poisonings. The poisoning must to the consumption of foods and water contaminated by OPs. After the Food Quality Protection Act of 1996, the US

Environmental Protection Agency put 40 OPs pesticides in the highest priority group and placed severe restrictions upon the use of three of them: chlorpyrifos, azinphos methyl and methyl parathion. Organophosphorus and carbamate pesticides primarily affect the nervous system by inhibiting acetylcholinesterase which main function is the break-down of the neurotransmitter acetylcholine. When acetylcholinesterase is inhibited, acetylcholine accumulates leading to an increase of the nerve impulse transmission then to nerve exhaustion and, ultimately, to general failure of the nervous system. When the nervous system fails, muscles do not receive the electrical input they require to move. The respiratory muscles are the most critical group affected and respiratory paralysis is often the immediate cause of death (Hart, 1993).

Some OPs pesticides are degraded by bacteria, such as *Pseudomonas, Flavobacterium, Altermonas* and *Bacillus,* harboring the organophosphorus hydrolase or organophosphorus acid anhydrolase activities whose are the enzymes involved in the cleavage of P-O bonds (see Amitai et al., 1998). However, these enzymes show limited capacity to cleave the thioether bond present in several OPs pesticides (Figure 1). On the other hand, the ligninolytic fungus *Phanerochaete chrysosporium* has been shown to be able to mineralize chlorpyrifos, phonofos and terbufos, and the contribution of the extracellular ligninolytic enzymes has been suggested (Bumpus et al., 1993).

In a previous work we demonstrated that *Caldariomyces fumago* chloroperoxidase was able to oxidize 7 of 10 different OPs pesticides assayed, although no oxidation was detected using other heme proteins such as lignin peroxidase, horseradish peroxidase, or cytochrome *c* (Hernández et al., 1998). A fungal laccase (polyphenol oxidase) produced by *Pleurotus ostreatus* has been reported as able to perform the oxidative degradation of two nerve agents, VX and RVX, which are organophosphorus compounds containing thioether bonds and are structurally similar to OPs pesticides (Amitai et al., 1998).

In this work, we present the identification of the enzymatic system involved in the degradation of OPs pesticides by white rot fungi. The pesticides used were those that are recalcitrant to the bacterial organophosphorus hydrolase and acid anhydrolase activities.

## MATERIALS AND METHODS.

**Chemicals.** 2,2'-azinobis(3-ethylbenzthiazoline-6-sulfonic acid) diammonium salt (ABTS), and sodium malonate were purchased from Aldrich (Milwaukee, WI). Organophosphorus pesticides (Figure 1): azinphos methyl (99%), DEF (94 %), phosmet (99 %) and terbufos (95 %) were obtained from Ultra Scientific (North Kingstown, RI). Buffer salts were obtained from J.T. Baker (Phillisburg, NJ.). Glucose, yeast extract and malt extract were purchased from Difco laboratories (Detroit, MI).

**Organisms and culture media.** *Bjerkandera adusta* 4312, 7308, 8258; *Pleurotus ostreatus* 7964, 7972, 7980, 7988, 7989 and 7992; *Phanerochaete chrysosporium* 3641, FC-322 and 3642; *Sporotichum pruinosum* 4521; *Coriolopsis gallica* 8260;

*Ganoderma. aplanatus* 8168 and *Trametes versicolor* 8272 were obtained from the University of Alberta Mold Herbarium (Edmonton, Canada). *Pleurotus ostreatus* IE8 (Chiapas) was obtained from the Ecology Institute (Xalapa, México). *Phanerochaete chrysosporium* ATCC 24725 was obtained from the American Type Culture Collection (Rockville Pike, MD). All fungi were maintained on potato dextrose agar plates (Difco). Inocula were prepared by homogenizing 1 $cm^2$ of mycelium from a colony grown on a potato-dextrose agar plate in 50 mL of GMY medium (Pickard et al., 1999a), using a Sorvall Omnimixer (Norwalk, CN) for 10 s. After 3 days incubation in an orbital shaker at 28°C and 200 rpm, the culture was homogenized again, and approximately 2 mg dw were used for inoculation of 100 mL of cereal-bran medium in 500 mL flasks (Pickard et al., 1999a). The cultures were incubated at 28°C and 200 rpm in an orbital shaker.

**Enzyme assays.** Lignin peroxidase was measured by the method of Tien and Kirk (1988) following the $H_2O_2$–dependent oxidation of veratryl alcohol to veratraldehyde at 25°C. Reaction mixtures contained 4 mM veratryl alcohol in 40 mM succinate buffer pH 3 and the reactions were started by the addition of 0.4 mM $H_2O_2$. Absorbance was monitored at 310 nm ($\varepsilon_{310}$ = 9300 $M^{-1}$ $cm^{-1}$ for veratraldehyde). Manganese peroxidase activity was estimated at 25°C by formation of oxidized manganic-malonate complex in the presence of $H_2O_2$ (Pickard et al., 1999a). The reaction mixture contained 1 mM manganous sulfate in 50 mM malonate buffer pH 4.5 and 5-50 μl of enzyme extract. Reactions were started by the addition of $H_2O_2$ to a final concentration of 0.1 mM and monitored at 270 nm ($\varepsilon_{270}$ = 11590 $M^{-1}$ $cm^{-1}$).

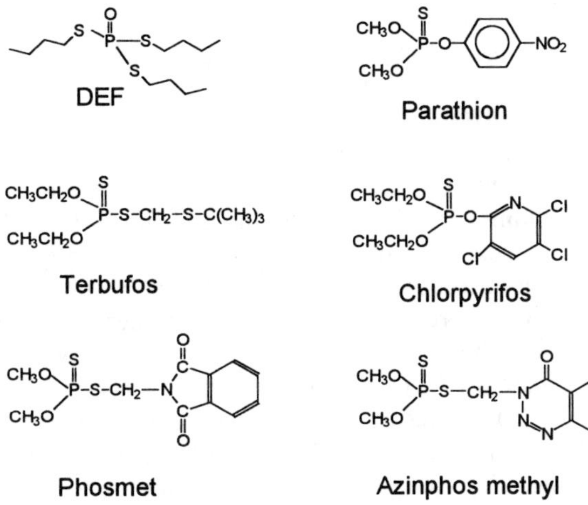


FIGURE 1. Structure of organophosphorus pesticides.

Laccase activity was determined by the oxidation of ABTS (Pickard et al., 1999b). The reaction mixture contained 1 mM ABTS in 0.1M sodium acetate buffer pH 4.5 with 5-50 μl of enzyme extract. The oxidation was followed at 436 nm ($\varepsilon_{436}$= 29300 $M^{-1}$ $cm^{-1}$ for ABTS).

***In vivo* degradation experiments.** Five-ml inoculum was prepared as described above and added to 50 ml of cereal-bran medium in a 250 ml flask. After 2-days incubation at 28°C and 200 rpm, 20 mM OP pesticides were added and the cultures incubated 4 more days. Then, 50 ml of tetrahydrofuran (THF) were added to the culture, the mixture was centrifuged and the supernatant analyzed by high performance liquid chromatography (HPLC, Perkin-Elmer, series 200) equipped with a 100 x 2.1 mm ODS Hypersil 5 μm column (Hewlett Packard). Samples were isocratically eluted with acetonitrile:water (45:55) at 0.3 ml/min and the OP pesticide concentration was monitored with a diode array detector (Perkin-Elmer, model 235C). Pesticide metabolization was estimated as the peak disappearance and quantified by using a standard curve.

**Microsomes preparation.** Fungal biomass was resuspended in a 50 mM Tris-HCl buffer pH 7.6 containing 25 mM $MgCl_2$ and grinded in four 30-seconds pulses a Sorvall Omnimixer. The mixture was then homogenized in a Milton Gauli homogenizer at 3,500 psi. The microsomal preparation was obtained by differential centrifugation as described in Cinti et al., 1972.

***In vitro* degradation experiments.** Enzymatic degradation of OPs pesticides was assayed with the microsomal preparations. Reactions were carried out in a 2-ml reaction mixture containing 2.0 mM OP pesticide, 25 mM $MgCl_2$ and 4.0 mM β-NADPH in 50 mM Tris-HCl buffer pH 7.6. The reaction was started by the addition of 0.2 to 0.28 mg of the microsomal protein and monitored by either HPLC or by spectrophotometric β-NADPH consumption at 340 nm.

## RESULTS AND DISCUSSION

Parathion metabolization by 17 different fungal strains was assayed for 96 hrs in cereal-bran medium (Table 1). From these results, it is clear that the capacity for OP pesticide degradation is widely distributed among the ligninolytic fungi although significant differences on this capacity can be found. Three strains showing the highest activity were selected for further studies: *Bjerkandera adusta* 8258, *Pleurotus ostreatus* 7989 and *Phanerocheate chrysosporium* 3641. With these strains, the degradation of OPs compounds (parathion, terbufos, azinphos-methyl, phosmet, and DEF) was tested, and the specific metabolization rate values are shown in Table 2. All the OPs pesticides tested were efficiently degraded by the three fungi. From the selected pesticides, only parathion is a phosphorothioate, while azinphos methyl, phosmet and terbufos are phosphorodithioates, and DEF are phosphorotrithioates. This is important because the presence of thioether bonds makes the pesticide recalcitrant to the bacterial enzyme, organophosphorus hydrolase.

In order to identify the the fungal degradation activity and to solve the controversy on the participation of extracellular peroxidases in the oxidation of OPs pesticides, several extracellular ligninolytic enzymes were evaluated *in vitro*. Extracellular fungal extracts containing lignin peroxidase, manganese-dependent peroxidase, laccase or veratryl alcohol oxidase activities were assayed for the oxidation of OPs pesticides (Table 3). No activity on pesticides could be found, in any of the extracts, under the different conditions tested. In addition, neither purified preparations of lignin peroxidase nor manganese-dependent peroxidase from *Phanerochaete chrysosporium* were able to perform the peroxidation of OPs pesticides.

**TABLE 1. Parathion degradation by 18 strains of ligninolytic fungi after 96 hrs incubation.**

| Fungi | Metabolization (%) |
|---|---|
| *B. adusta* 4312 | 49.0 (±4.2) |
| *B. adusta* 7308 | 49.0 (±5.5) |
| ***B. adusta* 8258 *** | **78.3 (±3.6)** |
| *G. aplanatus* 8168 | 0.0 |
| *P. ostreatus* 7964 | 44.5 (±4.1) |
| *P. ostreatus* 7980 | 44.5 (±4.7) |
| *P. ostreatus* 7988 | 18.4 (±4.0) |
| ***P. ostreatus* 7989 *** | **96.9 (±2.9)** |
| *P. ostreatus* 7972 | 32.3 (±3.1) |
| *P. ostreatus* 7992 | 26.0 (±2.5) |
| *P. ostreatus* IE8 | 48.5 (±2.3) |
| *P .chrysosporium* FC-322 | 10.5 (±3.0) |
| ***P.chrysosporium* 3641 *** | **55.0 (±2.1)** |
| *P. chrysosporium* 3642 | 28.6 (±2.5) |
| *P. chrysosporium* ATCC 24725 | 30.0 (±2.0) |
| *S. pruinosum* 4521 | 7.7 (±3.2) |
| *C. gallica* 8260 | 28.5 (±2.5) |
| *T. versicolor* 8272 | 9.5 (±3.4) |

*Fungi selected for further study.

Elsewhere, Amitai et al. (1998) reported the oxidative degradation of two nerve agents, VX and RVX. The chemical structure and the action mode of these toxic compounds were similar to those of OPs pesticides. They used a fungal laccase produced by *Pleurotus ostreatus* in the presence of ABTS as the radical mediator. We assayed fungal laccases pure preparations from *Pleurotus ostreatus*, *Trametes versicolor* and *Coriolopsis gallica* (Tinoco et al, 2001) either in the presence of ABTS or 1-hydroxybenzotriazole (HBT) and no oxidation of any OPs pesticide tested was found.

**TABLE 2. Specific metabolization rate (µmol g dry biomass$^{-1}$ h$^{-1}$) of five organophosphorus pesticides by three white-rot fungi strains.**

| Fungi | Pesticides | | | | |
|---|---|---|---|---|---|
| | Parathion | Azinphos methyl | Phosmet | Terbufos | DEF |
| *B. adusta* 8258 | 217 (±10) | 223 (±8) | 220 (±2) | 232 (±9) | 84 (±8) |
| *P. ostreatus* 7989 | 169 (± 4) | 106 (±6) | 103 (±4) | 589 (±5) | 45 (±5) |
| *P. chrysosporium* 3641 | 145 (± 6) | 203 (±8) | 45 (±3) | 174 (±7) | 25 (±2) |

Taken together, these results suggested that the degradation activity might be an intracellular function, thus microsomal fractions were obtained from *B. adusta* 8258 biomass and tested for the degradation of three OPs pesticides. The pesticides degradation rates were; 10 µmol mg-prot$^{-1}$ h$^{-1}$ for phosmet; 5.7 µmol mg-prot$^{-1}$ h$^{-1}$ for terbufos and 2.2 µmol mg prot$^{-1}$ h$^{-1}$ for azinphos-methyl. These rates are significantly higher than those found with human liver microsomes where from 0.1 to 1.1 µmol mg prot$^{-1}$ h$^{-1}$ have been reported (Butler and Murray, 1997). OPs pesticides degradation by the microsomal preparations required the addition of NADPH (FIGURE 2), suggesting the enzymatic activity involved might resemble the properties of cytochrome P450. In accordance, microsomal preparations from rat liver containing cytochrome P450 have been shown to be able to transform the phosphorothioate moieties of OPs pesticides to phosphate groups (Ma and Chambers, 1994; Whitehouse and Ecobiochon, 1975).

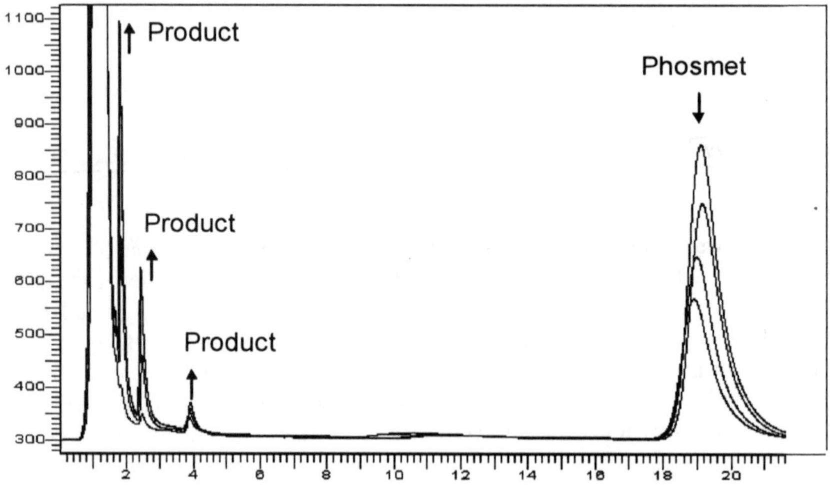


**FIGURE 2. Phosmet degradation by the microsomal preparation from *Pleurotus ostreatus* biomass with successive additions of β-NADPH.**

These preparations were also able to detoxify OPs pesticides by dearylation, as in the production of *p*-nitrophenol from parathion (Ma and Chambers, 1994). Paraoxon, diethylphosphorothioic acid, diethylphosphoric acid and *p*-nitrophenol were produced by NADPH-reconstituted cytochrome P450 systems (Yoshihara and Neal, 1977).

**TABLE 3. Ligninolytic enzyme activities in the extracellular concentrate from liquid cultures of three different white-rot fungi.**

| | | Specific activity (μmol $min^{-1}$mg $protein^{-1}$) | |
|---|---|---|---|
| White-rot fungi | Enzyme | Ligninolytic medium | No Ligninolytic medium |
| *B. adusta* 8258 | Laccase | 7 | Nd |
| | Mn-peroxidase | 35 | 12 |
| | Lignin peroxidase | 112 | 23 |
| *P. ostreatus* 7989 | Laccase | 250 | 6 |
| | Mn-peroxidase | 100 | Nd |
| | Lignin peroxidase | Nd | Nd |
| *P. chrysosporium* 3641 | Laccase | Nd | Nd |
| | Mn-peroxidase | 30 | 3 |
| | Lignin peroxidase | 4 | Nd |

Nd. Not detected.

## CONCLUSION

Ligninolytic fungi were able to metabolize *in vivo* several organophosphorus pesticides including phosphorodithioates and phosphorotrithioates which are know to be recalcitrant to bacterial detoxification activities. A microsomal cytochrome P450-like system is involved in a NADPH-dependent reaction. No oxidation of OPs pesticides could be detected with extracellular ligninolytic peroxidases.

## ACKNOWLEDGEMENTS

This work was funded by the National Council of Science and Technology of Mexico (grant 25376-A).

## REFERENCES

Amitai G., Adani R. Sod-Moriah G., Rabinovitz I., Vincze A., Leader H., Chefetz, B., Leibovitz-Persky L., Friesem D. and Hadar Y. 1998 "Oxidative biodegradation of phosphorothiolates by fungal laccase". *FEBS Letters*. 438:195-200.

Bumpus, J..A., Kakar, S.N. and Coleman, R.D. 1993. "Fungal degradation of organophosphorus insecticides". *Appl. Biochem. Biotechnol.* 39/40:715-726.

Butler, A.M. and Murray, M. 1997. "Biotransformation of parathion in human liver: participation of CYP3A4 and its inactivation during microsomal parathion oxidation". *Pharmacol. Exp. Ther*. 280:966-973.

Cinti, D.L., Moldeus, P. and Schenkman J.B. 1972. "Kinetic parameters of drug-metabolizing enzymes in $Ca^{2+}$-sedimented microsomes from rat liver". *Biochem. Pharmacol.* 21:3249-3256.

Grue, C.E., Fleming, W.J., Busby, D.G. and Hill, E.F.. 1983. Assessing hazards of organophosphate pesticides to wildlife. Pages 200-220 *in Transactions of the 48th North American Wildlife and Natural Resources Conference*. The Wildlife Management Institute, Washington, DC.

Hart, A.D.M. 1993. "Relationships between behavior and the inhibition of acetylcholinesterase in birds exposed to organophosphorus pesticides". *Enviro. Toxicol. Chem.* 12:321-336.

Hayes, W.J., Jr., and J. Wayland. 1975. Effects on wildlife. Chapter 11 *in Toxicology of pesticides*. Williams and Wilkins Company, Baltimore, MD.

Hernández J., Robledo N.R., Velasco L., Quintero R., Pickard M.A. and Vazquez-Duhalt R. 1998. "Chloroperoxidase-mediated oxidation of organophosphorus pesticides". *Pesticide Biochem. Physiol.* 61:87-94.

Ma, T. and J.E. Chambers. 1994. "Kinetic parameters of desulfuration and dearylation of parathion and chlorpyrifos by rat liver microsomes". *Food Chem. Toxicol*. 32 : 763-767

Pickard M.A., Roman R. and Vazquez-Duhalt R. 1999a. "High production of ligninolytic enzymes from white rot fungi cereal-bran liquid medium". *Can J. Microbiol*. 45:627-631

Pickard M.A., Roman R. and Vazquez-Duhalt R. 1999b. "Polycyclic aromatic hydrocarbon metabolism by white rot fungi and oxidation by *Coriolopsis gallica* UAMH 8260 laccase". *Applied Environ. Microbiol.* 65:3805-3809.

Tien, M., and Kirk, T.K. 1988. "Lignin peroxidase of *Phanerochaete chrysosporium*". *Methods Enzymol.* 161: 238-248.

Tinoco, R., Pickard, M.A., and Vazquez-Duhalt, R. 2001 "Kinetic differences of laccases from six Pleurotus ostreatus strains". *Lett. Appl. Microbiol.* 32:331-335

Yoshihara, S. and Neal, R.A. 1977. "Comparasion of the metabolism of parathion by a rat liver reconstituted mixed funtion oxidase enzyme system and by a system containing cumene hydroperoxide and purified liver cytochrome P450". *Drug Metab. Disp.* 5: 191-197.

Whitehouse, L.W. and Ecobichon, D.J. 1975. "Paraoxon formation and hydrolysis by mammalian liver". *Pestic. Biochem. Physiol.* 5 : 314-322.

# BIODEGRADATION OF SYNTHETIC PYRETHROID INSECTICIDES IN FORMULATED COMPOUNDS

***Russell J. Grant*** (University of York, UK)
W. Bernard Betts (Cell Analysis Ltd, UK)

**ABSTRACT**: Synthetic pyrethroids (SP) present in sheep dips were investigated in a series of cultures with previously isolated SP degrading bacteria. The experiments focused on the effect of culture temperature and presence of additional nutrients on bacterial cell growth and concomitant break down of SP. The goal was to begin to find a practical approach using biodegradation as a means of SP disposal in formulated insecticides. Results suggest that the addition of nutrients and more preferable growth substrates may increase the rate of SP degradation, and that temperature has a much lesser affect on the rate of break down per se. However, at low temperatures the cell growth was smallest and thus the degradation per cell was greatest.

## INTRODUCTION

Synthetic pyrethroids (SP) are used as the active ingredient in many insecticides, both domestic and industrial (Extoxnet PIP, 1994). Their usage is rapidly increasing as they replace the more mammalian toxic organophosphates (OP) in many insecticidal applications (Armstrong and Phillips, 1998. Stephens *et al.*, 1995). Some countries such as Britain have banned OP dip usage (MAFF, 1999). A major problem with this conversion is the environmental impact that SPs can have. The mammalian toxicity of SPs is much lower than the OPs (Tomlin, 1997), but to the aquatic environment SPs are over one hundred times more toxic (Pearce, 1997. Armstrong and Phillips, 1998). One major application of SPs is in the treatment of agricultural livestock such as cows and sheep. The treatment sites for these animals are often located close to a water source, and the disposal of leftover insecticide is usually by drain off into a field (Armstrong and Phillips, 1998). Thus the incidents of pollution have become much more serious, with the aquatic life from whole rivers being lost due to a minor volume of contamination. (Armstrong and Phillips, 1998. Virtue and Clayton, 1997).

**Objective.** The objective of these experiments was to examine how the elementary factors of temperature and nutrient availability affected bacterial growth and biodegradation of SPs when present as a sole compound or in a manufactured formulation. The particular aim of this work was to discover the best bacteriological approach for the disposal of SP-based animal dip treatments.

## MATERIALS AND METHODS:

The technical grade cypermethrin compound (72.6 % pure) was obtained from Zeneca Agrochemicals (Kent, UK). The Ecofleece® Non-OP sheep dip (contains 10 % cypermethrin) was donated by Bimeda (Liverpool, UK) and the

Bayticol® Scab & Tick dip (contains 6 % flumethrin) was donated by Bayer (Leverkusen, Germany). The mineral salts medium used was Bushnell Haas (BH) broth, composed of ($L^{-1}$ of deionized water) $K_2HPO_4$, 1 g; $KH_2PO_4$, 1 g; $NH_4NO_3$, 1 g; $CaCl_2$, 0.02 g; $MgSO_4$, 0.2 g; $FeCl_3$, 0.05 g; taken to pH7 with NaOH.

Nutrient broth (13 g $L^{-1}$), nutrient agar (28 g $L^{-1}$) and technical agar were obtained from Oxoid (Hampshire, UK).

The technical cypermethrin was prepared by making a 4 % solution in absolute alcohol.

**Isolation of SP degrading microorganisms.** The two organisms (Circle and White) profiled in Grant *et al.* (2000) were used as the SP degrading bacteria.

**Culture conditions.** 150 ml conical flasks with 50 mls of culture were agitated at 100 rpm. Starting numbers of approximately 2,000 cell $ml^{-1}$ Circle (*Pseudomonas sp.*) or 2,000 cell $ml^{-1}$ White (*Serratia sp.*) in BHB or NB were incubated at temperatures of 15, 25 or 37°C. 50 µl Bayticol® (≈60 mg $L^{-1}$ SP), 30 µl Ecofleece® (≈60 mg $L^{-1}$ SP), or 100 µl prepared cypermethrin (≈60 mg $L^{-1}$) were used as test compounds. Cultures using both organisms began with 1,000 cell $ml^{-1}$ of each microorganism. Control cultures using 100 µl absolute alcohol were also investigated.

**Surface tension measurement.** A surface and interfacial tension torsion balance (White Electrical Instruments OS, UK) was used to measure surface tension of the cultures and formulations.

**pH measurement.** A pH electrode (Jencons, USA) was used to measure the pH changes.

**Viable cell number count.** Cell numbers were monitored using serial dilutions of each culture, plated onto NA and incubated at 30 °C for 24 h.

**Estimation of SP present.** Using the bioassay shown in Grant (2000) (a direct toxicity assessment test using fruit flies) the concentration of SP was estimated. Grant (2000) showed that for these chemicals at the concentrations used the variation of result is less than 5 %.

## RESULTS AND DISCUSSION

**Controls.** The chemical controls with no bacteria present remained sterile throughout the experimental period. The control cultures with alcohol showed no significant differences in cell growth and SP break down (data not shown).

**Surface tension measurement.** The surface tension of each chemical culture at the beginning of the experiment is shown in Table 1. There were no changes of more than 0.001 N $m^{-1}$ over the whole experimental period.

**TABLE 1. Surface tensions of cultures.**

| Culture | Surface tension, N $m^{-1}$ |
|---|---|
| BHB | 0.072 |
| NB | 0.062 |
| Bayticol® | 0.044 |
| Ecofleece® | 0.033 |
| Cypermethrin | 0.072 |

BHB = Bushnell Haas broth. NB = Nutrient broth

**Cell growth.** In the 'BHB + chemical' cultures, growth with Ecofleece® and Bayticol® followed the pattern shown in Figure 1. With cypermethrin the growth was continuous, reaching around 1.00E +8 after 120 h.

Growth in the 'NB + chemical' cultures was similar (within ± ≈20 %) at each point in each culture and similar to the growth in Figure 2 except at 15 °C where the growth was lowest after 48 h, reaching around 1.00E +10.

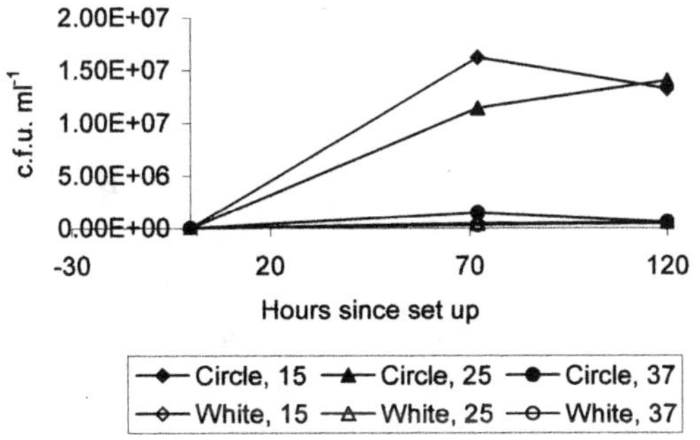


**FIGURE 1.Cell growth of 'Circle' and 'White' in BHB at three temperatures, °C.**

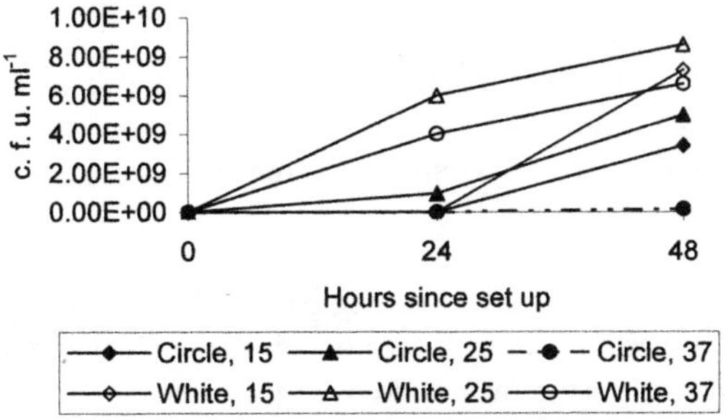


**FIGURE 2. Cell growth of 'Circle' and 'White' in NB at three temperatures, °C.**

**SP break down.** 5% error bars are shown on the SP break down figures. This is the amount of variation found with the SP measurement tool (Grant, 2000).

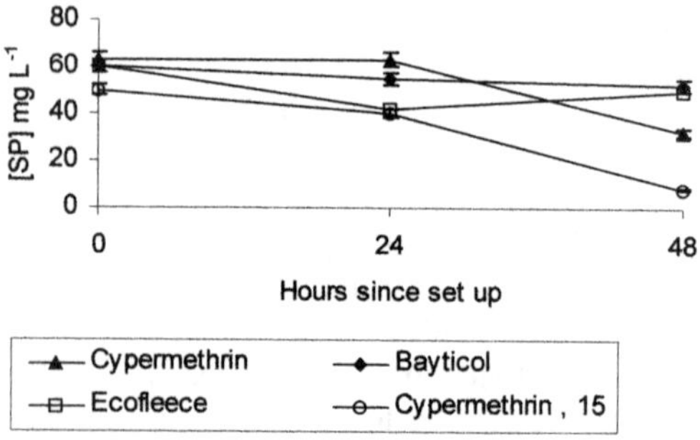


**FIGURE 3. Average natural break down of Bayticol, Ecofleece and cypermethrin in NB.**

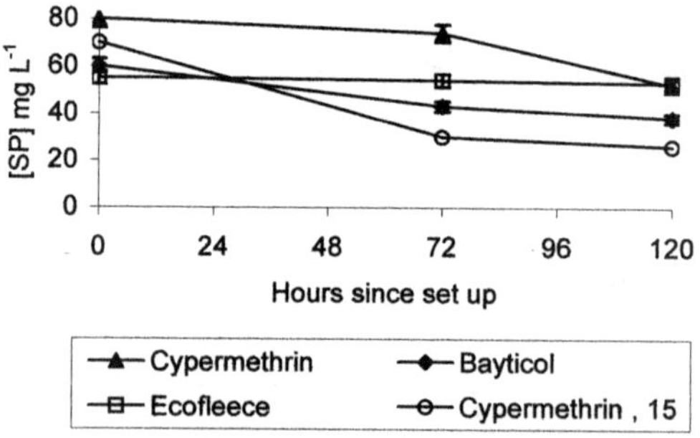


**FIGURE 4. Average natural break down of Bayticol, Ecofleece and cypermethrin in BHB.**

The combined bacteria cultures showed no significant differences (± ≈ 5 %) in cell growth or SP degradation to the single organism cultures at 25 °C. Cell growth of each organism was possible to measure due to the phenotypic differences in cell morphology.

**pH.** In the BHB cultures the starting pH of 7 only changed by no more than ± 0.1. In the NB cultures with Bayticol® and Ecofleece® the pH increased from 6.8 to 8, and with cypermethrin decreased to around 6. The small change in the BHB cultures is most likely due to the relatively low quantity of bacteria present. In the NB cultures with Bayticol® and Ecofleece® the pH increase is most likely because of the formulation break down. With NB + cypermethrin the pH decrease is likely to be due to the acidic products of the bacterial growth seen.

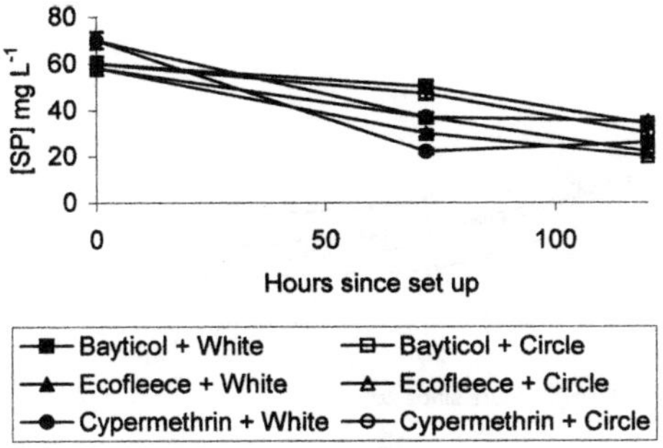


**FIGURE 5. Biodegradation of Bayticol, Ecofleece and cypermethrin in BHB at 15 °C.**

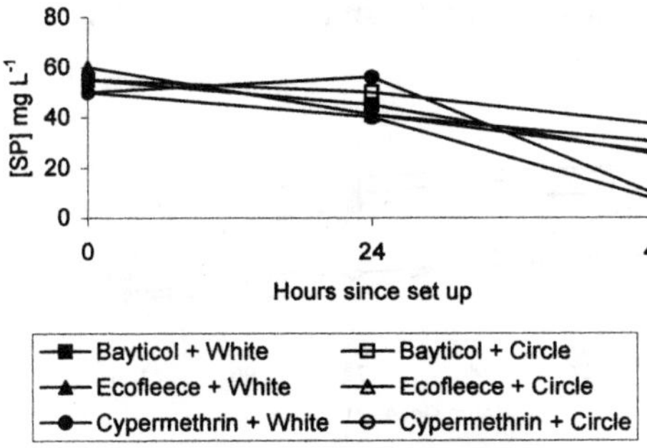


**FIGURE 6. Biodegradation of Bayticol, Ecofleece and cypermethrin in NB at 15 °C.**

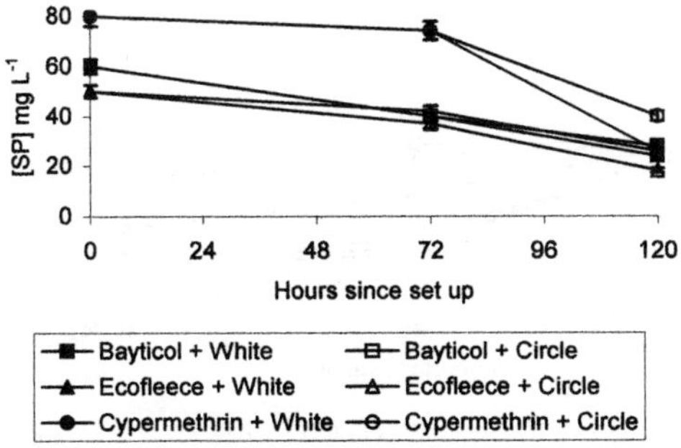


**FIGURE 7. Biodegradation of Bayticol, Ecofleece and cypermethrin in BHB at 25 °C.**

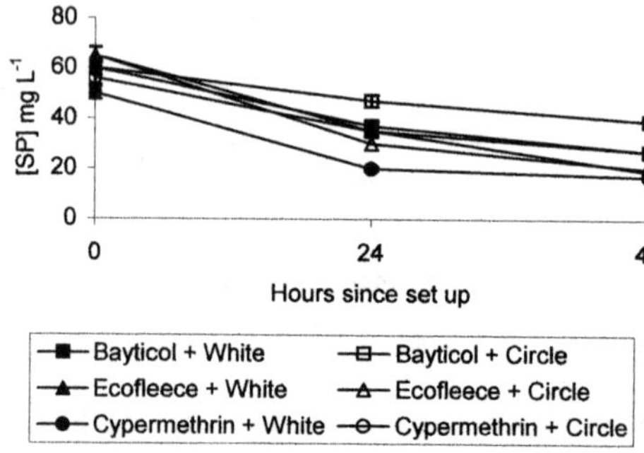


**FIGURE 8. Biodegradation of Bayticol, Ecofleece and cypermethrin in NB at 25 °C.**

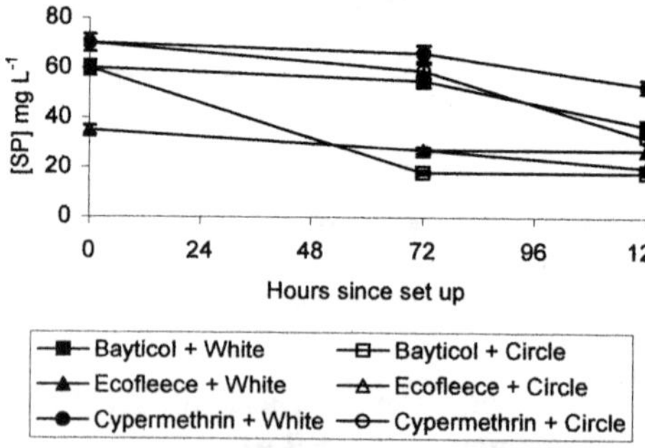


**FIGURE 9. Biodegradation of Bayticol, Ecofleece and cypermethrin in BHB at 37 °C.**

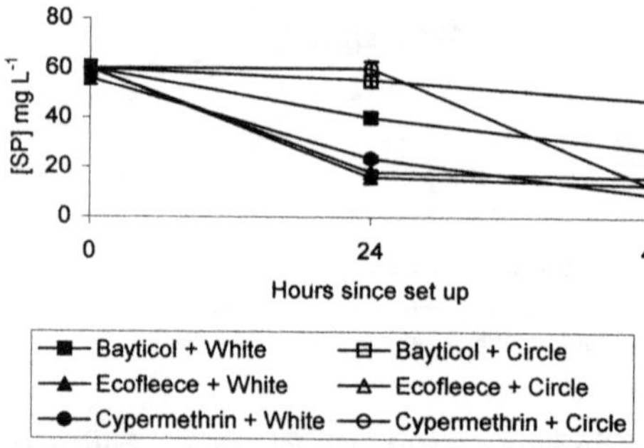


**FIGURE 10. Biodegradation of Bayticol, Ecofleece and cypermethrin in NB at 37 °C.**

All of the bacterial cultures degraded the SP above the rate of natural break down. The natural breakdown of SP at different temperatures only differed for the cypermethrin at 15 °C, shown in Figures 3 and 4. The most successful were the NB cultures, because although the BHB cultures proved equally as effective, this was over a greater time. The temperature appeared to play little part in the biodegradation, although with less growth at 15 °C the degradation per cell would be much greater.

The high growth in the NB cultures must be due to the bacteria using the more preferable substrates present, although the SPs were still degraded. This could be an indication to some sort of secondary metabolism of the bacteria were a secondary metabolite or product is degrading the SP, or a cometabolism system where the SP is broken down along with the more preferred substrates. To rule out uptake of SP by the bacteria other closely related species were tested, where uptake was not found to occur over the control culture break downs.

Degradation of the technical cypermethrin compound was greater than the SPs in the formulations. The sheep dip formulations may be making access to the SP difficult where as the technical compound is readily available.

This work has provided a basic study and provided some interesting data which will be looked at further. In particular the primary and secondary usage of the SPs with NB (at a lower concentration) will be examined to find out how the bacteria grow and degrade as the more preferable substrate runs out. The substrates will also be used to examine nutrition as a limiting factor, using SP and control cultures consisting of either water, water and sucrose, BHB, or BHB and sucrose.

## CONCLUSIONS

This piece of research demonstrates that a bacteriological approach to the break down of SPs is possible and that the availability of additional nutrients, which may be present in environmental contaminants, is important for the rate of breakdown providing that the temperature and possibly agitation are high enough.

## REFERENCES

Armstrong, A., and K. Phillips. 1998. "A Strategic Review of Sheep Dipping". *ADAS Report P170, UK Environment Agency*. Environment Agency, Station Road, Thirsk, York.

Extoxnet PIP. 1994. "Pyrethrins and Pyrethroids." http://ace.orst.edu/info/extoxnet/pips/pyrethri.htm. University of California at Davis, Environmental Toxicology Department.

Grant, R. J. 2000. "A Bioassay for the Measurement of Insecticide Concentration." Accepted for publication in *Arch. Environ. Contam. Toxicol. 2000*.

Grant, R. J., T. J. Daniell, and W. B. Betts. 2000. "Isolation and Identification of Synthetic Pyrethroid Degrading Bacteria." Submitted to *J. Appl. Microbiol. 2000.*

MAFF. 1999. "Government Announces Four point Plan on Organophosphates." MAFF News Release: 455/99, 20th December 1999. MAFF, Nobel House, 17 Smith Square, London.

Pearce, F. 1997. "Sheep Dips Poison River Life." *New Scientist. 2064* (153): 4.

Stephens, R., A. Spurgeon, I. A. Calvert, J. Beach, L. S. Levy, H. Berry, and J. M. Harrington. 1995. "Neuropsychological Effects of Long Term Exposure To Organophosphates in Sheep Dip." *Lancet*. (345): 1135-1139.

Tomlin, C. D. S. 1997. *The Pesticide Manual 11th ed.* British Crop Protection Council, London, UK.

Virtue, W. A., and J. W. Clayton. 1997. "Sheep Dip and Water Pollution." Sci. Total. Environ. (194): 207-217.

# BIOREMEDIATION OF HERBICIDES AND PESTICIDES (A Case Study)

***Paul Fikac*** (Jacobs Engineering Group, Houston, Texas)
Marty Fontenot (Syngenta Crop Protection Inc., St. Gabriel, Louisiana)
Bill Brown (Dunham Environmental Services, Houston, Texas)

**ABSTRACT:** An agrochemical production plant in Mexico was shutdown after 37 years of producing herbicides and pesticides. The site was contaminated from the surface disposal of contaminated process wastewater and land filling of product wastes and production residue. The goal was to clean up the site to meet the requirements of the Mexican Regulatory "Clean Company Certification". This certification was necessary for the sale or reuse of the site. The conduct of detailed solid waste segregation in conjunction with the application of known land farming techniques to bioremediate the herbicide/pesticide-contaminated soil resulted in achieving a clean closure of the site. The success of the bioremediation effort also eliminated the need for the construction of a lined treatment cell and the associated long-term maintenance and monitoring requirements. This project demonstrated that soils bioremediation technology could be used effectively for the clean up of agrochemical-contaminated sites when used in conjunction with detailed contaminated materials segregation and other remediation practices.

## INTRODUCTION

Site remediation is becoming increasingly important as the high cost of real estate necessitates the reuse of these industrial sites. The Mexico Procuraduria Federal de Proteccion al Ambiente (PROFEPA) requires the site receive a "Clean Company Certification" before it is reused. This paper discusses how soil bioremediation was used to clean up herbicide and pesticide contaminated soils at an Agrochemical Plant in San Luis Potosi, SLP, Mexico. Plant clean up was completed in late 2000 and was awarded a "Clean Company Certification". This was the first major remediation of its type in Mexico and received high visibility from the Regulatory Agencies.

**Objective.** The objective of the site remediation was to permanently treat pesticide/herbicide impacted soils to acceptable levels. Soils not treated to below the Preliminary Remediation Goals (PRGs) would be sent to the incinerator. The remaining soils with a Hazard Index of (Hazard Quotient) HQ>1.0 would be pretreated with nutrients and water and placed in a lined treatment cell for long-term anaerobic degradation. The treatment cell would be monitored and again re-evaluated after ten years.

**Site Description.** The 95-acre site is in the city of San Luis Potosi Industrial Zone in the San Luis Potosi Valley of the Mexican High Sierra Desert, at an elevation of 1800 meters and generally surrounded by volcanic mountains.

The soil to 2 meters below ground surface (bgs) is naturally fertile silty sediment of an old lakebed above a nonporous tuff of cemented sand, volcanic rock fragments, and talus outwash. Ground water is found at 100 meters below grade. The dry surface soil is normally crusted. When disturbed, it becomes a fine silty dust.

The annual evaporation rate is 79 inches per year with a rainfall average of 14 inches. A tuff layer is found at the depth of rainwater penetration. The highest rainfall is from May to September with little to no rain during February and March. Irrigation and other water uses are maintained with pumped groundwater or impounded storm water.

**Historical Background.** The site became impacted with herbicides and pesticides after formulation and packaging activities began in 1963 and continued until December 1998. Prior to 1986, process wastewater was disposed to a surface infiltration area. Solid wastes were disposed in an open pit landfill (landfill#1 Figure 1).

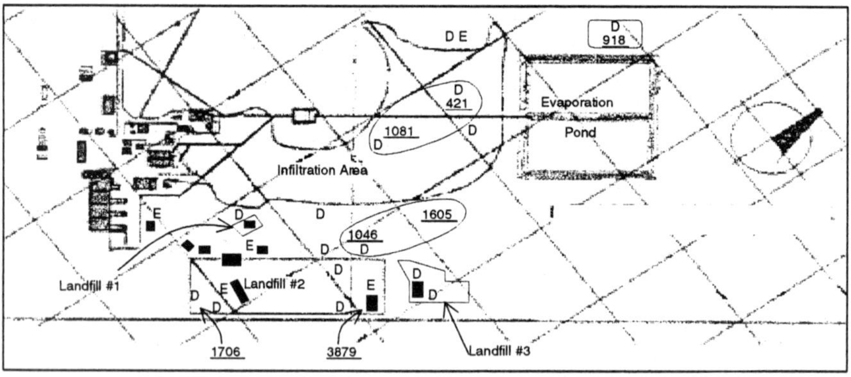


Legend: E - Endrin < PBG
D - Dicamba <PRG
Number - Atrazine >PRG
■ - Product/Packing Wastes

**FIGURE 1. General Agro Plant Layout, Pesticide / Herbicide Impacted Areas**

In 1986 an evaporation pond wastewater disposal system was constructed consisting of a plastic bottom liner overlain with one meter of compacted silty clay soil, and included a leakage monitoring system. The old wastewater infiltration area was excavated and backfilled with compacted clean soil from an onsite barrow area. This onsite barrow area, Landfill #2 (Figure 1), was lined with re-compacted tuff, backfilled with the excavated contaminated soil and capped with 0.8-meter of clean soil. Landfill #2 also received surface dumped construction demolition debris and other buried production wastes. A third pit, Landfill #3 (Figure 1), was also used for the burial of waste product and burned waste material-ashes.

A voluntary environmental audit was conducted in 1997 under the guidance of the Mexican PROFEPA. It was estimated that up to 11,000 tons of waste pesticide/herbicide product residue was disposed to the surface and into the landfills. The audit concluded that a site clean up was required.

## METHODS

A detailed Site Investigation and a Risk Assessment were conducted in 1998 to identify the Contaminants of Concern and establish the industrial level PRGs. The Mexican PROFEPA approved the PRGs in Table 1.

**Table 1. Preliminary Remediation Goals (PRG's)**

| Chemical | PRG (mg/kg) Industrial | Chemical | PRG (mg/kg) Industrial |
|---|---|---|---|
| Alpha-Bhc | 5.5 C | Dicamba | 27,000 |
| Atrazine | 110 C | Dithiocarbamates | 94C |
| Chlordane | 72 C | 2,4-D | 11,000 |
| Copper | 75,600 | Endrin Total | 321 |

C=Risk of $10^{-5}$, All other concentration values are based on the Hazard Index HQ <1.0

For multiple contaminants, a Hazard Index of HQ<1.0 for potential exposures would also have to be achieved.

$$HQ = (C_1/PRG_1 + C_2/PRG_2 +\text{---}+ C_N/PRG_N)$$
$$C = \text{contaminant concentration in the soil}$$

The chlorinated pesticides Endrin and Heptachlor can be treated by incineration, thermally desorbed, or biotreated anaerobically. Herbicides are generally biodegradable. Atrazine is easily biodegraded and has a half-life of 7 to 21 days. Dicamba and 2-4D have longer half-lifes, however, they had no values above their PRGs. The Risk Assessment identified the impacted areas and the recommended remediation. Figure 1 above shows the impacted areas and the type and level of contamination.

1. Insitu Thermal Desorption and incineration were used to clean up the Endrin contaminated soils in the physical plant area and are not included in this paper.
2. Insitu bioremediation was selected to treat the soils in the old wastewater infiltration area and adjacent areas contaminated with Atrazine, Dicamba, 2,4-D and Endrin. Since only Atrazine contamination was above the PRG, the soils would be biotreated to a level of HQ <1.0, inclusive of any Endrin present, and left in place.
3. Pretreatment with nutrients and water with placement in a lined treatment cell was recommended for the soil from the evaporation pond and Landfills #1, 2, & 3. Over 11,000 tons of known product wastes were disposed in these areas. This would allow for long-term anaerobic biodegradation of the disposed pesticides and herbicides. The treatment cell would be monitored for ten years and the soils reassessed for final disposition. Earlier investigative sampling in these areas found only two sample areas with Atrazine above its PRGs and several sample points with HQs>1.0.

## RESULTS AND DISCUSSION

The site remediation was scheduled in two phases.

**Phase 1.** This phase consisted in the thermal desorption, excavation and incineration of the impacted soils in the plant area. The soils in the wastewater infiltration and adjacent surface contaminated areas were bioremediated. A lined treatment cell was designed and planned for construction in the east corner of the site. Only the bioremediation of the infiltration area and the adjacent surface soils are discussed in this paper.

The old wastewater infiltration area and adjacent impacted areas (Figure 2) were cleared to prepare for treatment. Water was applied daily for one week to enhance natural bioremediation and to eliminate dust problems. The soil was aerated by plowing to a depth of 16 inches.

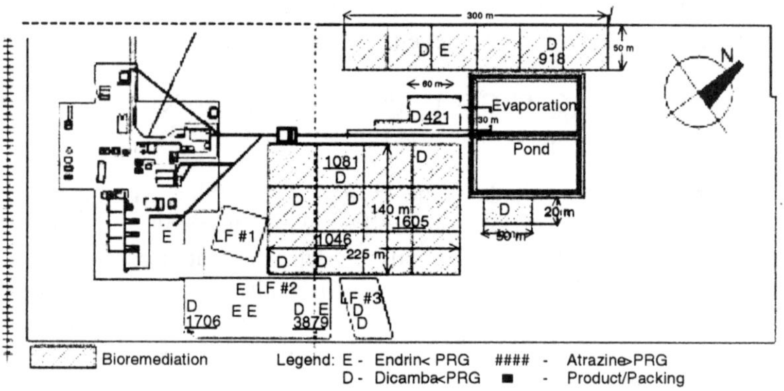


**FIGURE 2. Phase 1 Bioremediation**

The target contaminants for bioremediation were Atrazine, Dicamba, and 2,4-D. Atrazine, with a PRG of 110 kg/mg, had soil concentrations ranging from 420 to 1685 kg/mg. Dicamba and 2,4-D had soil concentrations below their PRGs of 27,000 mg/kg and 11,000 mg/kg, respectively. The naturally fertile desert soil needed no nutrients. The area was wetted with one inch of water every four days. The soil was plowed just before watering. After seven weeks, confirmatory soil samples were collected from the 50-meter square grid shown in Figure 2. Soils from three 50-meter square grids were composited to make one soil sample. The results were non-detect. Because of the possibility of dilution with composite sampling, the soils were sampled again with two random point samples collected from each 50-meter grid. The results were again non-detect. Four point samples were taken at points of the elevated Atrazine concentrations during the 1998 investigative sampling. These results were also non-detect. A partial "Clean Company Certification" was received for this biotreated area.

**Phase 2.** This phase originally required excavating the contaminated soils from the three landfills and the evaporation pond, preconditioning them with nutrients and water and placing the soils in the constructed treatment cell.

The Mexican PROFEPA did not permit construction of the lined treatment cell in the east corner of the site. Instead, an alternate site overlapping the existing Landfills # 1 and #2 was negotiated. This required the excavation and temporary storage of the Landfills #1 and #2 until the treatment cell could be constructed. This became an opportunity to use soils segregation and bioremediation to potentially reduce the quantity of soil to be placed into the treatment cell.

The 1998 investigative soils sampling program found the landfill soils had only Atrazine with known concentrations above its PRGs. Treating for Atrazine would reduce the quantity of soil going to the treatment cell. The Mexican PROFEPA approved eliminating the placement of landfill soils into the treatment cell if the soil Hazard Index was HQ <1.0.

A new landfill soils excavation and temporary storage work plan was prepared that segregated the landfill soils and materials into three categories.

1. Potentially clean soils consisting of the known clean soils used for landfill cover and the soils around the periphery of the landfills or not in direct contact with the buried wastes. These Category 1 soils would be stockpiled, sampled and analyzed to determine if they could be eliminated from placement in the treatment cell.
2. Potentially contaminated soils consisting of soils without any visible waste product but from areas of the landfills that had investigative sample analysis results below the pesticide PRGs or contained only herbicide contamination and a probable HQ >1.0. The Category 2 soils would be stockpiled, sampled, and analyzed to determine the nutrients required for preconditioning before placing into the treatment cell.
3. Known highly contaminated soils and materials consisting of soils containing visible waste product, pure waste product, waste product packaging and soils with pesticide contaminant analysis results significantly above the PRGs. These Category 3 soils and materials would be segregated, packaged, and shipped to an approved offsite incinerator.

Discussion of the new work plan determined that significant amounts of water would be needed for dust control during soil excavation and placement. Adding the nutrients to the Category 2 soils, when placing them in the temporary storage area, would save future time and work. Nutrients could be added using the 1998 investigative sampling results as the basis for nutrient addition. The soils would be placed in 16-inch lifts to allow for the addition of water and nutrients, and to mix the soils for later placement into the treatment cell. The tractor and plow could also be used to enhance bioremediation. The soils in the temporary soil storage area would be sampled after three weeks of biotreatment for possible elimination from the treatment cell. The target compound for the bioremediation was Atrazine. Eliminating the Atrazine would bring much of the soil HQ to<1.0. The fertilizer will also help to anaerobically degrade any Endrin after placement in the treatment cell. The nutrient calculations were based on a locally available balanced 17-17-17 fertilizer. A general application of 800 pounds per acre for a 20-inch soil thickness was recommended. The Atrazine hotspot soils in landfill

#2 required 3500 pounds per acre for phosphorous needs. The general fertilizer rate was applied to help degrade any residual herbicides.

The temporary soil storage areas were constructed in the Phase 1 bioremediated area, as shown in Figures 2 and 3. It was near the existing landfills and the irrigation header was in place. The existing evaporation pond was also used as a temporary soil storage area.

The landfill soils were excavated and placed in the temporary soil storage areas as shown in Figure 3, Ex Situ Bioremediation Areas.

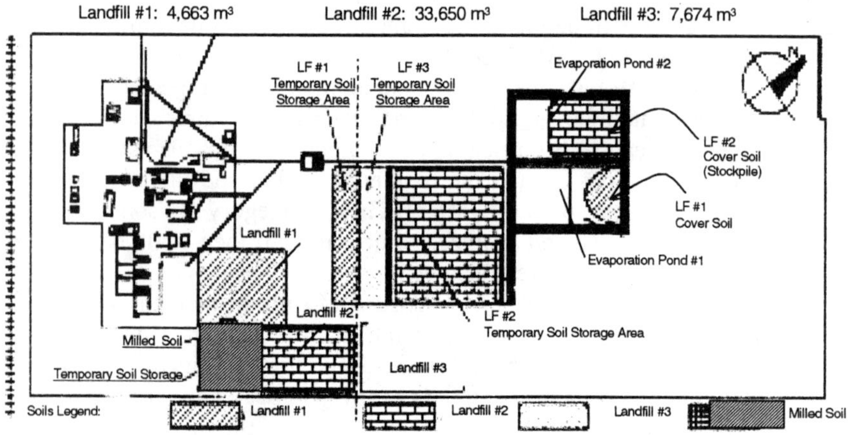


**FIGURE 3. Phase 2, Exsitu Bioremediation Areas, Excavated Landfill #1, 2 and 3 Soils**

The Category 1-landfill cover soils were stockpiled, sampled, and analyzed. The results all had an HQ<1.0 except for one sample from the Landfill #1 cover soil, which had Endrin at three times the PRG level. The sample area soil was spread and an apparent bag of pesticide residue was found and removed. Consequently, the entire stockpile of Landfill #1 Category 1 soil was spread to 1-foot thick to preclude any other bags of pesticide remaining in the soil. Fertilizer was added and the soils were land farmed for seven weeks. Follow up sampling and analysis found the Endrin to be below the PRG limits and the herbicides were none detect.

The residue sludge and clay liner in the evaporation ponds were also sampled. The analytical results indicted that most of the contaminants of concern were below the detection limits and that all soil samples had an HQ<<1.0

The Category 2 landfill soils were excavated and placed in the temporary soil storage areas in lifts of approximately 15 inches. The soils excavation was continuously monitored to assure that no visible waste product was included with the soils. All visible trash and packaging wastes were hand picked and packaged for incineration. Excavation of the Category 2 soils was terminated when visible waste products and significant packaging wastes were encountered. Fertilizer was applied to the placed Category 2 soils at a rate of 800-1000 pounds per acre surface area for each lift. The soils were land farmed with a minimum of two waterings and three tillings between succeeding lifts. The land farming activities continued for at least 10 days after conclusion of the category 2 soils excavation before sampling. Composite samples were taken from each 30-meter grid square to the depth of the placed soil. The analytical results found all the Category 2 soils had an HQ<1.0.

Deletion of the lined treatment cell was approved by the PROFEPA at the conclusion of the Category 2 soil treatment and analysis. At that time less than 8,000 cubic meters of the potentially contaminated Category 3 soils and waste remained to be processed. It would be more economical to package and incinerate any additional product waste or impacted soils that could not be treated below the HQ<1.0.

The Category 3 soils and waste product or packaging materials were manually or mechanically separated. All visible waste products were manually repackaged and shipped to incineration. Soils with mostly product packaging wastes and other mixed debris were milled and mechanically separated. The separated waste materials were sent to incineration. The separated Category 3 soils were placed to a thickness of 1-foot on the bottom of the Landfill #1 and #2 excavations. Fertilizer was applied at 3500 pounds per acre. The soil was land farmed with two waterings and three tillings per week for three weeks. Composite soil samples were collected on each approximate 25-meter square grid. The sample results had most soil with an HQ<1.0. However, some of the initial milled soil sample results had an HQ>1.0 so more fertilizer was applied at 3500 pounds per acre and land farming continued for three more weeks. Composite samples were then taken on a 12-meter square grid. These soil analysis results had an HQ<1.0 except for one sample area with Endrin above the PRG. The soils from that sample area were packaged and sent to incineration.

**Closure.** The now successfully treated landfill soils were used to backfill the empty landfills. The evaporation pond was demolished. The entire area was graded to its natural contours. The site was issued a "Clean Company Certification" by the PROFEPA.

**Sampling and Analysis.** All soil sampling collection and analysis followed the USEPA SW846 Standard Methods for Sampling and Analysis. Methods 8141 and 8151 were used for analyzing for the Herbicides. Method 8081 was used for analyzing for the Pesticides. QA/QC samples were collected and submitted in accordance with the SW846 standards. The Instituio Tecnologico Estudios Superiores de Monterrey (ITESM) laboratory in Monterrey Mexico conducted the sample analysis. The client QA samples were sent to a laboratory in the United States for comparative analysis.

## CONCLUSIONS

Initially, 55,000 $m^3$ of pesticide/herbicide-impacted soils were destined for placement in a lined treatment cell and long term monitoring and future assessment. Approximately 2,500 $m^3$ of highly impacted soils, product wastes and packaging waste was identified for incineration.

The final tally at site closure was no soils in a treatment cell and only approximately 1400 $m^3$ of impacted soils and wastes were sent to incineration. What happened to the 11,000 tons (8000 $m^3$) of product waste disposed on site? Natural degradation over time, enhanced bioremediation during clean-up and very stringent segregation of pesticide/herbicide-impacted soils and wastes resulted in a significant soil volume reduction.

The chlorinated pesticides were not expected to respond to any bioremediation, yet despite the observation that hundreds of drums of heptachlor liquid disappeared as rust destroyed the drums, significant heptachlor contamination was not observed. The pesticides released in the infiltration area were not found in the soil placed in the landfill. The pesticides and herbicides released to the evaporation pond were below PRGs and mostly non-detected. It is assumed that these areas received sufficient water and that the natural subsurface anaerobic biodegradation processes aided in destroying these compounds during the time they lie buried or immersed in water. The Endrin, not very soluble in water, was generally concentrated in the landfill waste product areas in plastic lined packaging and was repackaged and incinerated. The soluble herbicides, i.e., Atrazine and Dicamba were more dispersed in the soils but were successfully treated by enhanced bioremediation, thereby reducing the concentration levels to an HQ<1.0.

The success of this project capitalized on the past natural degradation of the buried pesticides, and the use of enhanced bioremediation to eliminate the treatable herbicides. An additional contributing factor to the success of the project was the close communications with the PROFEPA that allowed for critical changes in the soils treatment process. The PROFEPA rapidly approved requested changes and representatives were present during task initiation and sampling. These timely approvals to work plan changes resulted in the acceptance of bioremediation for the landfill soils and elimination of the treatment cell. This provided for closure of the site without a need for long term monitoring.

## REFERENCES

H.C. Finklea and M.M. Fontenot Jr. "Accelerated Bioremediation of Triazine Contaminated Soils: A Practical Case Study." Ciba-Giegy Corporation, St. Gabriel, Louisiana, Copyright – 1995 Soil Science Society of America, American Society of Agronomy, and Crop Science Society, 677 S. Segoe Rd., Madison, WI. 537111 USA. *Bioremediation: Science and Applications.* SSSA Special Publication 43

David W. Hill and Perry L. McCarty. 1996, Anaerobic *Degradation of selected chlorinated Hydrocarbon Pesticides. Paper presented at the 39th Annual Conference of the Water Pollution Control Federation, Kansas City, Missouri.*

# REMEDIATION OF SOIL PAH: COMPARISON OF BIOSTIMULATION AND BIOAUGMENTATION

**Catherine Nestler** (Applied Research Associates, Inc., Vicksburg, MS USA)
Lance D. Hansen (US Army Engineer Research and Development Center, Vicksburg, MS USA), David Ringelberg (Cold Regions Research Environmental Laboratory, Hanover, NH USA) and Jeffrey W. Talley, (US Army Engineer Research and Development Center, Vicksburg, MS USA),

**Abstract:** A pilot-scale, landfarm treatability study was designed to demonstrate improved bioremediation of PAHs using bioaugmentation in combination with bulking agents and slow release nitrogen (biostimulation). Three galvanized steel water troughs were used to contain the contaminated soil and the various amendments in the pilot facility. Pan 1 was left untreated. Pan 2 was amended with bulking agent and slow-release organic nutrient. Pan 3 was amended with the bulking agent, nutrients, and bioaugmented with a known biosurfactant producer, a *Pseudomonas* sp. All pans were mixed weekly using a mechanical rotary cultivator and soil moisture was maintained at 50-80% of the field moisture capacity (FMC). Initial PAH concentrations were 10,500, 7,200 and 7,300 mg/kg for Pan 1, 2, and 3, respectively. After 16 months, only 12% degradation was observed in Pan 1 (control). An initial lag was observed for Pan 2 and Pan 3. Following this, Pan 3 exhibited a higher degradation rate that was maintained through month 7. At 16 months, Pans 2 and 3 showed 86% and 87% degradation of total PAH, respectively. Contaminant reduction extended into the 4-ring homologues (fluoranthene, pyrene) and the BaP TEQ compounds (chrysene, benzo(a)anthracene). The results suggest landfarming is possible for sites with high concentrations of PAHs (>10,000 mg/kg) by using these alternative treatment strategies.

## INTRODUCTION

The POPILE, Inc. Superfund site is a former wood-treatment facility located in El Dorado, Arkansas. The primary contaminants found at the site include pentachlorophenol (PCP) and creosote compounds associated with wood treatment, including polycyclic aromatic hydrocarbons (PAHs). The process area soils had unusually high concentrations of PAHs (>10,000 mg/kg). Despite the high concentration of PAHs and PCP (>1200 mg/kg) in the soil, a low level of microbiological activity was maintained. During experimental landfarming treatment, the microbial biomass increased, corresponding to a decrease in the PAH concentrations (Hansen et al., 2000). The current study, also using the POPILE soil, focused on increasing the rate and the extent of the PAH biodegradation by employing biostimulation and bioaugmentation processes.

PAHs occur naturally as components of incompletely burned fossil fuels but they are also manufactured. A few of these are used in medicines, dyes, and pesticides, but most are found in coal tar, roofing tar, and creosote, a commonly used wood preservative. The persistence of PAHs in the environment, coupled with their hydrophobicity, gives them a high potential for bioaccumulation. PAHs are considered both mutagenic and carcinogenic. The different PAH homologues each have different toxic potencies that vary widely. Because PAHs generally occur in mixtures, toxic equivalency factors (TEFs) were proposed by the US EPA (1984), similar to those used in the risk assessment of mixtures of polychlorinated biphenyls (PCBs). The TEFs were modified by Nisbet and LaGoy (1992) to reflect updated toxicity information on the PAHs. Benzo(a)pyrene (BaP), whose concentration is commonly used to set clean-up goals, was retained as the reference carcinogen and given the value of 1.00. Sixteen other PAHs were ranked in comparison to BaP carcinogenicity. There are seven PAHs with the greatest environmental risk. These are benzo(a)anthracene, chrysene, benzo(b)fluoranthene, benzo(k)fluoranthene, benzo(a)pyrene, indeno(1,2,3-cd)pyrene, and dibenzo(a,h)anthracene. All of these compounds are found in the POPILE soil.

Traditional landfarming, as a form of applied bioremediation, is the cultivation of contaminated soil at properly engineered sites to stimulate the naturally occurring microorganisms to degrade organic contaminants (USACE 1996). The degradation takes place in two phases; an initial rapid phase that lasts weeks or months and involves the most soluble, and bioavailable, hydrocarbons, and a second phase of much slower degradation (years). The initial phase results in a significant reduction in total PAHs, but leaves the most recalcitrant and toxic homologues largely untouched. The second phase can remove some of these higher molecular weight PAHs (Brown et al. 1995, Pollard et al. 1994).

Contaminated soils are most often deficient in soil nutrients, such as nitrogen, high in carbon sources (usually the contaminant), and high in clay content-factors that inhibit contaminant degradation by the indigenous microbial populations. The chances for successful bioremediation are improved by conditioning the soil with the addition of bulking agents, nutrients and co-factors that also enhance the metabolic capabilities of the natural microbial communities. This process, called biostimulation is one method to improve PAH degradation (Pritchard et al., 1999).

Because they are hydrophobic and neutral in charge, PAHs are strongly adsorbed onto soil particles, especially clays (Luthy et al., 1997) and organic matter. The second method to improve PAH degradation is, therefore, to enhance their availability to the soil microbial communities. Surfactants increase the solubility of non-polar organic contaminants, such as the PAHs. The increased solubility significantly enhances PAH bioavailability and, therefore, PAH degradation rates (Deziel et al., 1996). There are bacteria, notably *Pseudomonas* sp., which produce their own surfactants. These biosurfactants desorb the PAHs, are non-toxic, biodegradable, non-competitive, and a self-sustaining resource. In addition, they have the advantage of "checks and balances", that is, surfactant isn't produced in greater quantities than the rate at which the PAHs can be degraded. Therefore, PAH availability will not greatly exceed its biodegradation rate, reducing the potential for contaminant losses through leaching.

Bioaugmentation involves the addition of specially adapted bacterial strains into the contaminated soil in order to enhance the initial degradation phase. When the added strains are known to be surfactant-producers, the opportunity exists to develop a reasonable and effective approach for predictably reducing concentrations of PAHs to below regulatory levels.

The objective of this project was to demonstrate improved bioremediation of PAHs using bioaugmentation plus biostimulation over biostimulation alone, and compare the efficiency and effectiveness of these processes in the removal of PAHs.

## MATERIALS AND METHODS

Three galvanized steel water troughs, 3 x 1 x 0.6 m (10'x3'x2'), seated on wooden pallets, were used to contain the contaminated soil and the amendments. The bottom of each pan was reinforced with aluminum sheeting. Three rotary cultivators were used to mix and aerate the soil and the amendments. The test soil was collected from the POPILE, Inc., wood treatment facility in El Dorado, Arkansas. The physical and chemical characterization was reported by Hansen et al. (2000). The soil was kept at 50-80% of the field moisture capacity (FMC). The ground rice hulls used as the bulking agent were purchased from Rice Hull Specialty Products, Inc., Stuttgart, AK. The rice hulls have a particle size of 30/80 and a bulk density of 19-24 lbs/ft$^3$. Slow-release nitrogen, in the form of pelletized, dried blood (12-0-0), was purchased from Dragon Corporation, Roanoke, VA. The nitrogen pellets have a bulk density of 8-10 lbs/ft$^3$. The carrier for the microorganisms was medium vermiculite, AF2, obtained from Strong-Lite Products; Corp. Bioaugmentation consisted of *Pseudomonas* strain #64 (a biosurfactant producer). The vermiculite-bacteria combination was produced according to Pritchard et al. (1999). Cell density at the time of application was $10^6$ cells/ml.

Pan 1, the control, was left untreated. Pan 2, demonstrating biostimulation alone, was amended with the bulking agent and slow-release nitrogen. Pan 3 was amended with the bulking agent and nutrients, and bioaugmented with the biosurfactant producer. The rice hulls were applied at 1:1 volume with the soil. The nitrogen was applied at a rate of 2% of the soil (w:w). The soil and amendments were thoroughly mixed, the soil moisture level checked, and water was added to bring it up to 80% of the FMC. The soil was mixed again to distribute the added water. The vermiculite-cultures for bioaugmentation were added to Pan 3, also at 2% w/w concentration with the soil, and the pan was thoroughly mixed. All the pans were sampled (T=0) then raked to fill the sampling holes.

For sampling, each pan was divided into five replicate collection zones. Each replicate sample was a composite of five randomly selected sub-samples. The sub-samples were collected as cores, combined in a single sample jar and manually homogenized. Analysis was performed as scheduled, and the remainder of each sample was archived at 4° C. The pans were tilled after each sampling event. Water was added before tilling the pans to maintain the moisture content of the soil.

Soil moisture was analyzed by oven drying at 105°C for 24 hrs. PAH, nitrogen, phosphate and total organic carbon analyses were performed by the Environmental Chemistry Branch of the ERDC on both treated and untreated soil. PAH concentrations were determined using SW846 EPA Method 8270c for GC/MS after extraction by Method 3540c. The nitrogen and phosphate analysis was performed using the Lachat 8000 Flow Injection Analyzer (FIA). The preparation methods were modified versions of EPA-600/4-79-020(1983 revision), 365.1 and 351.2, respectively. Soil pH was determined for a soil-distilled water slurry (1:1, wt/vol) using a Cole-Parmer® pH meter. Statistical significance was established for all data at a 95% confidence interval using n=5.

Microbiological characterization of the contaminated soil consisted of biomass and community composition analysis of the indigenous microbiota. Two grams (wet wt) of the soil sample were subjected to a modified Bligh-Dyer organic solvent extraction to quantitatively recover bacterial membrane lipid biomarkers (ester-linked phospholipid fatty acids, PLFA) as outlined in White and Ringelberg (1998). Biomass was estimated from the total concentration of membrane lipids and ester-linked PLFA (Balkwill, 1988). Similarities between single PLFA profiles were evaluated by application of a hierarchical cluster analysis. Correlation between PLFA and other study variables were assessed by Spearman rank order correlation statistics. Both cluster and correlation analyses were performed using the Statistica software package, v.5.0 (Statsoft Inc., Tulsa, OK).

## RESULTS AND DISCUSSION

The total soil PAH concentration at 16 months in Pans 1, 2 and 3 was 9,000, 900, and 900 mg/kg, respectively, down from an initial concentration of 10,500, 7,300 and 7,200 mg/kg for Pans 1, 2 and 3 (Figure 1). Using normalized concentrations to account for the addition of the bulking material, this corresponds to a reduction of 12%, 86%, and 87% for the control, biostimulation and bioaugmentation treatments, respectively. The final concentrations were not significantly different between Pans 2 and 3, but the difference was significant between Pan 1 and the others ($p=0.05$). An initial apparent increase in the concentration of soil PAH lasted 4 months in Pans 2 and 3, and 5 months in Pan 1. This time lag in degradation is usually attributed to acclimation of the bacteria. The PAH degradation from Month 4-16 in Pans 2 and 3 followed first-order decay curves with kinetic coefficients of 0.025 ($r^2 = 0.959$) and 0.032 ($r^2=0.934$), respectively. PAH concentrations were significantly different between Pan 2 and 3 for Months 4-8. The PAH half-life in Pan 2 is 7 months, in Pan 3 it is 6 months. No significant reduction in total PAH was observed in Pan 1 after 16 months of treatment. There was no reduction in any of the 4-ring compounds or the BaP TEQ compounds in Pan 1.

Table 1 shows Pan 3 (bioaugmentation) total PAH reduction over time. Similar reductions were observed in Pan 2. Blank spaces in the table indicate that reduction had not begun at that point. For example, there was no reduction in pyrene by 6 months. This was followed by a rapid decrease in concentration so that at 12 months pyrene was reduced by 82%.

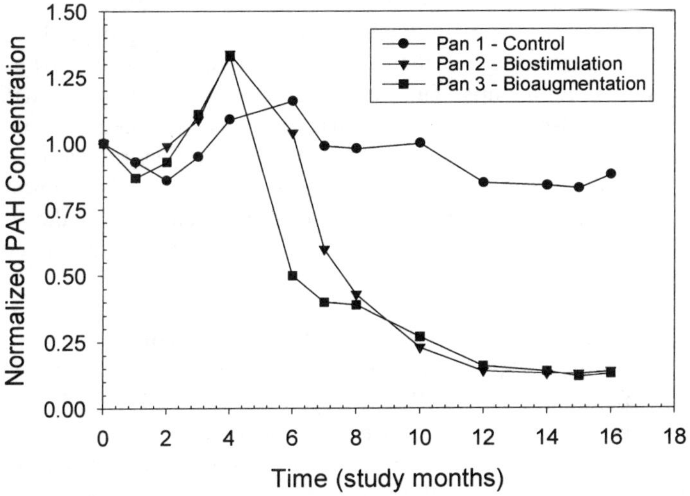


**FIGURE 1. The normalized PAH concentration after 16 months in control soil compared to soil being treated by biostimulation or bioaugmentation**

**TABLE 1. The reduction (%) from initial concentrations (mg/kg) of PAH homologues, including the BaP toxic equivalent compounds (indicated in bold), and total PAH at two midpoints and final analysis.**

| PAH Homologue | Initial Concentration | 6 months | 12 months | 16 months |
|---|---|---|---|---|
| Naphthalene (2-ring) | 14 | 64% | 100% | 100% |
| 2-methylnaphthalene (2-ring) | 33 | 78% | 82% | 100% |
| Acenaphthylene | 14 | 68% | 100% | 100% |
| Acenaphthene (3-ring) | 685 | 90% | 99% | 100% |
| Fluorene (3-ring) | 782 | 92% | 95% | 97% |
| Phenanthrene (3-ring) | 2172 | 88% | 95% | 97% |
| Anthracene (3-ring) | 941 | 25% | 60% | 72% |
| Fluoranthene (4-ring) | 1288 | 33% | 89% | 92% |
| Pyrene (4-ring) | 744 | | 82% | 84% |
| **Chrysene (4-ring)** | 190 | | **74%** | **78%** |
| **Benzo(a)anthracene (4-ring)** | 175 | **11%** | **82%** | **85%** |
| **Benzo(b)fluoranthene (5-ring)** | 61 | | | |
| **Benzo(k)fluoranthene (5-ring)** | 56 | | | |
| **Benzo(a)pyrene (5-ring)** | 46 | | | |
| **Indeno(1,2,3-c,d)pyrene(6-ring)** | 22 | | | |
| **Dibenzo(a,h) anthracene(5-ring)** | 4 | **100%** | | |
| Benzo(g,h,I)perylene (6-ring) | 12 | | | |
| **Total PAH** | 7239 | 50% | 84% | 87% |

The apparent PAH degradation pattern is lower molecular weight homologues degraded first, followed by consecutively larger molecules, with methyl substitution, as in 2-methylnaphthalene, also slowing degradation. These observations are supported by the work of Leblond et al. (2001) on PAH structure-activity relationships. The pattern appears to break down with pyrene and chrysene (both 4-ring compounds), which are not reduced in concentration until 7 months, several months after the beginning of degradation for several larger, higher molecular weight compounds. It was also observed that degradation into the next higher ring number didn't require that the lower ring compounds completely degrade. Work is in progress to attempt a correlation between structure and degradation.

The degradation of the BaP Toxic Equivalent homologues in Pan 3 is illustrated in Figure 2. Pan 2 showed a similar pattern, to a lesser extent, and there was no degradation of the BaP compounds in Pan 1. Each of the PAH homologues in the treated pans showed an apparent increase in concentration before degradation commenced. We believe this is a result of desorption by the biosurfactant produced by both indigenous and added populations.

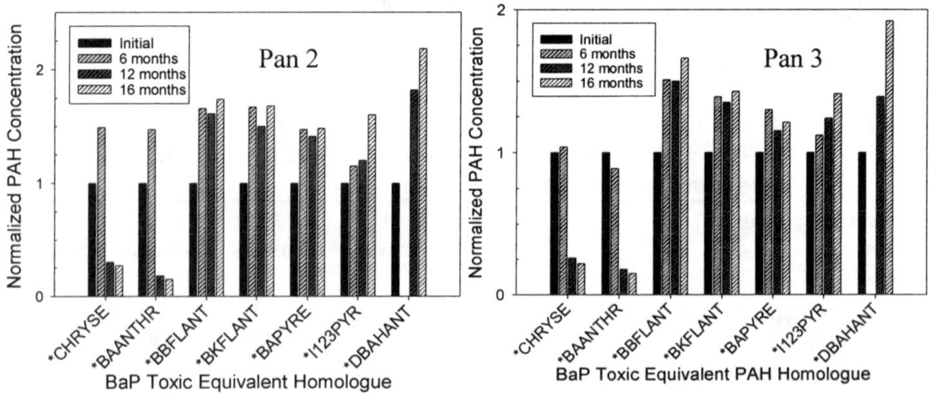


**FIGURE 2. Degradation of the BaP toxic equivalent PAHs over the 16-month study shown as normalized concentrations.**

Pan 1 showed distinct differences in community composition from Pans 2 and 3. In Pan 1, Gram-positive bacterial biomarkers average between 15-25% throughout the 10 months of the study, whereas in Pans 2 and 3 the relative abundance averaged 6%. The low relative abundance of Gram-positive bacterial biomarkers in Pans 2 and 3 is offset by increases in the relative abundances of Gram-negative bacterial biomarkers. The PLFA pattern detected in both of these pans strongly suggests the enhancement of a Gram-negative Pseudomonad community. This result is certainly far from being unexpected as: (1) the addition of nitrogen often preferentially stimulates the more opportunistic and less selective Gram-negative bacterial population and (2) *Pseudomonas* sp. are known

PAH and phenol degraders. A hierarchical cluster analysis of PLFA (arcsin transformed molar percentages) from Pans 1, 2 and 3 throughout the 10 months, showed that Pan 1 populations are distinct from Pans 2 and 3. The bacterial communities within Pans 2 and 3 appeared to evolve in a similar manner.

The biomass in the control pan remained relatively constant at 2.54 x $10^7$ cells/gm. The biomass in Pans 2 and 3 exhibited similar growth, increasing from 1.5 x $10^8$ to >3 x $10^8$ cells/gm with a decrease in both pans around 3 months. PCR analysis of both pans suggests that the decrease was due to selection of PAH-degrading and surfactant-producing populations (Perkins et al., 2001).

## CONCLUSIONS

Both biostimulation and bioaugmentation improved the rate of PAH degradation over traditional landfarming and natural attenuation. Bioaugmentation had an initial rapid phase of degradation (Months 4-8) that surpassed that of Pan 2 (biostimulation). Efforts should be made to determine how to sustain this phase. A re-application of the biosurfactant producer, and/or re-application of nitrogen are two parameters that require investigation. Bioaugmentation attempted with biostimulation shows promise in the remediation of very highly contaminated soils that until now have been considered unsuitable for landfarming.

## ACKNOWLEDGEMENTS

This work was supported by the Strategic Environmental Research and Development Program (SERDP), CU 720. The vermiculite-bacteria amendment was supplied by Drs. H. Pritchard and J. Jones-Meehan of the Naval Research Laboratory, Washington, DC. Technical assistance provided by Mr. Karl Konecny, and Mr. Samuel Tucker is gratefully acknowledged.

## REFERENCES

Balkwill, D.L., Leach, F.R., Wilson, J.T., McNabb, J.F., White, D.C. 1988. Equivalence of microbial biomass measures based on membrane lipid and cell wall components, adenosine triphosphate, and direct counts in subsurface aquifer sediments. Microbial. Ecol. 16, 73-84.

Brown, R.A., Hicks, P.M., Hicks, R.J., and Leahy M.C. (1995). Postremediation Bioremediation, p. 77-84. In R.E. Hinchee, J.T. Wilson, and D.C. Downey (eds.), *Intrinsic Bioremediation*, Battelle Press, Columbus, OH.

Deziel,E., Paquette,G., Villemur,R., Lepine,F., Bisaillon, J-G. 1996. Biosurfactant Production by a Soil *Pseudomonas* Strain Growing on Polycyclic Aromatic Hydrocarbons. Appl. Environ.Microbiol. 62, 1908-1912.

Hansen, L.D., Nestler, C., and Ringelberg, D. 2000. "Bioremediation of PAH/PCP Contaminated Soils from POPILE Wood Treatment Facility". In G.B. Wickramanayake, A.R. Gavaskar, J.T. Gibbs and J.L. Means (Eds.), *Proceedings of the Second International Conference on*

Leblond, J.D., Schultz, T.W., and Sayler, G.S. 2001. Observations on the preferential biodegradation of selected components of polyaromatic hydrocarbon mixtures. Chemosphere 42, 333-343.

Luthy,R.G., Aiken, G.R., Brusseau, M.L., Cunningham, S.D., Gschwend, P.M., Pignatello, J.J., Reinhard, M., Traina, S.J., Weber, W.J., Jr., Westall, J.C. 1997. Sequestration of hydrophobic organic contaminants by geosorbents, Environ. Sci. Technol. 31, 3341-3347.

Nisbet, C., LaGoy, P. 1992. Toxic equivalency factors (TEFs) for polycyclic aromatic hydrocarbons (PAHs). Reg. Toxicol. Pharmacol. 16, 290-300.

Perkins, E., Hansen, L.D., Nestler, C.C., Byrnes, J. 2001. Changes in abundance of in-situ aromatic degrading bacteria during a pilot scale landfarming of a polycyclic aromatic hydrocarbon contaminated soil, In, Proceedings of the Ninth International Symposium on Microbial Ecology, Amsterdam.

Pollard, S.J., Hrudey, S.E., and Fedorak, P.M. (1994). Bioremediation of petroleum- and creosote-contaminated soils: A review of constraints. Waste Mange. Res. 12,173-194.

Pritchard, P.H., J. Jones-Meehan, J.G. Mueller and W. Straube. 1999. "Bioremediation of high molecular PAHs; Application of techniques in bioaugmentation and bioavailability enhancement". In "Novel Approaches for Bioremediation of Organic Pollution", R. Fass, Y. Flashner, and S. Reuveny, (eds.), Kluwer Academic/Plenum Publishers, New York, pp. 157-169.

U.S. Army Corps of Engineers. 1996. Bioremediation using landfarming systems. Engineering and Design. ETL 1110-1-176. Washington, D.C.

U.S. EPA. 1984. Health effects assessment for polycyclic aromatic hydrocarbons (PAH). EPA/549-1-86-013. Cincinnati, OH.

White, D.C., Ringelberg, D.B. 1998. Signature lipid biomarker analysis. In: Burlage, R.S., Atlas, R., Stahl, D., Geesey, G., Sayler, G. (Eds.) Techniques in Microbial Ecology. Oxford University Press, Inc., New York, pp. 255-272.

# REMEDIATION OF SEDIMENTS CONTAMINATED WITH PAHs AND HEAVY METALS USING A SEQUENTIAL BIOSLURRY REACTOR

***Bumhan Bae*** (Kyungwon University, Sungnam, Kyunggi, Rep. of Korea)
Kyung-Suk Cho and Tae Young Kim (Ewha University, Seoul, Rep. of Korea)
Sung-Jae Lee and Kyoohong Park (ChoongAng University, Seoul, Rep. of Korea)
Youn-Kyoo Choung (Yonsei University, Seoul, Rep. of Korea)

**ABSTRACTS**: A lab-scale sequential slurry reactor was developed to remediate sediments contaminated with both polycyclic aromatic hydrocarbons (PAHs) and heavy metals. The sediment was first treated with an aerobic bioslurry reactor containing a PAH-degrading mixed culture and subsequently treated with a sulfur oxidizing bacterium, *Thiobacillus thiooxidans* AZ11 in a second bioleaching slurry reactor. The results showed that 82.6% of naphthalene and 92.6% of phenanthrene, in solid phase, were removed in 21 days. However, significant fractions of PAHs, 40.0% of naphthalene and 10.3% of phenanthrene were volatilized. No detectable levels of PAHs were found in the liquid phase after 3 days and the highest MPN counts occurred at Day 3 and decreased thereafter. These results suggest that biodegradation was limited by the dissolution of PAHs from sediments.

The optimum culture conditions for *T. thiooxidans* AZ11 were studied. With the decrease in the initial pH, the time to reach a steady state was shortened. The optimum sulfur/sediment ratio was determined to be $\geq$ 0.04 W/W, and solid content $\geq$ 30% was inhibitory due to the buffering capacity of the sediment. Under the optimum conditions, the PAH treated sediment was treated in a bioleaching slurry reactor. Results showed that 94% of 3500 mg/kg Zn, 70% of 420 mg/kg Cu, 54% of 105 mg/kg Pb, and 46% of 8.5mg/kg Cd were removed from the sediment in 8 days.

## INTRODUCTION

Polycyclic aromatic hydrocarbons (PAHs) and heavy metals are among the most frequently reported contaminants in sediment. PAHs are released into the environment through combustion of fossil fuel, oil spills, and natural processes (Francis, B.M., 1994). Industrial and domestic use of metals causes extensive release of heavy metals through abrasion, corrosion, and dissolution (U.S. EPA, 1997). The accumulation of these pollutants in sediments poses substantial risks both to human and aquatic ecosystems. Due to the highly toxic nature of these pollutants and persistence in the natural environment, many *in-situ* and off-site sediment remediation technologies have been developed. However, when these two types of pollutants co-exist, in the case of sediments of a storm water detention basin, a suitable remedial option is hard to find. In addition, dredged contaminated sediments are usually disposed in a landfill, which is very costly, and there is only a limited landfill space available in Korea.

A sequential bioslurry reactor, consisting of a conventional slurry reactor and a bioleaching reactor, was used to treat sediments of a storm water detention basin contaminated with both PAHs and heavy metals. A bioslurry reactor is one of the effective remedial options frequently used for the treatment of contaminated sediments (Hughes *et al.*, 1997). It has the potential to treat wide range of organics including pesticides, fuels, creosote, PCP, PCBs, and some halogenated volatile organics. However, the presence of elevated levels of heavy metals can be highly toxic to PAH degraders, and PAHs and other culture conditions may inhibit activity of a sulfur oxidizer used in the bioleaching process. These conditions should be identified for the successful operation of the sequential slurry reactor.

The objectives of this study were to test the feasibility of treating mixed contaminants using a sequential bioslurry reactor, to obtain optimum culture conditions for the best performance, and to estimate design parameters for a pilot-plant design. The final goal of the project is to recycle the contaminated sediment that is otherwise disposed in a landfill, as soil amendments.

## MATERIALS AND METHODS

**Slurry Reactors.** Two lab-scale slurry reactors (36L each) made of plexiglass (Figure 1) were used for the treatment of site sediments, and a 1.2 L scaled-down glass slurry reactor was employed to identify the PAH mass balance. Aeration, 2 liters per minute (Lpm) for the PAH degrading reactor and 36 Lpm for the bioleaching reactor, provided oxygen and mixing. Additional mixing was provided using an agitator with three impellers to prevent solids settling. The 1.2 L reactor was equipped with an additional XAD-2 resin column to collect volatilized PAHs. The XAD-2 resin in the column was regularly changed and the collected PAHs in the spent resin was extracted using dichloromethane.

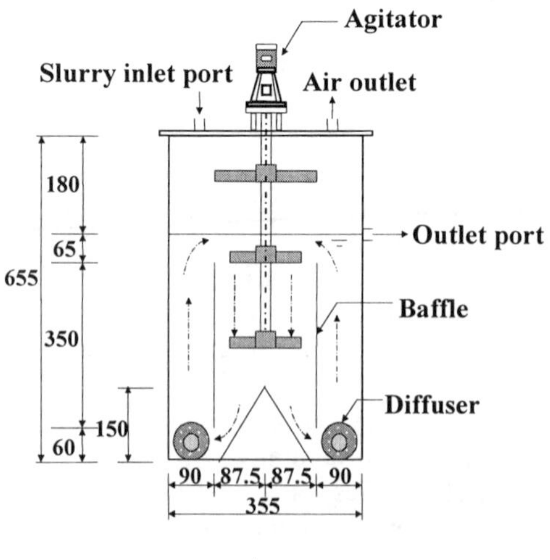


**FIGURE 1. Schematics of a laboratory scale slurry reactor (unit:mm)**

The reactor is similar to an air-lift reactor comprised of two risers and a down-flow chamber. Measurement of gas holdup of each riser showed that flows were equally divided between the two risers under various superficial gas velocities ($U_{Gr}$) induced by aeration (Figure 2a). The tracer study showed that mixing time was 4, 9, and 23 seconds under the condition of aeration with agitation, agitation only, and aeration only, respectively (Figure 2b). As a result, solid settling was not observed during the reactor operation.

$$\varepsilon_G = (H_d - H_L) / H_d$$

where, $\varepsilon_G$ = gas holdup, $H_d$=aerated liquid height, $H_L$=clear liquid height

For the PAH degrading reactor, a mixed PAH degrading microbial consortium isolated from soil was inoculated at the start of experiments. Minimal nutrients and 10mM phosphate buffer (pH 7.0) was provided every 3 days. For the bioleaching reactor, *T. thiooxidans* AZ11 was inoculated with 10g/L of sulfur. All of the reactors were operated at room temperature (25±5°C).

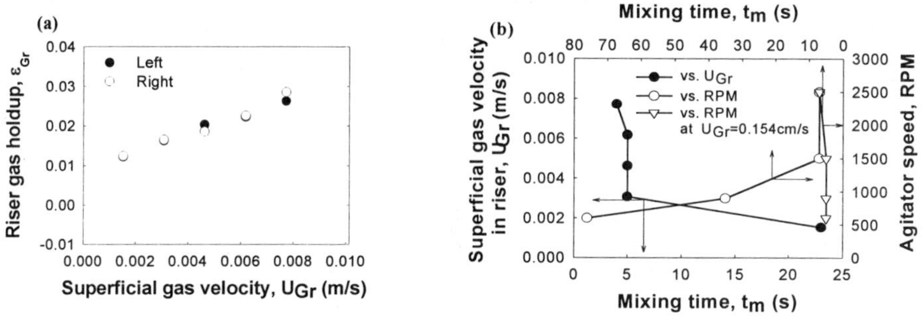


**FIGURE 2. Mixing characteristics of the bioslurry reactor used in the experiment. (a) gas holdup. (b) mixing time**

**Sediment Sample.** A sediment sample of a storm water detention pond at Incheon city, Korea, was collected using a vacuum dredger and was physically treated to remove particles larger than 177 μm using a hydrocyclone and flotation column. The physical and chemical characteristics of the sediment sample are summarized in Table 1. As the PAH concentrations were less than 10 mg/L, naphthalene (99%, Acros Co.) and phenanthrene (98+%, Acros Co.) dissolved in pentane were added to the sediment and left for 3 days before the test.

**TABLE 1. The characteristics of sediment treated in a sequential slurry reactor.**

| Contaminants | Concentration (mg/kg) | | |
|---|---|---|---|
| Naphthalene | 100.5[b](3.17[c]) | Particle size | <177 μm |
| Phenanthrene | 171.9[b](7.89[c]) | Total solids | 11.6 % |
| Zn | 3500.0 | Volatile solids | 9.54 % |
| Cu | 420.0 | COD | 7191 mg/L |
| Pb | 105.0 | Total nitrogen | 1.52 mg/L |
| Cd | 8.5 | Total phosphorous | 5.25 mg/L |
| TPHs[a] | 542.6 | | |

[a] Total petroleum hydrocarbons; [b] After PAHs amendment; [c] Before PAHs amendment

**Analysis**. Reactor conditions including temperature, pH, solid contents, dissolved oxygen, and COD were monitored every 3 days. PAHs in sediment and TPH were extracted, concentrated, and cleaned for GC-FID (Finnigan 9001) analysis following the method in SW-846 (U.S. EPA, 1996). Naphthalene-$d_8$ (98+%, Acros Co.) and phenanthrene-$d_{10}$ (99%, Aldrich Co.) were used as internal standards. A capillary DB-5 (J&W Scientific, Inc.) was used for the analysis of PAH and TPH. Method detection limits for naphthalene and phenanthrene were 0.803 mg/Kg and 0.688 mg/Kg respectively. A 96-well plate MPN method was used to monitor PAH-degrading microorganisms (Venosa *et al*., 1996). The concentration of heavy metals was analyzed by an atomic-absorption spectrometer (AA100, Perkin Elmer, USA) after digestion with 37% HCl and 65% $HNO_3$.

## RESULTS AND DISCUSSION

PAHs in the solid phase decreased rapidly until Day 12. However, much of the PAH removal was due to volatilization as shown in Figures 3a and b. Park *et al*. (1990) reported more than 20% of 2-3 ring PAHs volatilized during the initial phase of a slurry reactor. Considering the short 3-day adsorption period, the absence of aging effects might be another reason for the high volatilization (Alexander, 1999).

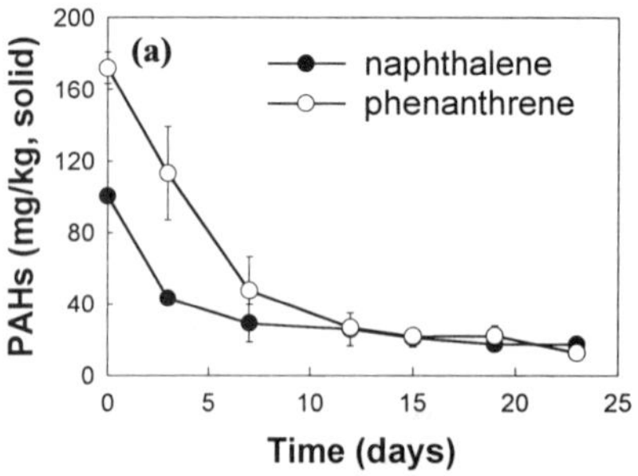


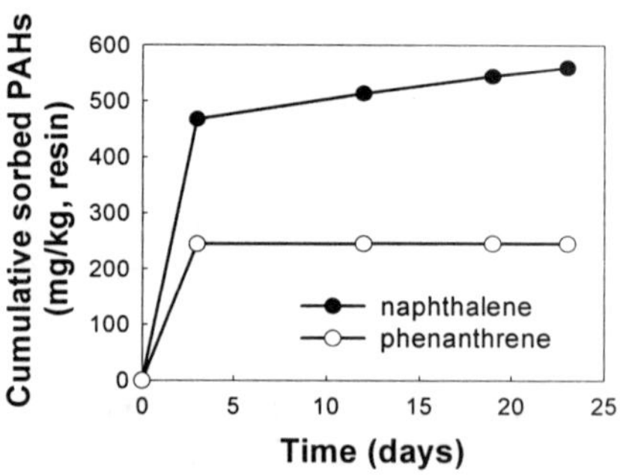


**FIGURE 3. Time profiles of PAHs in the solid phase and in the XAD-2 resin column. (a) PAHs in sediment. (b) Cumulative PAHs in XAD-2 resin.**

The aqueous phase PAH concentrations were detected only in the initial phase. The MPN counts of PAH-degrading microorganisms decreased with 3 days of lag as the PAH concentration in the aqueous phase decreased (Figure 4). The high MPN counts ($10^{11}$-$10^{12}$) imply the PAH degraders were still active. However, it seems that further growth of the PAH degraders was limited by the low aqueous PAH concentrations.

Mass balance of two PAHs is presented in Table 2. In the calculation of mass balance, all of the disappeared PAHs were assumed to be biodegraded as we did not use $^{14}C$ radio-labeled compounds. The result showed that biodegradation of less volatile phenanthrene is two times higher than that of naphthalene. In addition, 46.5% of TPH in solid phase was removed in 11 days.

**TABLE 2. The fate of naphthalene and phenanthrene in the bioslurry reactor.**

| Compound | Volatilized | In Solids | Biodegraded |
|---|---|---|---|
| Naphthalene (%) | 40.0 | 17.4 | 42.6 |
| Phenanthrene (%) | 10.3 | 7.4 | 82.3 |

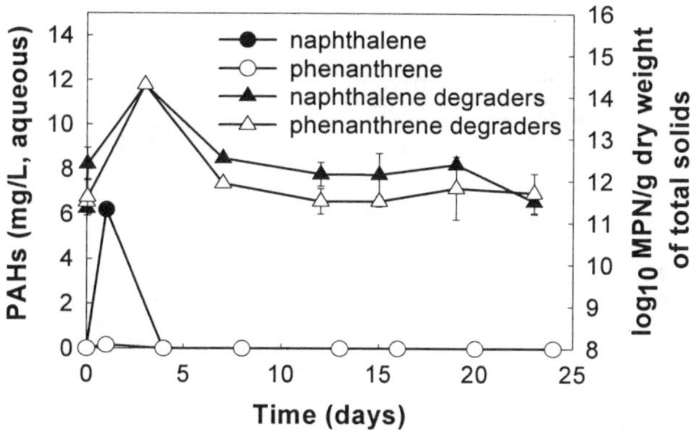


**FIGURE 4. Concentration profiles of aqueous phase PAHs and PAH degrading microbial MPN counts in the slurry reactor.**

After the PAHs were treated, the sediment was transferred to a bioleaching reactor for the treatment of heavy metals. In many studys, *T. ferrooxidans* has been widely used for heavy metal removal from sludges, however, *T. ferrooxidans* has been shown to be effective at low solids concentration ranges of 1.3-4.0% due to the inhibitory effects of organics in solids (Ryu *et al.*, 1998). As the sediment contains higher than 10% of solids. *T. ferrooxidans* was not suitable. To overcome the limitation, a sulfur oxidizing species, *T. thidoxidans* AZ11 that was isolated from soils and can tolerate higher concentrations of organics and other inhibitory substances, was used in this study. The electron micrograph of the *T. thidoxidans* AZ11 was shown in Figure 5.

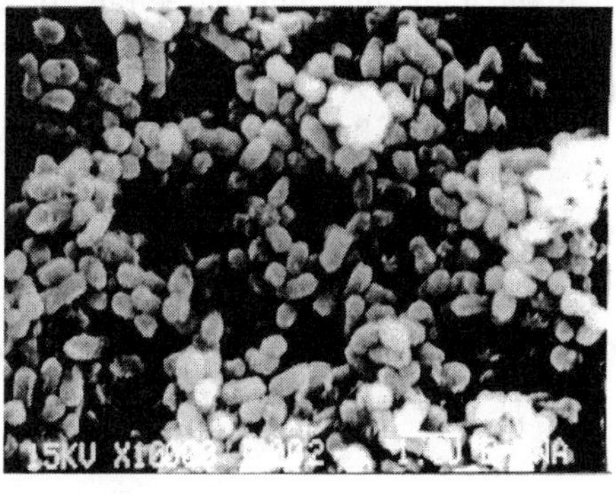

**FIGURE 5. Scanning electron micrograph of *T. thiooxidans* AZ11.**

The bioreaching reactor was operated with the inoculation of *T. thidoxidans* AZ11 and the addition of elemental sulfur 10 g/L. After 2 days of lag time, the liquid pH, which is an indicator of sulfur oxidation by *T. Thidoxidans* AZ11, started to decrease and reached to as low as 2.0 g/L at Day 10 with the corresponding increase in ORP up to 470 mV at Day 10 as shown in Figure 6a. No significant inhibition due to the sediment was not observed.

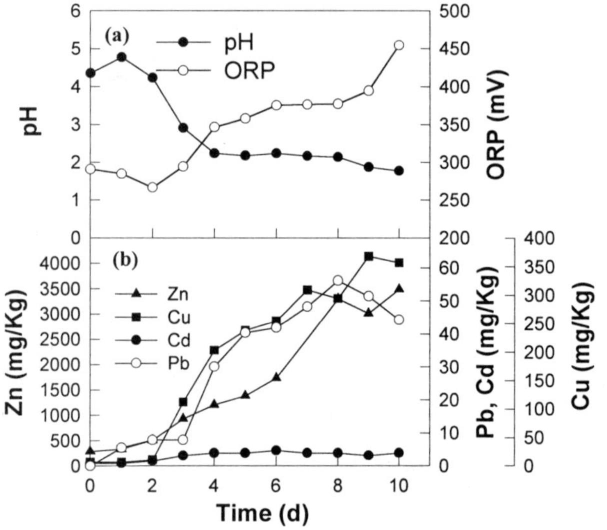


**FIGURE 6. Time profiles of pH, ORP and heavy metal concentrations in the aqueous phase during the bioleaching process. (a) pH and ORP. (b) heavy metal concentrations in the aqueous phase.**

The changes in the liquid phase concentration of heavy metals leached from the sediment solids are presented in Figure 6b. The leaching of heavy metal started as the decrease in the reactor pH (after 2 days of inoculation). The heavy metal concentrations in the liquid phase increased rapidly with the decrease in the reactor pH except for Cd. Cadmium was the least leachable compound among the heavy metals. Results showed that 94% of 3500 mg/kg Zn, 70% of 420 mg/kg Cu, 54% of 105 mg/kg Pb, and 46% of 8.5mg/kg Cd were removed in 8 days.

## ACKNOWLEDGEMENTS

This research was conducted as a part of G7 project (1999) funded by the National Institute of Environmental Research, the Republic of Korea.

## REFERENCES

Alexander, M. 1999. Biodegradation and Bioremediation, 2nd Edition. Academic Press, San Diego.

Francis, B.M. 1994. *Toxic Substances in the Environment*, John Wiley & Sons, Inc. New York.

Hughes, J.B., Beckles, D.M., Chandra, S.D., and Ward, C.H. 1997. "Utilization of Bioremediation Processes for the Treatment of PAH-contaminated Sediments." *J. of Ind. Microbiol. & Biotechnol.* 18:152-160.

Park, K.S., Sims, R.C., Dupont, R.R., Doncette, W.J., and Matthews, J.E. 1999. "Fate of PAH Compounds in Two Soil Types: Influence of Volatilization, Abiotic Loss, and Biological Activity." *Environ. Toxicol. Chem.*, 9:187-195.

Ryu, H. W., Kim, Y. J., Cho, K. S., Kang, K. S., and Choi, H. M. 1998. "Effect of Sludge Concentration on Removal of Heavy Metals from Digested Sludge by *Thiobacillus ferrooxidans.*" *Kor. J. Biotechnol. Bioeng.* 13:279-283.

Sreekrishnan, T. R. and Tyagi, R. D. 1996. "A Comparative Study of the Cost of Leaching out Heavy Metal from Sewage Sludges." *Proc. Biochem.*, 31:31-41.

U.S. EPA. 1996. *Test Methods for Evaluating Solid Wastes,* 3rd Edition.

U.S. EPA. 1997. *The Incidence and Severity of Sediments Contamination In Surface Waters of the United States*. EPA 823-R-97-006.

Venosa, A.D., Suidan, M.T., Strohmeier, K.L., Haines, J.R., Eberhart, B.L., King, D., and Holder, E. 1996. "Bioremediation of an Experimental Oil Spill on the Shoreline of Delaware Bay." *Enviorn. Sci. Technol.* 30:1764-1775.

# SURFACE TREATMENT OF POLLUTED SEDIMENT IN AN ENERGY PLANTATION

***Hans Breteler***, Rik Duijn, Peter Goedbloed
(Oostwaardhoeve Co., Slootdorp, the Netherlands)
Joop Harmsen (Alterra, Wageningen-UR, Wageningen, the Netherlands)

**ABSTRACT:** Soils and sediments polluted with mineral oil and PAHs can be bioremediated by landfarming. The landfarming procedure may be used to convert polluted sediment into suitable building material. Landfarming, however, requires many years and a large area of land. Because land is very expensive in the Netherlands, a combination of bioremediation with another profitable use of the land becomes mandatory. The prospect of landfarming, combined with the cultivation of an energy crop (willow), is investigated commercially and technically on an experimental farm by Oostwaardhoeve Co., which has been in business for 5 years, on approximately 15 ha. This paper discusses the state of the art of bioremediation, crop performance, required techniques, and the risks for the environment. In combination with energy crops and with the installation of a power station, this type of bioremediation becomes attached to spatial planning and landscaping. The prospects for the Oostwaardhoeve concept are discussed.

## INTRODUCTION

Soils and sediments polluted with mineral oil and PAHs (polycyclic aromatic hydrocarbons) can be bioremediated by landfarming. To achieve contaminant concentrations that allow a limited or multifunctional reuse of the cleaned soil, the landfarming procedure can be adopted. However, to yield sufficient results, this procedure requires many years. In a densely-populated country like the Netherlands, the use of land is expensive and a combination of bioremediation with another profitable use of the land helps to reduce costs and to make better use of this limited resource (Harmsen *et al.*, 1999).

The prospect of landfarming, combined with the cultivation of an energy crop, is investigated on the Oostwaardhoeve, an experimental farm in the Northwestern part of the Netherlands. The energy crop (willow) yields biomass fuel, which is to converted into power and heat in a power plant by combustion or gasification. The Oostwaardhoeve Co. is a company in which a contractor of soil and sediment works and Wageningen University and Research Centre co-operate to combine practical experience and research and development (R&D), respectively. Oostwaardhoeve Co. is involved in field-scale research with the objective to implement the results from research projects on a smaller scale or from the laboratory to full-scale applications. The Oostwaardhoeve Co. has been in business since 1996 and has approximately 15 ha in use to clean polluted sediments in combination with a willow plantation.

**Objective.** This paper discusses the state of the art of landfarming (as investigated on the Oostwaardhoeve) and examines the prospects and the future development of this concept.

## STATE OF THE ART

The activities of Oostwaardhoeve Co. are strictly regulated by Dutch environmental legislation, which requires regular checks of soils, sediments and water, the very careful transport and application of the polluted sediments, and thorough bookkeeping. A permit was granted to landfarm sediments with a PAH content of at least 40 mg/kg dry sediment, and/or a mineral oil content of at least 100-500 mg/kg dry material, depending on sediment composition. Also, the permit sets a limit to the total amount of sediment that may be accepted annually. To date approximately 15 ha have been treated with a total volume of approximately 100,000 $m^3$ of sediment (dredged sludge).

On this scale, the degradation of polluted sediments initially occurs erratically, due to the heterogeneity in the dehydration of the sediments, heterogeneity in physical and chemical characteristics of the sediments, and heterogeneity inherent in the process of applying the sediments on the field. The logistics from dredging to application of sediment on the field is an important problem. The sediment has been brought to the Oostwaardhoeve by truck, and placed in a transitional depository to temporarily store some of the sediments.

The proper application of sediments on a hectare scale requires a thorough macro-filtration of the sediment, so that objects that may be encountered in the sediments removed from the bottom of waterways, like bicycles, safes, car wheels, *etc.* are removed before the sediment is transported to the landfarm. Application of sediment in an even layer on the land requires experience and technical skill to tune the pumping equipment to the acceptable variation of quality and quantity of the applied sediment. At this site, it appeared easier to apply sediments to an existing willow crop than to establish a new crop on a layer of fresh sediment. On the other hand, the existing crop presented a barrier towards the even distribution of sediment. On the average, approximately 1.0 m of sediment was applied on the land, eventually leaving approximately 0.5 m of matured sediment. With this concept, 10 to 20 years of landfarming are available to clean the sediments to the extent that the remaining material fulfils the requirements for reuse.

In 1998 the first experiments on a field scale were started. Results of PAH and mineral oil degradation (Table 1) confirm that several years or even decades may be necessary to clean the sediment to be eventually suitable for agricultural use (Harmsen *et al.*, 1997). Legislation in the Netherlands has given an opening for earlier application of the cleaned sediment, if the produced 'clean' soil is to be used as a building material under controlled circumstances (dikes, road construction, *etc.*). The concentration limits for building material are 40 (PAHs) and 500 (mineral oil) mg/kg dry material (corrected to standard soil with 10% organic matter). Depending on the degradation rate, the final disposition of the applied sediments may differ for the Oostwaardhoeve:

- The developed soil stays on the Oostwaardhoeve to be reused for agricultural purposes.
- The developed soil will be removed and used elsewhere as building material.

**TABLE 1. Contents of PAH and mineral oil (mg/kg dry sediment, PAH sum of 10 National List) on experimental fields of the Oostwaardhoeve. Experiments started 1-2 y before November 2000.**

| | *PAH* | | *Mineral oil* | |
|---|---|---|---|---|
| **Sediment (source)** | **Start** | **November 2000** | **Start** | **November 2000** |
| Callantsoog | 40 | 38 | 80 | 80 |
| Biograp | 8 | 8 | 1180 | 320 |
| Amstelveen | 5.5 | 2.6 | 1840 | 1070 |
| Wieringerwaard | 20 | 14 | 200 | 200 |
| Schiphol | 8 | 4 | 1330 | 500 |
| Sassenheim | 10 | 7 | 4000 | 770 |
| Nobis | 53 | 50 | 860 | 490 |

The initial limiting factor for the degradation of pollutants in sediments is the availability of oxygen. In the neighbourhood of the willows, the sediments were completely aerobic. However, tree roots were not yet present in the driving tracks on the fields, where oxygen is present in large cracks, but not in individual soil aggregates.

The growth and yield of willow trees are stimulated by the application of fresh sediment. Yield levels usually approximate 10 tons of dry matter/ha/y. Exceptions occur both on the low (<5) and the high (>12 tons DM/ha/y) side. The stimulation is probably mainly caused by the extra supply of water, and possibly also by nutrients and organic matter present in the sediment.

The most commonly used willow (*Salix viminalis* L.) clones on the Oostwaardhoeve, Orm and Rap, also perform well when sediments are applied. Little information, however, is available on the most suitable genotypes for biomass production combined with landfarming. Nor is there information about the effect of the choice of the genotype of willow, or of other crops, on the bio-remediation process. Due to the non-polar nature of the prevailing contaminants (mineral oil, PAHs), their concentration in the aqueous soil solution is low. Therefore, accumulation in the crop and environmental problems caused by the eventual conversion of the wood in an energy plant are not likely. Similarly, contamination of the subsoil and of groundwater by these components may be excluded. Recently, pot experiments have been started to learn whether land-farming of salt-containing polluted sediments, combined with an energy crop, is feasible. In the Netherlands, high volumes of salty sediments, usually polluted, are available.

The willow cultivation on sediments requires new techniques, *e.g.* for planting, crop protection, fertilisation, and harvest. Over the last 5 years, new equipment was used and sometimes conventional machinery had to be adapted. One of the main problems was the low carrying capacity of freshly-deposited sediments. The use of caterpillar tracks on 'dry land' machinery and the construction of earthen elevations around the willow field, from which the crop could be treated, are possible solutions. To keep a positive economic balance as well as a favourable energy balance, the amount of energy to be used in the passive concept of landfarming has to be limited.

Between various lots of sediments, major differences occur in the rate of water loss and, consequently, the start of crack formation and the cracking pattern. Water loss, aeration around the cracks, and the establishment of an active micro-flora in the soil and in the rhizosphere of the energy crop, may affect the crop yield and the biodegradation of the pollutants in the sediments. At Oostwaardhoeve, research is carried out to select the best combination of sediment characteristics and crop genotype to optimise biodegradation and biomass yield. In the spring of 2000 twenty different willow genotypes were planted in the framework of such an investigation. Another agronomic aspect that varies widely between various lots of sediments is the infestation with weed seeds. Moreover, weed control on wet sediments requires the design of specific techniques and equipment.

Although the response in the public media, and environmental and agricultural circles is usually positive, not all inhabitants of the area around the Oostwaardhoeve are supporting our trials. Despite various types of information, explanation and invitations to visit the experimental field, Oostwaardhoeve Co. has incurred various law suits, mainly based on allegations about the risks for people and for the environment. Not in my backyard! To extend our activities even more emphasis shall be placed on communication and information.

## PROSPECTS AND FUTURE DEVELOPMENT

Land, including farmland, is scarce and expensive in the Netherlands. The production of biomass from energy crops in the Netherlands is too expensive to compete with other biomass fuels, because the costs of the use of land contribute for approximately 50% to the crop production costs. However, due to the income associated with the acceptance of polluted sediments, biomass fuel at the Oostwaardhoeve is produced at a much lower price. Despite the green tax system and the ambition of the Dutch and the EU governments to increase the use of sustainable energy, with emphasis on biomass, there is not much biomass-based power generation in the Netherlands. Consequently, the price for premium biomass fuel, derived from energy crops, is low and unstable. Therefore, a power plant to produce green heat, power and $CO_2$ is foreseen at the premises of the Oostwaardhoeve. This power plant will also save the costs and energy consumption for long distance transport from the plantation to the power plant.

Moreover, the whole cycle from the start of an energy plantation to the generation of energy can be demonstrated at the Oostwaardhoeve, combined with the cycle from polluted sediment to useful building material.

The current permit of the Oostwaardhoeve allows 35 ha of energy crops, combined with the bioremediation of sediments. By its size, and in combination with power generation and economic activities that use the energy on site, our future bioremediation efforts become attached to spatial planning and landscaping (Figure 1).

Finally, the landscaping of the Oostwaardhoeve will stimulate the conservation of nature values (water, land, crops, fauna, biodiversity) and enable the recreational use of the facility.

**FIGURE 1. Artist impression of the Oostwaardhoeve in 2010 with energy crops, landfarming, an energy plant, greenhouses and a traffic canal.**

## REFERENCES

Harmsen, J., H.J.J. Wieggers, J.J.H. Van de Akker, O.M. Van Dijk-Hooyer, A. Van den Toorn and A.J. Zweers, 1997. "Intensive and extensive treatment of dredged sediments on landfarms". *Proceedings In Situ and On-Site Bioremediation*: Volume 2, 153-158. Battelle Press.

Harmsen, J., A. Van den Toorn, D. Boels, B. Vermeulen, W. Ma, J. Van der Waarde, R. Duijn & R. Kampf, 1999. "Growing of biomass to stimulate bioremediation in technical and economical perspective". In: A.Leeson and B.C. Alleman (eds), *Phytoremediation and Innovative Strategies for Specialized Remedial Applications*, 39-44. Battelle Press.

# RAPESEED OIL FOR EXTRACTION OF LIPOPHILIC POLLUTANTS FROM SOIL PARTICLES.

***Jessica H. Schuur*** and Bo Mattiasson
(Dept. of Biotechnology, Lund University, Lund, Sweden)

**ABSTRACT:** The aim of this study was to enhance the efficiency of bioslurry reactors by studying the possibility of a pre- or postbioslurry treatment. Because of the lipophilic character of many persistent pollutants the use of a non-toxic lipophilic extraction medium seemed suitable. Therefore, a new method was tested for the separation of creosote constituents from their soil matrix using vegetable oil.

In shake flask experiments soil slurry was mixed with rapeseed oil and a surfactant (Brij 30) under varying conditions. The soil, taken from an old gasworks site in Stockholm (Sweden) mainly polluted with creosote, was sampled for analyses of 16 U.S. EPA polycyclic aromatic hydrocarbons (PAHs). The experiments have shown the pretreatment with rapeseed oil to be effective. The amount of oil used seemed to play a crucial role in the extraction process. In case of oil surplus, up to 87 % of high-molecular-weight (HMW) PAHs diffused into the vegetable oil, which formed a layer of small droplets on top of the water phase. Less oil caused inhibition of creosote transfer to the oil phase. The time for extraction was also studied. When multiple extractions were performed time was of minor importance. For example, two extractions during one day each gave a better separation (77 %) of pyrene than one extraction during eight days (67 %). The results obtained using the new method were compared with those of biodegradation during a bioslurry treatment of 30 days, indicating a maximum degradation of 55 % for HMW PAH. This two-liquid-phase (TLP) bioslurry treatment might become a more environmentally friendly alternative to the present chemical/physical pretreatment methods.

## INTRODUCTION

Polycyclic aromatic hydrocarbons (PAHs) are recalcitrant hydrophobic organic compounds that are released into the environment on a large scale. Some of them are toxic, carcinogenic or mutagenic (Mehlman et al., 1997) and therefore identified as priority pollutants by the United States Environmental Protection Agency (U.S. EPA) (Keith and Teillard, 1979). PAHs are formed during incomplete combustion of e.g. vehicle fuels or pyrolysis of organic material like during coal gasification. Coal gasification and wood preservation activities have caused the contamination of many places by the spillage of coal-tar oil and creosote.

The relationship between environmental persistence and increasing number of benzene rings is consistent with the results of various studies correlating biodegradation rates and molecule size. For example, reported half-lifes in soil and sediment of phenanthrene (3 rings) may range from 16 to 126 days while for benzo(a)pyrene (5 rings) they might range from 229 to over 1,400 days (Shuttleworth and Cerniglia, 1995). This suggests that complete bioremediation of HMW PAHs has a greater chance in long-term treatments like landfarming, composting, and fungal degradation, than during short-term treatments, like bioslurry. But by introducing a pretreatment the soil can be cleaned before the actual biodegradation has started.

The objective of this study is development of an environmentally friendly soil treatment process with the possibility of soil recycling. Recently Villemur (Villemur et al., 2000) published the first article on a two-liquid-phase (TLP) slurry bioreactor system showing a rapid transfer of HMW PAHs into non-biodegradable silicone oil. Up to 431 mg/L of benzo(a)pyrene was dissolved in silicone oil while its water solubility is 0.0038 mg/L (Cerniglia, 1992), proving the potential in this technique.

## MATERIALS AND METHODS

**Soil.** The soil used for this study was taken from a gasworks site in Stockholm, which was used for the production of town gas between 1893-1972. These activities caused soil pollution with creosote, cyanide and heavy metals. Typically, the concentration of the 16 U.S. EPA PAHs ranged from a few hundred to a few thousand mg/kg (dry wt). In this study soil with 3,000 mg/kg was used. Comparison of the initial concentration of 3- and 4-ringed PAHs in the soil described in this report with the initial values from about 1.5 years earlier revealed high degradation rates. Therefore the PAHs left in the soil at the beginning of these experiments can be described as having been relatively hard to biodegrade. The soil had an organic content of 5%. After excavation the soil was wet sieved, mixed thoroughly and stored in darkness at a maximum of 7 °C. Only the 0-2 mm fraction was used in the experiments.

**Bioslurry reactors.** Biodegradation studies were done at laboratory scale using 1 litre bioslurry reactors (with exception of shake-flask experiments in Figures 1 a,b) with 75-150 g of soil (dry wt), demineralised water, and nutrient mixtures Bio-D® and Soil Activator®(Medina Agricultural Products Company Inc., Hondo, USA). The indigenous microflora was supplemented with an inoculum isolated from another gasworks-site. Oxygen was supplied by filtered pressurised air (~0.5 L/s) during one month, mixing the slurry simultaneously. During reactor operation the temperature was maintained at 20 ± 2 °C and the pH at 7.8. Additional nutrients were supplied when necessary. To cause inhibition of the microflora in the reference reactor sodium azide was tried, but did not seem to work in this aerated system. Therefore the pH was brought down to 3.

**Shake flasks.** The experiments were done in baffled shake flasks of 500 ml containing 28-30 g soil and 150 ml water. For inhibition of the microflora, 0.5% $NaN_3$ was used. The surfactant Brij 30 (polyoxyethylene 4 lauryl ether; $M_w$= 362 $g.mol^{-1}$; critical micelle concentration (CMC) = 0.055 mM) was added for emulsification of the vegetable oil and minimising attachment of oil to glass surfaces. In all flasks containing vegetable oil a final concentration of 10 CMC was used.

**Sampling and Analysis.** In order to take representative samples, the whole reactor content was centrifuged and the supernatant was discarded. The slurry was mixed thoroughly and triplicate samples were taken. Soil samples were then mixed with acetone and extracted during stirring for one hour at 300 r.p.m at room temperature (Noordkamp et al., 1997). After sedimentation the acetone extracts were transferred into a glass syringe and filtered through a 0.50 μm polytetrafluoroethene (PTFE) membrane (Advantec MFS Inc., Pleasanton, USA). Extracts (except for results in Table 1) were analysed by HPLC (Varian 9100 Autosampler, Varian 9010 Solvent Delivery System, Varian 9050 Variable Wavelength UV-VIS Detector) with a Supelcosil LC-PAH column (25 cm * 3.0 mm Supelco) with a LC18-guard column. All injected samples were analysed at a wavelength of 254 nm. Samples in Table 1 were extracted using accelerated solvent extraction (ASE) (Lundstedt et al., 2000) and analysed by GC/MS.

## RESULTS AND DISCUSSION

It was found that HMW PAHs were not sufficiently degraded within the time frame of the 30-days bioslurry treatment, but were easily released from the soil particles by the vegetable oil. The amount of oil used had a great impact on the release, as well as the extraction time and frequency.

**Bioslurry treatment.** Biological degradation was studied through bioslurry treatment during 30 days. The results from these experiments show biodegradation rates of the 3-, 4-, 5- and 6-ringed PAHs being 57 %, 54 %, 44 % and 44 % respectively (Table 1). Higher biodegradation rates can hardly be expected within the time frame due to aging and aerobic half-lifes of the pollutants.

**TABLE 1. PAH concentrations in soil from gasworks site before and after 30 days bioslurry treatment and efficiency.**

| Total PAH (nr) | Initial (mg/kg) | After 30 days bioslurry treatment (mg/kg) | Removal efficiency (%) |
|---|---|---|---|
| 3 rings (5) | 452 | 196 | 57 |
| 4 rings (4) | 1262 | 586 | 54 |
| 5 rings (4) | 654 | 367 | 44 |
| 6 rings (2) | 359 | 202 | 44 |

**Soil extraction with vegetable oil.** The ability to use vegetable oil for the extraction of creosote constituents from polluted soil was studied in a TLP system using shake flasks. In order to get good mixing properties of oil with water and soil particles, surfactant Brij 30 was added to the slurry. The lipophilic core of surfactant micelles should theoretically hold a higher content of PAHs than "empty" micelles. Rapeseed oil and sunflower oil were both tested, but the results turned out to be similar. Rapeseed oil was chosen for continuation because it is a local product. Extraction parameters studied were time, oil/soil ratio, and extraction frequency.

When 28-30 grams of soil were extracted with 7.5 ml oil, phenanthrene was released in the same extent after 12 days as biodegraded during the same period (Figure 1a). But for pyrene, which has a water solubility that is a factor 10 lower than for phenanthrene, more than double was released compared to biodegradation (Figure 1b). The release of pyrene was 87% of the initial value. Extraction with 1.5 ml oil resulted in inhibition of the coal-tar transfer. Hardly any oil phase was formed on top of the water, even after centrifugation. This shows the importance of the right oil/soil ratio. The autoclaved reference indicated that biological activity was responsible for the degradation. PAH-release from the autoclaved soil was partly caused by the high temperature and pressure in the autoclaving process itself and/or partly because of lower extraction efficiency after this process.

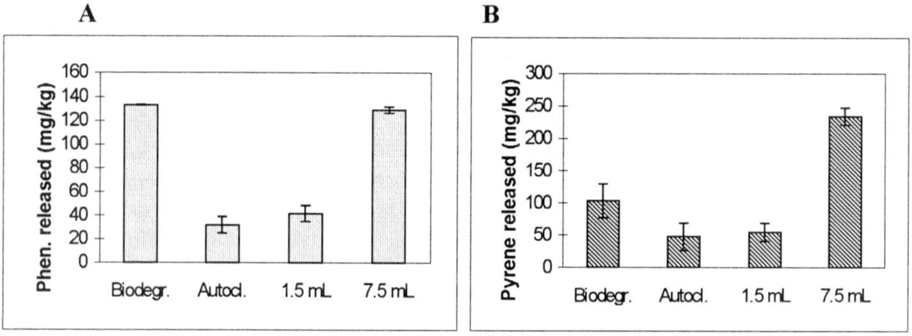


**FIGURE 1. (A) Phenanthrene released in mg/kg after autoclaving, 12 days of biodegradation in shake flasks or treatment with 1.5 or 7.5 ml rapeseed oil. (B) Pyrene released in mg/kg after autoclaving, 12 days of biodegradation in shake flasks or treatment with 1.5 or 7.5 ml rapeseed oil. The error bars show the standard deviation of three samples. 28-30 g of soil was used per flask.**

When studying the time needed for the diffusion process to be completed, it was found that treatment for 12 days might not be sufficient. After a contact time of 30 days, 88% of the initial phenanthrene and 83% pyrene was released (Figure 2 a,b). For phenanthrene, a significant release is seen after a 12 days contact time, while for pyrene the diffusion could be static after less than 8 days. This should be compared with a 30 days bioslurry treatment during which 60 % and 55 % is biodegraded, respectively.

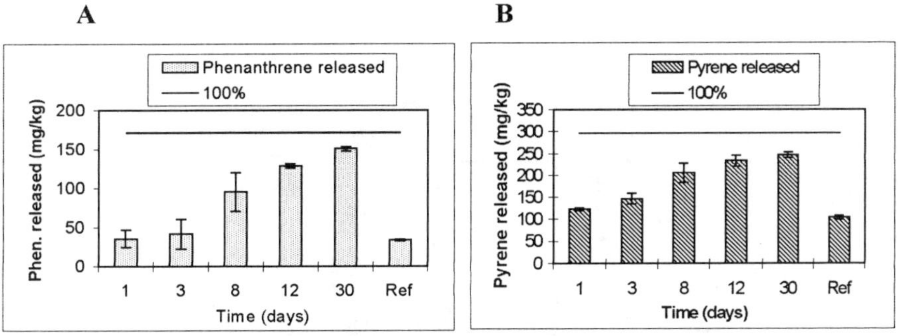


**FIGURE 2. (a) Phenanthrene released during 30 days using 7.5 mL rapeseed oil. (b) Pyrene released during 30 days using 7.5 mL rapeseed oil. The error bars show the standard deviation of three samples. 28-30 g of soil was used per flask.**

During this extraction not only PAHs will enter the oil phase, but even other lipophilic substances. For this heavily polluted soil, the amount of extraction medium seemed crucial. In figure 3 the release of pyrene is shown using between 3.0 and 7.5 mL oil. It was found that for this soil 6.0 mL of rapeseed oil gave the highest release of pyrene within 8 days.

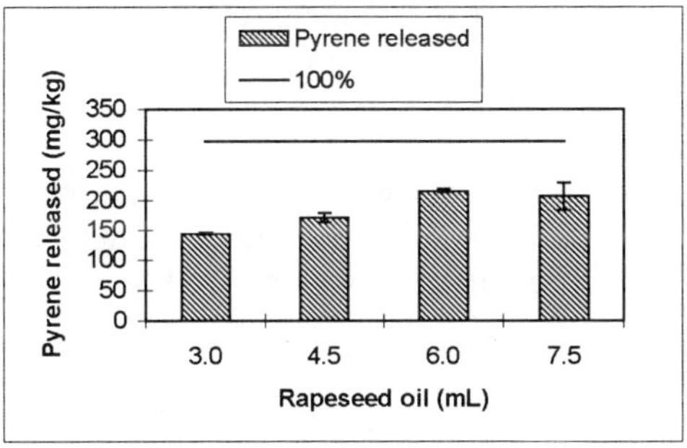


**FIGURE 3. Pyrene released after 8 days using different volumes of rapeseed oil. The error bars show the standard deviation of three samples. 28-30 g of soil was used per flask.**

When multiple treatments were performed (Figure 4) the time factor did not seem to be that important for the hydrocarbon release. When one extraction was performed during one day only 36% of pyrene was released, while a second extraction during one day yielded more than double (77%) the amount. A third extraction yielded another 5%. When two extractions during four days each were performed 87% was released.

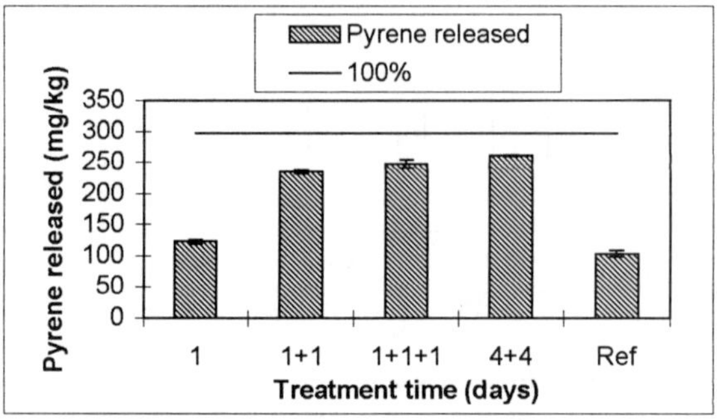


**FIGURE 4. Pyrene released after 1, 2 or 3 oil treatments with 7.5 ml during 1 or 4 days at the time. The reference was treated ones during 8 days. The error bars show the standard deviation of three samples. 28-30 g of soil was used per flask.**

In general a high separation efficiency could be seen especially for HMW PAHs. Even higher diffusion rates should be possible to achieve because some oil is left on the soil particles after treatment. Earlier experiments have shown neither Brij 30 nor rapeseed oil to be toxic to the microorganisms present in the soil or to the added inoculum, which is promising for a combined TLP-bioslurry system.

## ACKNOWLEDGEMENTS

This study was carried out within the Swedish national research program COLDREM (Soil remediation in a cold climate). The authors wish to acknowledge MISTRA (Foundation for Strategic Environmental Research) for the financial support.

## REFERENCES

Cerniglia C.E. 1992. "Biodegradation of polycyclic aromatic hydrocarbons." *Biodegradation*. 3: 351-368.

Keith L.H., W.A. Teillard. 1979. "Priority pollutants". *Environmental Science and Technology*. 13 (4): 416-423.

Lundstedt, S., B. van Bavel, P. Haglund, M. Tysklind, L. Öberg. 2000. "Pressurised liquid extraction of polycyclic aromatic hydrocarbons from contaminated soils". *Journal of Chromatography A*. 883:151-162.

Mehlman, M.A., M.M. Mumtaz, O. Faroon, C.T. De Rosa. 1997. "Health effects of polycyclic aromatic hydrocarbons." *Journal of Clean Technology, Environmental Toxicology and Occupational Medicine*. 6 (1): 1-22.

Noordkamp, E.R., J.T.C. Grotenhuis, W.H. Rulkens. 1997. "Selection of an efficient extraction method for the determination of polycyclic aromatic hydrocarbons in contaminated soil and sediment." *Chemosphere*. 35(9):1907-1917.

Shuttleworth K.L., C.E. Cerniglia. 1995. "Environmental aspects of PAH biodegradation." *Applied Biochemistry and Biotechnology*. 54: 291-302.

Villemur, R., E. Déziel, A. Benachenhou, J. Marcoux, E. Gauthier, F. Lépine, R. Beaudet, Y. Comeau. 2000. "Two-Liquid-Phase Slurry Bioreactors to Enhance the Degradation of High-Molecular-Weight Polycyclic Aromatic Hydrocarbons in Soil." *Biotechnology Progress*. 16: 966-972.

# ENHANCEMENT OF PAHs DEGRADATION BY NONIONIC SURFACTANTS IN COMPOSTING SYSTEM

***M. Fang*** (University of Massachusetts, USA;
Hong Kong Baptist University, Hong Kong SAR, P.R. China)
C.K. Wan, and J.W.C. Wong (Hong Kong Baptist University,
Hong Kong SAR, P.R. China)

**ABSTRACT:** The effect of a nonionic surfactant, Tween 80, on the biodegradation of PAHs in acidic loamy soil spiked with 400 mg $kg^{-1}$ each of phenanthrene, anthracene, and pyrene was assessed in a bench-scale composting system. PAH-spiked soil was mixed with pig manure at a ratio of 25% (w/w dry wt). Tween 80 was applied at a rate of 20, 40, or 100 mg $kg^{-1}$ (w/w dry wt) to the composting mass at Day 0 or 14 of the composting period. Tween 80 enhanced the removal of anthracene and pyrene, whereas no significant improvement of phenanthrene removal was noted. The effect of Tween 80 on biodegradation of PAHs was also time-dependent with significantly higher degradation for treatment added at Day 14 than Day 0, especially at a Tween 80 concentration of 100 mg $kg^{-1}$. The highest removal efficiency of PAHs in the spiked soil was obtained in the treatments with Tween 80 application rate of 20 to 40 mg $kg^{-1}$ applied at Day 14 of composting process. In addition, maturation test of composting by $C/N_{organic}$ and seed germination showed that addition of Tween 80, regardless of its concentration or time of addition, did not affect organic matter decomposition during the composting process. The results support the use of Tween 80 to enhance the degradation of pyrene and anthracene during composting treatment but its general use for PAH contaminated sites still requires further studies on PAHs of higher ring number.

## INTRODUCTION

Polycyclic aromatic hydrocarbons (PAHs) are a group of ubiquitous contaminants with a high potential hazard to human health. Bioremediation for cleaning up PAH-contaminated sites is the currently favored approach, because microorganisms can totally degrade or partially transform many PAHs. Composting is a potential cost-effective alternative for decontamination of PAH-contaminated soil (Joyce et al., 1998; Potter et al., 1999). Nevertheless, biodegradation of PAHs in soil is limited by their poor availability to the microorganisms, which is due to their hydrophobicity, low aqueous solubility, and strong adsorptive capacity to organic matter (Volkering et al. 1998). Surfactants with their hydrophobic and hydrophilic moieties would increase the desorption of PAHs from the solid phase by solubilization or emulsification and as a result promote biodegradation. It is well documented that surfactants can facilitate the release of hydrophobic organic compounds from soil and thus enhance the overall performance of bioremediation (Abdul et al., 1992; Tiehm, 1994). Madsen and Kristensen (1997) showed that surfactants that are mineralized at moderate rates may be more applicable for increasing the availability of PAHs in soil. However, Laha and Luthy (1992) reported that solubilization of sorbed hydrophobic PAH compounds from soil following the addition of nonionic surfactants may not experience a concomitant enhancement of soil bioremediation. Although surfactants can increase the apparent mobility of PAHs, their beneficial effect on PAH degradation has not been satisfactorily addressed.

Composting has been adopted effectively in our previous research to remediate PAH-contaminated soil except at a high starting concentration of PAHs, which caused a low efficiency of PAH removal. This is likely due to slow release of PAHs from the soil matrix to the aqueous phase. It leads us to think about addition of surfactants in composting mass for stimulating PAH degradation during composting. A nonionic surfactant, Tween 80, was selected in the present experiment because it has a high HLB (hydrophile/lipophile balance) number, weaker sorption to charged sites, and relatively low toxicity to bacteria (Laha and Luthy, 1992). However, Tween 80 is considered biodegradable and may act as a substrate for microorganisms, and as a result, the high concentration of Tween 80 in composting mass may compete with microbial biodegradation of PAH compounds. Apart from that, the time of surfactant addition during the composting process also is an important aspect needed to be considered, since sorption of PAHs onto organic matter is time-dependent. Therefore, the objectives of the present experiment were to study the feasibility of surfactant addition in enhancing biodegradation of PAHs in soil spiked with PAHs during composting process and to optimize the Tween 80 amendment by testing the effect of Tween 80 concentrations and timing of addition on the biodegradation of PAHs.

## MATERIALS AND METHODS

**Preparation of Composting Piles.** Three PAH compounds, i.e. phenanthrene, anthracene, and pyrene, each at 400 mg $kg^{-1}$ (w/w dry wt) were mixed thoroughly with a sandy loam (texture: 71% sand, 19% silt and 10% clay; organic matter: 2.31%) collected from the Kadoorie Farm and Garden located in Tai Po, Hong Kong. After one month of incubation at room temperature, this PAH spiked soil was amended with pig manure using sawdust as a bulking agent at a ratio 3:1 (w/w dry wt) to obtain a C/N ratio of about 30. Polyoxyethelene (20) sorbitan monooleate known as Tween 80 (Sigma Chemical Co.) at concentrations of 20, 40, and 100 mg $kg^{-1}$ (w/w dry wt) were added separately to the manure amended soil at Day 0 and Day 14 of the composting period, respectively. Four kilograms $\pm$ 0.1 kg of the mixtures were prepared and composted in bench-scale home-made composting reactors connected with an on line computer control system to provide sufficient air (0.4 L $min^{-1}$ $kg^{-1}$) to the composting mass. There were duplicate samples for each treatment. Periodically, about 100 g of sample was removed from each reactor at days 0, 7, 14, 21, 28, 35, and 49 for chemical testing.

**Analytical Procedures.** Water soluble carbon and soluble $NH_4$-N was measured on the water extracts at a rate of solid:double distilled water = 1:10 (w/v) using a SHIMADZU TOC-5000A Total Organic Analyzer and the indophenol blue method, respectively. Total soluble organic N was extracted by Kjeldahl digestion method on soluble extract followed by soluble $NH_4$-N determination using the methods described before.

Seed germination and root length tests were performed on water extracts obtained by shaking fresh samples at 1:2 solid:deionized water (w/v) for 1 hour. Suspensions were centrifuged at 10000 rpm for 20 min before filtering through Whatman #1 filter paper. Ten cress (*Lepidium sativum, L.*) seeds were placed in a sterilized plastic petri-dish with 6 mL of each test solution and incubated at 25°C in the dark for 72 h. Treatments were evaluated by counting the number of germinated seeds and the length of the root radical. The responses were calculated by a germination index obtained by multiplying the percent germination by the percent root growth as related to control.

About 5 g of sample was Soxhlet extracted with 150 mL of acetone/dichloromethane (DCM) for 16 hours followed by an alkaline hydrolysis step to extract PAHs sorbed onto the organic matrix using a method described by Loser and Holm (1999). The extract was evaporated in a rotary evaporator under reduced pressure to near dryness, and then dissolved in 1mL of hexane before clean-up. The extracts were analyzed by gas chromatography/mass spectroscopy (GC/MS) according to USEPA Methods 8270C. An HP-5MS fused silica capillary column (30 mm x 0.25 mm, 0.25 μm film thickness, Supelco) was used and the column outlet was located in the MS ion source. Helium was used as the GC carrier gas. Following injection (1 μL), the GC column was held at 50°C for 1 min and the temperature-programmed to 300°C at 4°C/min. The MS was operated in the selected ion monitoring (SIM) mode. Identification of the target compounds was based on their GC retention time relative to those of the internal standards. The minimum detectable level was approximately 0.01 $\mu g\ g^{-1}$.

**Statistical Analysis.** All data were analyzed by a SAS statistical package through an IBM compatible personal computer. The means and standard deviations of duplicate treatments were calculated. The ANOVA and Least Significant Difference test were used to compare the means of all different treatments within the same sampling interval for all parameters.

## RESULTS AND DISCUSSION

**Effect of Surfactants on the Biodegradation of PAHs.** Complete biodegradation of phenanthrene was observed after 21 days of the composting process for all the treatments except for the treatment receiving Tween 80 at 100 mg $kg^{-1}$ at the beginning of composting (Figures 1a and b). This indicates that phenanthrene was readily available to microbial degradation even without surfactant addition. The degradation rate of phenanthrene in the treatment with 100 mg $kg^{-1}$ of Tween 80 addition was even slower than the control receiving no surfactant addition. This indicated that the high concentration of Tween 80 might encourage degradative microorganisms to use Tween 80 as an energy source for growth. There was no discernible beneficial effect in the addition of various concentrations of Tween 80 on the biodegradation of phenanthrene at different time intervals.

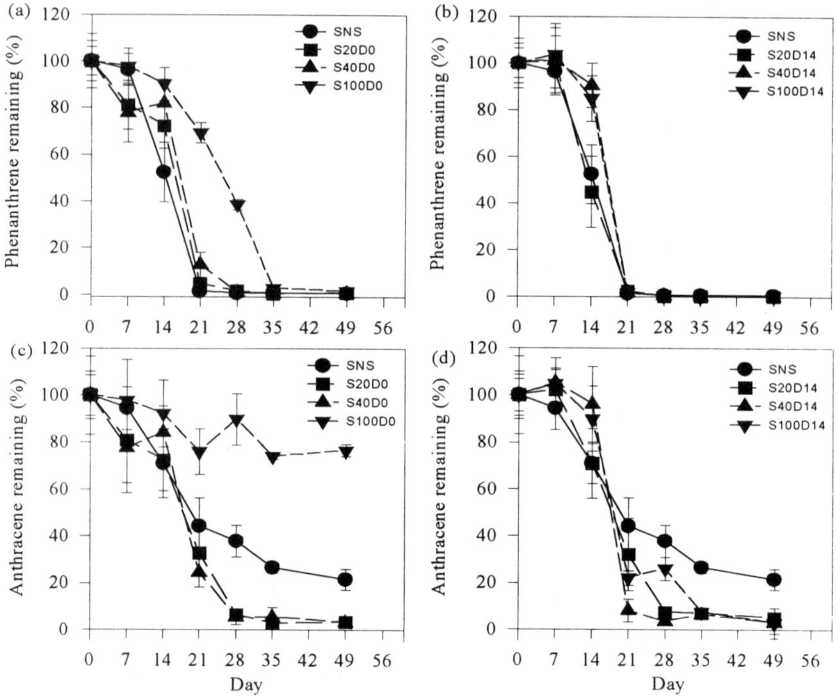


**FIGURE 1. Percentage of phenanthrene (a, b) and anthracene (c, d) remaining after composting of PAHs contaminated soil with the addition of Tween 80 at Day 0 (a, c) and Day 14 (b, d). (SNS=without surfactant, S20, S40 and S100 = surfactant added at a concentration of 20, 40 and 100 mg $kg^{-1}$, respectively).**

After the first 14 days of composting, addition of Tween 80 had a significantly high efficiency in anthracene removal except for those with 100 mg $kg^{-1}$ of Tween 80 added at Day 0 (Figure 1 c). The poor degradation efficiency in this treatment might be due to the preferential use of the more available Tween 80 by degradative microorganisms rather than using the less soluble anthracene. Tiehm (1994) also found that sodium dodecyl sulfate (SDS) inhibited PAH degradation because this surfactant was preferred as a growth substrate. On the other hand, the high concentration of Tween 80 added may inhibit degradation of PAHs at a concentration above its critical micelle concentration of 13-15 mg $L^{-1}$ for Tween 80 (Page et al., 1999). The addition of Tween 80 at 20-40 mg $kg^{-1}$ clearly demonstrated its beneficial effect in enhancing the biodegradation of anthracene during composting treatment as compared to the control. Since the solubility of anthracene (0.073 mg $L^{-1}$) is lower than that of phenanthrene (1.29 mg

$L^{-1}$), addition of Tween 80 is expected to accelerate the mass transfer of anthracene from soil to water during composting. Addition of Tween 80 at Day 14 as compared to its counterpart with Tween 80 added at Day 0 caused further enhancement in degradation of anthracene, especially at 100 mg $kg^{-1}$ with a dramatic drop in anthracene from 80% to 20% (Figure 1d).

A continuous removal of pyrene was noted in the control without the addition of Tween 80, with a total removal of about 45-50% (Figure 2), which was comparatively lower than that of phenanthrene and anthracene. Similar to anthracene, treatment with 100 mg $kg^{-1}$ of Tween 80 added at Day 0 exhibited the lowest biodegradation of only 20% pyrene being removed. Those receiving Tween 80 at a concentration range of 20-40 mg $kg^{-1}$ showed a complete biodegradation of pyrene after 35 days of composting. The removal of pyrene was significantly improved by adding Tween 80 at Day 14 instead of adding right at the beginning of the composting process at each particular concentration of Tween 80, although the overall rate were about the same. This effect was more obvious for soil receiving 100 mg $kg^{-1}$ Tween 80 with pyrene removal rate increasing from 20% to over 90%. The reasons of such a phenomenon are not clearly understood. The delay in the addition of Tween 80 would allow time for the building up of PAH degradative microbial population in the soil and hence the addition of Tween 80 would increase the availability of PAHs to become available for microbial utilization (Deschenes et al., 1996). In addition, the high temperature of composting mass might enhance the effectiveness of Tween 80 added at Day 14 by improving cell adhesion or attachment at surface-$H_2O$-NAPL (non-aqueous-phase liquids) interfaces resulting in an increased bioavailability of PAHs (Macur and Inskeep, 1999).

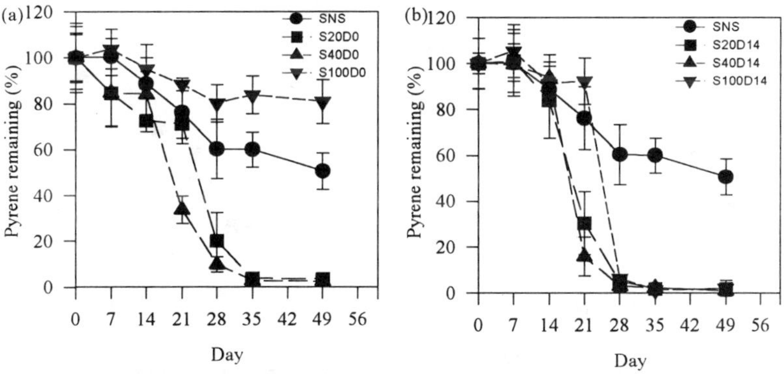


**FIGURE 2. Percentage of pyrene remaining after composting process of PAHs contaminated soil with the addition of Tween 80 at (a) day 0 and (b) day 14. (Refer to Figure 1 for legend explanation)**

In the present study, the use of Tween 80 to enhance the removal of PAHs is proven to be feasible under thermophilic conditions. However, the loss of PAHs can be explained by the vaporization loss, organic matter adsorption, degradation activity of indigenous microorganisms or a combination of any of these processes. Vaporization loss of PAHs was negligible as indicated by the undetectable amount of PAHs in the exhaust air from all the reactors. Wild et al. (1990) also reported that volatilisation loss of PAHs from soil is likely to be important for naphthalene, but not for the higher ringed compounds. Another possible loss of PAHs is the amount of sorption of PAHs on the organic matter of pig manure so that it was not available in the Soxhlet extraction. Loser and Holm (1999) included alkaline hydrolysis in the traditional Soxhlet extraction in order to set free the proportion sorbed to the organic matter in a PAH-contaminated wood. They concluded that the disappearance of PAHs was mainly due to biodegradation and less than 5% due to sorption onto organic soil matrix. Therefore, microbial degradation in soil would likely be the major factor contributing to removal of PAHs in the present composting system with the addition of Tween 80.

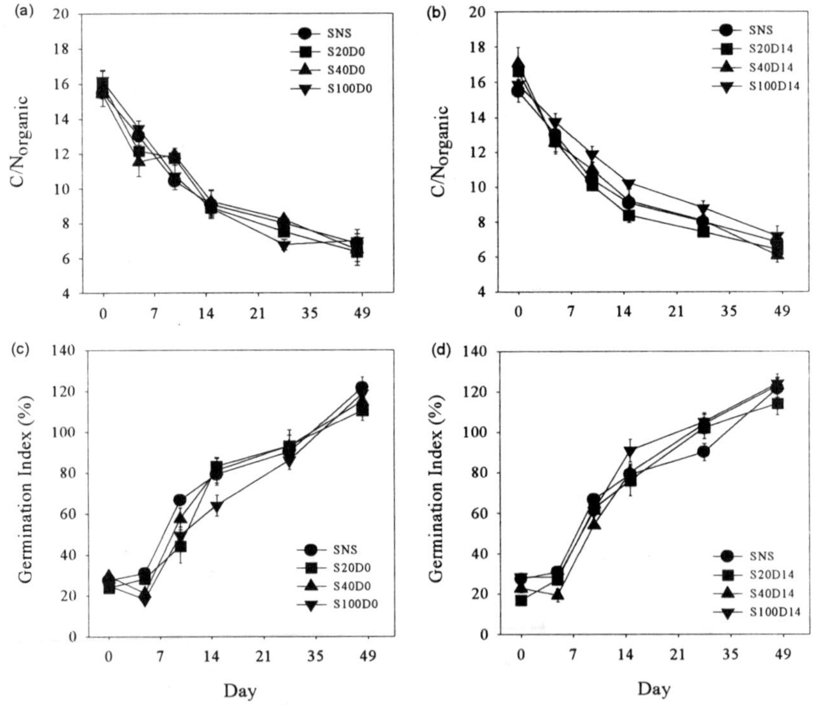


**FIGURE 3. Changes in C/N$_{organic}$ (a, b) and germination index (c, d) during composting process of PAHs contaminated soil with the addition of Tween 80 at Day 0 (a, c)and Day 14 (b, d). (Refer to Figure 1 for legend explanation)**

**Changes in C/N$_{organic}$ and Germination Index (GI).** Figure 3 shows the changes in the C/N$_{organic}$ and germination index for all treatments. The C/N$_{organic}$ decreased gradually from about 16 to 6 for all of the treatments and the decrease became slow down after Day 14. No significant difference was noted among the various treatments. $C/N_{organic}$ is used as an indictor for compost maturation because composting reaction is a biochemical decomposition of organic matter by microorganisms whose metabolic activity mainly occurred in the water-soluble phase. The values of $C/N_{organic}$ in water extracts of well-maturated compost is within the range of 5 to 6, which are similar to those in microbial cells (Chanyasak and Kubota, 1981). Hence, the present results indicated that addition of Tween 80 at different concentrations and period did not affect microbial activity during composting process and all treatments reached maturation phase after 49 days of composting. The initial GI values in all treatments were below 30%, and then GI increased significantly and reached over 100% at around 35 to 42 days. An increase in concentrations of Tween 80 addition did not cause a reduction of GI regardless of the time of surfactant addition during the composting process. The results demonstrated that Tween 80 did not contain phytotoxic substances that would affect the maturity of compost, which supported the use of Tween 80 as a surfactant to enhance the availability of PAHs during composting treatment.

## CONCLUSIONS

The present study demonstrated that addition of Tween 80 increased the removal of phenanthrene, anthracene and pyrene during composting treatment of soil spiked with these three PAH compounds. The enhancement effect was more obvious for PAHs with low solubility or high ringed structure, i.e., pyrene > anthracene > phenanthrene. Among the three concentrations of Tween 80 used, a concentration ranged from 20-40 mg $kg^{-1}$ demonstrated a better efficiency in biodegradation of PAH compounds. Moreover, addition of Tween 80 at Day 14 instead of Day 0 had a better biodegradation of PAHs, especially at the high concentration of Tween 80. It indicates that the effect of the surfactant on PAH removal was time-dependent. Addition of Tween 80, on the other hand, did not affect the decomposition of organic matter during composting and all treatment reached maturation after 49 days of composting as indicated by $C/N_{organic}$ and germination index. The present results confirmed the positive role of Tween 80 in increasing PAH biodegradation through enhancing the availability of PAHs. However, its effectiveness on the degradation of PAHs with higher ring number and aged soil samples warrant further studies.

## ACKNOWLEDGMENTS

The work described in this paper was fully supported by a grant from the Research Grant Council of the Hong Kong Special Administrative Region, P.R. China (Project No. HKBU 2043/97M).

## REFERENCES

Abdul, S. A., T. L. Gibson, C. C. Ang, J. C Smith, and R. E. Soyczynski. 1992. "In situ surfactant washing of polychlorinated biphenyls and oils from a contaminated site." *Ground Water 30*: 219-231.

Chanyasak, V., and H. Kubota. 1981. "Carbon/organic nitrogen ratio in water extract as measure of composting degradation." *J. Ferment. Technol. 59*: 215-219.

Deschenes, L., P. Lafrance, J. P. Villeneuve, and R. Samson. 1996. "Adding sodium dodecyl sulfate and *Pseudomonas aeruginosa* UG2 biosurfactants inhibits polycyclic aromatic hydrocarbon biodegradation in a weathered creosote-contaminated soil." *Appl. Microbiol. Biotechnol.46*: 638-646.

Joyce, J. F., C. Sato, R. Cardenas, and R.Y. Surampalli. 1998. "Composting of polycyclic aromatic hydrocarbons in simulated municipal solid waste." *Water Environ. Res. 70*: 356-361.

Laha, S., and R. G. Luthy. 1992. "Effects of non-ionic surfactants on the solubilization and mineralization of phenanthrene in soil-water systems." *Biotechnol. Bioeng. 40*: 1367-1380.

Loser, C., and U. Holm. 1999. "Composting of wood containing polycyclic aromatic hydrocarbons (PAHs)." *Compost Sci. Util., 7:* 16-32.

Macur R. E. and W. P. Inskeep. 1999. "Effects of a nonionic surfactant on biodegradation of phenanthrene and hexadecane in soil." *Environ. Toxicity. Chem. 18*: 1927-1931.

Madsen, T., and P. Kristensen. 1997. "Effects of bacterial inoculation and nonionic surfactants on degradation of polycyclic aromatic hydrocarbons in soil." *Environ. Toxic. Chem., 16*: 631-637.

Page, C A., J. S. Bonner, S. A. Kanga, M. A. Mills, and R. L. Autenrieth. 1999. "Biosurfactant solution of PAHs." *Environ. Eng. Sci. 16*: 465-473.

Potter, C. L., J. A. Glaser, L. W. Chang, J. R. Meier, M. A. Dosani, and R. F. Herrmann. 1999. "Degradation of polynuclear aromatic hydrocarbons under bench-scale compost conditions." *Environ. Sci. Technol. 33*: 1717-1725.

Tiehm, A. 1994. "Degradation of polycyclic aromatic hydrocarbons in the presence of synthetic surfactants." *Appl. Environ. Microbiol. 60*: 258-263.

Volkering, F., A. M. Breure, and W. H. Rulkens. 1998. "Microbiological aspects of surfactant use for biological soil remediation." *Biodegradation. 8*: 401-417.

Wild, S. R., K. S. Waterhouse, S. P. McGrath, K. C. Jones. 1990. "Organic contaminants in an agricultural soil with a known history of sewage sludge amendments: polynuclear aromatic hydrocarbons." *Environ. Sci. Technol., 24*: 1706-1711.

# CASE STUDY: SUCCESSFUL BIOTREATMENT OF PAH-CONTAMINATED SOIL

***Michael Profit,*** Gary Clemons, Tim Eggert, and Tim Turner, CDM Federal Programs Corp., Atlanta, GA, USA; Al W. Bourquin, Camp Dresser & McKee Inc., Denver, CO, USA; and Jon Bornholm, U.S. Environmental Protection Agency, Atlanta, GA, USA.

**ABSTRACT:** Soils contaminated with creosote and diesel fuels at the Benfield Industries Site in Waynesville, North Carolina were remediated using a modified biopile treatment technology. Soil was excavated, amended with straw and manure, and formed into windrow configurations in a land treatment unit (LTU). An excavator was used to lift, mix, aerate, and redistribute the soil in the windrows.

Confirmatory sampling was performed for baseline conditions and then monthly to document the progress of biological treatment. Similarly, the LTU was sampled initially for baseline biological monitoring concentrations conditions, and then either weekly or monthly until treatment was complete.

Soil was classified as successfully treated when analytical results demonstrated that remediation goals had been met or that biodegradation had ceased to reduce polycyclic aromatic hydrocarbon (PAH) levels. If PAHs were above the remediation goals, soils were sampled for Toxicity Characteristic Leaching Procedure (TCLP) analysis. Such soils were considered successfully treated if TCLP extract results contained contaminant concentrations that were at or below their respective groundwater remediation levels and North Carolina groundwater quality standards. The remediation goal for benzo(a)pyrene was not changed. Soil that failed to meet the benzo(a)pyrene target was buried by at least one foot of clean fill.

## INTRODUCTION

Recent studies have shown that microbial degradation of PAHs is the major process that results in the decontamination of sediment and surface soils. PAH compounds can be totally degraded (i.e., mineralized) or transformed by either a community of microorganisms or by a single microorganism. Molecular oxygen is usually required to initiate the transformation. Generally, two- and three-ring PAHs are able to serve as sole sources of carbon and energy for growth.

Biodegradation of mixtures of PAHs containing four or more rings, as found in creosote, is well documented (Sims et al. 1986). The higher ring PAHs are thought to be biodegraded by a cometabolic process in which the two- and three-ring PAHs, which are structural analogs of the higher ring PAHs, serve as cosubstrates. Biodegradation of higher ring PAHs, which are not able to support growth by themselves, has been demonstrated in the laboratory using naphthalene and phenanthrene as cosubstrates. As a result, higher ring PAHs when present in mixtures, such as those found in creosote, are biodegraded to a far greater extent than would occur if present individually in environmental matrices.

This paper reports the results of the full-scale bioremediation conducted in two phases. Phase 1 was conducted from May through September 1998 by CDM Federal Programs and its subcontractor, OHM, under EPA contract 68-W9-0056. This work ceased when CDM Federal's EPA contract expired. Phase 2 was conducted by CMC, Inc. under EPA contract 68-S7-4006 between June 1999 and August 2000.

**Site Description and Project Background**. The Benfield Industries Site is the location of a former chemical product formulating and packaging facility. In April 1982, a fire destroyed the majority of the structures. Subsequent investigations by EPA and the North Carolina Department of Environment, Health and Natural Resources resulted in the Site being added to Superfund's National Priorities List.

The Site had widespread soil contamination consisting principally of PAHs and lesser occurrences of pentachlorophenol (PCP). The Record of Decision (ROD) required treatment or removal of soils contaminated above the cleanup goals for the Site's Contaminants of Concern (COC) shown in Table 1. The ROD was amended following a soil bio-treatability test that failed to achieve the cleanup goals for several COCs, but showed through TCLP testing that groundwater protection criteria had been met. These findings were the basis of the ROD revision which allowed for biotreated soils to remain as backfill on the Site as long as exhaustive biodegradation was achieved and TCLP extract testing demonstrated all COCs were below the water quality criteria.

**TABLE 1. Range of concentrations and remediation goals for chemicals of concern.**

| Compound | Range (mg/kg) | Remediation goal (mg/kg) |
|---|---|---|
| Benzo(a)anthracene | 0.072 - 94 | 0.8 |
| Benzo(a)pyrene | 0.041 - 14 | 0.3 |
| Benzo(b and/or k)fluoranthene | 0.042 - 75 | 1.6 |
| Chrysene | 0.040 - 77 | 1.6 |
| Indeno(1,2,3-cd)pyrene | 0.11 - 5.1 | 2.8 |
| Naphthalene | 0.043 - 120 | 10 |
| Pentachlorophenol | 0.074 - 19 | 1 |

## MATERIALS AND METHODS

Excavated soils were first stockpiled and allowed to drain free of water. They were then processed through a mechanical screen to remove cobble and debris greater than 4 inches in size. The resulting large particle fraction was cleaned by pressure washing and the small particle fraction was prepared for biotreatment in the LTU.

**LTU Construction.** All biotreatment occurred within an LTU that was rectangular in shape and approximately 4 acres in size. Its main features were a 60-mil high-density polyethylene (HDPE) liner that was placed over a prepared 4-inch sand layer. A 12-inch sand layer was placed on top of the liner. The outer boundaries of the LTU were small dykes that were also covered by the HDPE liner. This design prevented inflows of surface run-off during rainfall and contained all liquids within the system. Excess water was periodically discharged to the Town of Waynesville's waste water treatment plant.

**Windrow Construction**. Screened material less than 4 inches was transported by dump truck to the LTU where the windrows were constructed. A CAT 330 excavator mixed the soil and blended in straw and manure to make the soil more friable. After blending, the excavator shaped the soil into the final windrow configuration: 45 feet wide at the base, 25 feet wide at the top, and 10 feet high. As additional soil was delivered to the windrow, the excavator tracked along the length of the LTU area continuing windrow construction. When soil placement was complete, three windrows (180, 340, and 400 feet in length) were constructed in the LTU area.

In order to maintain the proper moisture content and not interfere with windrow aeration, a portable irrigation system was installed in the LTU. The system consisted of a supply pump, flexible garden hose, and lawn sprinklers. A pump was placed next to the LTU sump to draw water from the sump for irrigation. If the sump was dry, additional water was supplied from either the temporary storage pools located on the Site or the municipal water system.

**Operation**. Operation of the windrows included continuous mixing and aeration of the soil utilizing the excavator with the air-bucket attachment. The excavator was used to lift, mix, aerate, and redistribute the soil in the windrows. The objectives of both the soil configuration and mixing techniques were to maintain optimal conditions for the degradation of the target PAH compounds.

The optimal operating parameters, and typical corrective actions to maintain LTU operating conditions, are summarized in Table 2. These operating parameters and the ranges presented either are those generally defined as optimal for enhanced biological treatment of available organics (Cookson 1995) or were specifically determined in the Kiber pilot study as optimal for treatment of PAHs at the Site.

**PAH Degradation Sampling**. Confirmatory sampling was performed for baseline conditions and then monthly to document the progress of biological treatment. One ten-point composite confirmatory sample was collected for every 500 cubic yards of soil in the LTU. Based on the dimensions of the biopiles, this corresponded to approximately 50 linear feet of windrow. Samples were analyzed for the target PAHs and PCP using US EPA Method 3550/8270.

**TABLE 2. LTU Operating Parameters**

| Parameter | Operating Range | Typical Corrective Actions |
|---|---|---|
| Oxygen | 5% preferred, or at a minimum maintained at > 1% by volume in soil gas | Increase mixing |
| Moisture content | 40% <target <70% of field capacity | Increase mixing to dry or irrigate to increase moisture |
| pH | 6.0 su <target <8.0 su | Add lime to increase pH or ammonium sulfate to decrease pH |
| Ammonia-nitrogen | >20 mg/kg | Add manure or fertilizer |
| Phosphate-phosphorous | >10 mg/kg | Add manure or fertilizer |
| Microbial population | > 1000 CFU/g | Add manure |

**Biological Monitoring Sampling**. The LTU was sampled for baseline biological monitoring concentrations conditions and then either weekly or at a minimum monthly until treatment was complete. Four composite samples were collected for biological monitoring analyses. The samples were analyzed for total petroleum hydrocarbons, total organic carbon, total Kjeildahl nitrogen, ammonia-nitrogen, nitrate-nitrogen, phosphate-phosphorous, bacterial population density, moisture content, and soil-gas oxygen.

**Successful Treatment.** After pressure washing, the large particle soil fraction (> 4 inches) underwent TCLP analysis. Results were compared to the groundwater remediation levels established in the ROD and to North Carolina groundwater standards (NCAC 15-2L.0202). It was classified as successfully treated when analytical results demonstrated that it contained target contaminant concentrations that were at or below their respective goals.

The fine particle soil fraction was classified as successfully treated when analytical results demonstrated that the remediation goals had been met or that biodegradation had ceased to reduce PAH levels. This was demonstrated by the results of three successive weekly sampling events on the same portion of treated soil that had been maintained under optimum conditions for biotreatment. Such soils were sampled for TCLP analysis and considered successfully treated if TCLP extracts contained contaminant concentrations that were at or below their respective goals (i.e., groundwater remediation levels and North Carolina groundwater standards).

## RESULTS AND DISCUSSION

As stated previously, the work was conducted in two phases. The Phase 2 work was performed essentially the same as Phase 1 with one major exception: treatment times were extended much beyond those used in Phase 1. The results show similar trends but with important differences.

**Phase 1.** The treatability study conducted before Phase 1 construction showed that total PAH concentrations were reduced by approximately 90 percent and target PAHs (benzo(a)anthracene, benzo(a)pyrene, benzo(b)fluoranthene, benzo(k)fluoranthene, chrysene, indeno(1,2,3-cd)pyrene, and naphthalene) were reduced on the order of 80 percent in less than 50 days (Figure 1). Despite this reduction, not all target PAHs were reduced to or below the remedial goals set in the ROD (Table 3). The PAHs that were not destroyed exhibited very low leachability. Based on this finding, it was assumed that the PAHs that remained were not readily soluble or available for microbial degradation.

**FIGURE 1**
**Reactor 4-1: Treatability Study**
**Total and Target PAHs**

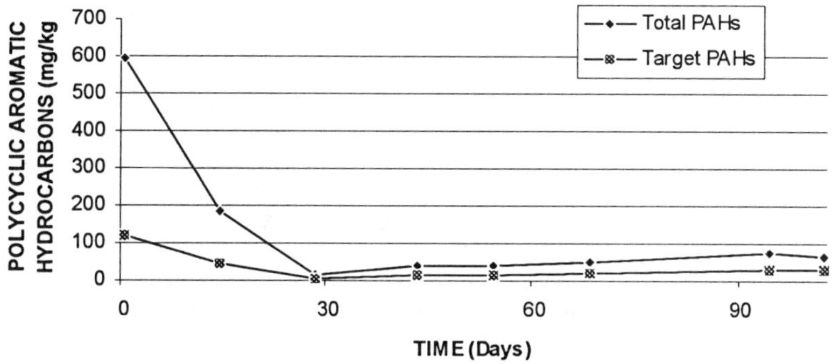


**TABLE 3. Reduction of chemicals of concern in treatability study, treatment duration: 102 days**

| Compound | Goal (mg/kg) | Start (mg/kg) | Finish (mg/kg) |
|---|---|---|---|
| Benzo(a)anthracene | 0.8 | 17.0 | 2.6 |
| Benzo(a)pyrene | 0.3 | 7.1 | 3.7 |
| Benzo(b)fluoranthene | 1.6 | 12.0 | 5.2 |
| Benzo(k)fluoranthene | 1.6 | 3.5 | 0.02 |
| Chrysene | 1.6 | 25.0 | 3.8 |
| Indeno(1,2,3-cd)pyrene | 2.8 | 2.1 | 3.4 |
| Naphthalene | 10 | 53.0 | 9.4 |
| Total target PAHs | NA | 120 | 28.1 |

Phase 1 field work was conducted from May through September 1998. The windrows were aerated continuously over this period. Weekly sampling data included bacterial counts (PAH degraders) to verify maintenance of specified biological conditions. Oxygen concentrations at depths greater than 1 foot diminished to zero within one hour due to the high biological activity. However, in the outer 1-foot zone, oxygen concentrations remained aerobic between mixing intervals. Table 4 shows the average reduction in COCs in Phase 1 cells over this period. Figure 2 depicts the degradation of these compounds in Cell C6.

**TABLE 4. Reduction of chemicals of concern in Phase 1, average treatment duration: 106 days**

| Compound | Goal (mg/kg) | Start (mg/kg) | Finish (mg/kg) |
|---|---|---|---|
| Benzo(a)anthracene | 0.8 | 9.2 | 1.5 |
| Benzo(a)pyrene | 0.3 | 3.8 | 2.8 |
| Benzo(b)fluoranthene | 1.6 | 4.6 | 3.0 |
| Benzo(k)fluoranthene | 1.6 | 4.7 | 1.9 |
| Chrysene | 1.6 | 10.6 | 3.4 |
| Indeno(1,2,3-cd)pyrene | 2.8 | 0.9 | 0.7 |
| Naphthalene | 10 | 6.9 | 0.8 |
| Total target PAHs | NA | 40.1 | 14.1 |

**FIGURE 2**
**Phase 1: Cell C6**
**Total and Target PAHs**

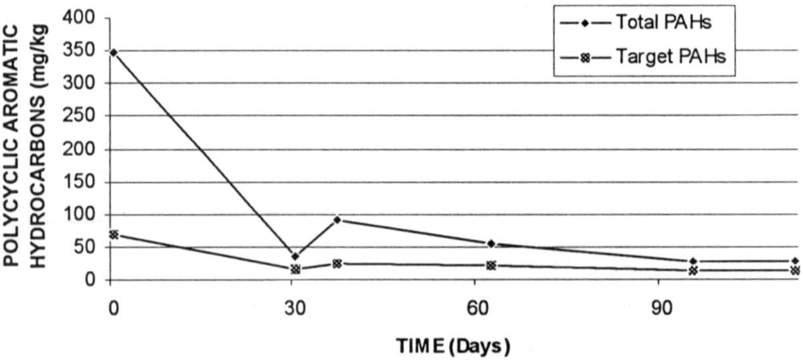


The pattern shown in Figure 2 is similar to what was observed in the treatability study. Most of the degradation occurred in the first 30 to 40 days, after which, the general trend was downward but at a slower rate. Degradation of target PAHs leveled out after 90 days. Consistent with the treatability study, target PAHs were above remedial goals and showed no further evidence of degradation after about100 days.

Much of the target PAHs is comprised of heavy molecular weight PAHs (4-6 ring compounds, particularly benzo(a)anthracene, benzo(a)pyrene, and chrysene). Although above the remedial goals, the soil fraction was classified as having met the definition of exhaustive biotreatment because analytical results demonstrated that biodegradation had ceased to reduce PAH levels. Such soils were considered successfully treated when TCLP extracts showed contaminant concentrations that were at or below their respective groundwater remediation level and North Carolina groundwater standard and thus protective of groundwater.

**Phase 2**. Phase 2 of the soil remediation was conducted by CMC Inc. between June 1999 and August 2000. The operations were essentially identical to Phase 1 operations; the only difference was in the duration of treatment (i.e., mean duration 106 days in Phase 1 vs. 255 days in Phase 2). This significantly longer treatment period improved biodegradation results. Table 5 shows the average reduction in COCs in Phase 2 cells over this period. Figure 3 shows the results for Cell A1A.

**TABLE 5. Reduction of chemicals of concern in Phase 2, average treatment duration: 255 days**

| Compound | Goal (mg/kg) | Start (mg/kg) | Finish (mg/kg) |
|---|---|---|---|
| Benzo(a)anthracene | 0.8 | 5.3 | 0.4 |
| Benzo(a)pyrene | 0.3 | 2.9 | 1.0 |
| Benzo(b)fluoranthene | 1.6 | 3.1 | 1.1 |
| Benzo(k)fluoranthene | 1.6 | 3.1 | 0.4 |
| Chrysene | 1.6 | 5.8 | 0.7 |
| Indeno(1,2,3-cd)pyrene | 2.8 | 0.5 | 0.4 |
| Naphthalene | 10 | 0.9 | ND |
| Total target PAHs | NA | 19.5 | 3.5 |

Note that the degradation of total and target PAHs was similar to the treatability study and Cell C6 from Phase 1. The majority of the degradation was complete after about 100 days. The major difference between the treatability study/Phase 1 results and Phase 2 was the improved degradation of recalcitrant compounds in Phase 2. On average, soil remediation goals were met for all compounds except benzo(a)pyrene. This contrasts with the treatability study/ Phase 1 results where the remediation goals were largely unmet. It is speculated that prolonged mixing gradually achieved solubilization of recalcitrant compounds making them susceptible to microbial action.

FIGURE 3
Phase 2: Cell F1
Total and Target PAHs

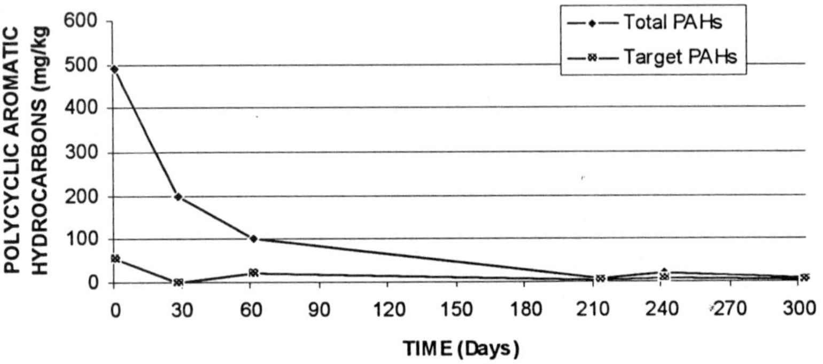


## ACKNOWLEDGMENT

The authors wish to thank C. Leggett of CMC Inc. for providing data obtained from the Phase 2 work.

## REFERENCES

Cerniglia, C.E.. 1992. "Biodegradation of Polycyclic Aromatic Hydrocarbons" *Biodegradation*, 3:351- 368.

Cookson, Jr., J.T.. 1995. *Bioremediation Engineering - Design and Application*, New York, NY: McGraw-Hill.

Sims, R.C., J.L. Sims, D.L. Sorensen, and L.L. Hastings. 1986. *Waste/Soil Treatability Studies for Four Complex Industrial Wastes: Methodologies and Results, Vol.1, Literature Assessment, Waste/Soil Characterization, Loading Rate Selection*, EPA/600/6-86/003a. Robert S. Kerr Environmental Research Laboratory, Office of Research and Development, U.S. Environmental Protection Agency, Ada, Oklahoma.

# LONG-TERM PASSIVE PCB/PAH BIOREMEDIATION FOLLOWING ACTIVE LAND TREATMENT

***John R. Smith*** (Alcoa, Pittsburgh, Pennsylvania)
Janice V. Fleckenstein (Pittsburgh, Pennsylvania)
Maria Mitraka (Serres, Greece)
Upal Ghosh (Stanford University, Stanford, CA)
A. Scott Weber (State University of New York at Buffalo, Buffalo, NY)
James N. Jensen (State University of New York at Buffalo, Buffalo, NY)
Robert Green (Alcoa, Massena, New York)

**ABSTRACT**: To demonstrate that biostabilization of PCBs and PAHs in soil-sludge-sediment is a technically sound, environmentally protective, and a low-cost bioremediation approach, Alcoa is in the midst of a multi-year pilot-scale engineered land-treatment based program to demonstrate the efficacy of a sequential active-passive bioremediation process. Four land treatment units (LTUs) were loaded with PAH- and PCB-impacted sludge, subjected to 2–3 months of active biotreatment, and then covered with soil to simulate a closed land treatment unit subject only to intrinsic (i.e., naturally occurring) bioremediation. PCB and PAH concentrations in soil and in water leachate generated from the soil were measured over time to document chemical losses in the soil and soil leachate characteristics. In the LTUs, 2–3 months of active biotreatment reduced total PAH concentrations in soil by 55–80% and total PCB concentrations by 27–40%. Long-term monitoring (5–6 years) of the pilot-scale units indicates that subsequent intrinsic bioremediation has yielded further contaminant destruction. Auxiliary measurements also indicate that observed contaminant reductions are the result of biodegradation with minimal losses to volatilization and transport.

## INTRODUCTION

Alcoa is in the midst of a multi-year pilot-scale study to demonstrate the efficacy of a sequential active-passive bioremediation process to achieve risk-based endpoints for soil-sludge-sediment materials containing PCBs and PAHs. Biostabilization refers to short-term active bioremediation to provide significant reductions of the more soluble, desorbable, and biodegradable lower molecular weight organic compounds as they rapidly desorb from the solid phase matrix. Following active bioremediation, longer-term intrinsic bioremediation serves to biodegrade the less soluble and slower desorbable organics remaining after the active phase, as they slowly desorb from the solid phase matrix.

This concept is supported by research that has shown that bioavailability is a primary controlling factor influencing both the rate and extent of organic biodegradation (Alexander, 1997; Loehr and Webster, 1997). Here, bioavailability is defined as accessibility of organics to bacteria present in the solid phase matrix macropore water. Bioavailability is largely a function of the sorption and dissolution mechanisms associated with the organics, solids, and NAPLs present in the solid phase matrix. With low bioavailability as the primary factor influencing biodegradation rate and extent, and assuming that sorption is reversible, given sufficient time frames, the potential exists for biodegradation to achieve significant concentration reductions of even the most highly sorptive compounds, including PAHs and PCBs.

Information pertinent to both the active and passive bioremediation phases of the study to date are presented below.

## METHODS AND MATERIALS

This Alcoa pilot-scale Land Treatment Unit (LTU) study has involved extensive research, planning, sampling, analyses, and documentation. Much detail about the study has been documented in previously published reports (Alcoa, 1995; Alcoa, 1998a; Alcoa, 1998b; Alcoa, 2000). Only a summary of the study details are presented here, and the information presented focuses on the target compounds: PAHs and PCBs.

**LTU Construction and Loading.** Four pilot-scale LTUs were located at Alcoa's Massena, New York facility. Constructed in 1994/1995, each LTU represents 2–3 months of active biotreatment followed by 5–6 years of intrinsic bioremediation. All four LTUs are of similar construction: a 1.2 by 9.1 meter cell containing ~7.5 cm of uncompacted sand resting on ~15 cm of compacted soil. The compacted soil is underlain by filter fabric, a drain net, and an impermeable high-density polyethylene bottom liner, all resting on a clay base. The portion of the LTU length available for biotreatment is ~7 m; a berm and sump occupy the remainder of the cell. Each LTU was constructed with a 2% slope leading toward a sump with a ~946 L capacity.

Each LTU was loaded with de-watered lagoon sludge from one or more of four distinct settling-treatment lagoons. For each of the four source lagoons, the sludge characteristics were unique as each lagoon received different materials for settling-treatment; but in each lagoon, PAHs and PCBs coexisted within the sludge. No supplemental bacteria (i.e., bioaugmentation) or supplemental organics were added. After loading, the sand-sludge layer was rototilled and mixed by shoveling until visual inspection indicated homogenization of the sand-sludge layer; then, the homogenized sand-sludge (soil) layer was sampled to determine the initial contaminant concentrations in the untreated soil, and active biotreatment commenced.

**Operation and Maintenance.** Each LTU was subjected to 2–3 months of active biotreatment. Active biotreatment operational activities included daily soil layer mixing and periodic nutrient addition (phosphorus and nitrogen), irrigation, dust control, sump water management, storm water management, and soil temperature management.

Immediately following the active biotreatment phase the intrinsic bioremediation phase commenced. The LTUs were covered to simulate a closed land treatment unit subject only to intrinsic (i.e., naturally occuring) bioremediation. Initially, cover was provided with an HDPE cover; subsequently, permanent coverage was established with a permeable geofabric layer and 3–6 inches of topsoil. Once applied, the topsoil was watered, fertilized, and resident vegetation was allowed to take over. Automatic pumps were placed in the LTU sumps to remove (for storage and eventual treatment) any excess water; all other operation procedures were terminated.

**Sampling and Analysis.** LTU soil was sampled and analyzed for PAHs and PCBs 3-4 times during the active biotreatment period. Three samples per LTU were analyzed for PAHs and PCBs with the following exceptions: single sample analyses for LTU-3 and extend PCB analyses for LTU-1 (6-8 soil samples per sampling event). For each LTU, replicate soil samples were obtained from the same well-mixed composite of 10 LTU grab samples. For LTUs 1, 2, and 4, a single soil sample was subjected to a modified version of the Toxicity Characteristic Leaching Procedure (TCLP) to produce a water leachate sample for PAH and PCB analysis. Other parameters evaluated included: nutrients, moisture, soil gases, oil & grease, total organic carbon (TOC), microbial assays, soil toxicity, a number of inorganics, and a number of other SVOCs and VOCs.

During the intrinsic bioremediation phase, LTU soil was sampled and analyzed for PAHs and PCBs approximately once per year. For each LTU, PAH and PCB analyses were executed on three replicate soil samples each created as a unique composite from 12 separate LTU grab samples. For each LTU, a single soil sample was subjected to a modified version of the Toxicity Characteristic Leaching Procedure (TCLP) to produce a water leachate sample for PAH and PCB analysis. Additionally, the following parameters were monitored one or more times during the intrinsic bioremediation phase: nutrients, moisture, soil gases, oil & grease, TOC, microbial assays, soil toxicity, and contaminant concentrations in resident vegetation.

All PAH analyses were conducted according to EPA standard protocols, SW-846:8310 or SW-846:8270, by one of the following laboratories: Aquatech Laboratories, Lancaster Laboratories, Alcoa Massena Laboratories, or Center Analytical. All PCB congener analyses were conducted according to laboratory developed in-house standard protocols by one of the following laboratories: Aquatech Laboratories, Northeast Analytical, or Alcoa Analysis and Testing Services. During the active biotreatment phase, multiple laboratories were used to analyze split/replicate samples for QA/QC purposes. During the intrinsic bioremediation phase, although different laboratories were utilized over the course of the study, all of the soil and water leachate samples for any single sampling event were consistently analyzed by a single laboratory for PAHs and a single laboratory for PCB congeners.

## RESULTS AND DISCUSSION

All soil sample results reported, except active biotreatment results for LTU-3, are average values from three to eight replicate samples; LTU-3 active biotreatment results are single sample analytical results. All water leachate results reported are single sample analytical results. All summations and averages include compounds reported at below detection limits equated to zero. The reported 90%

confidence interval on the mean was computed using the 2-tailed Student-t distribution.

**PAH Reductions in Soil.** PAH biodegradation from the soil was substantial during the active bioremediation phase and continued in the intrinsic bioremediation phase though at a much slower rate. Active biotreatment reduced total PAH concentrations in soil by greater than 55% for all LTUs. Long-term (5-6 years) intrinsic bioremediation continued reduction of PAHs yielding cumulative percent reductions of 93% or greater for all LTUs. Total PAH concentrations in soil over time are reported in Table 1 and graphically represented in Figure 1. Figure 1 legend entries include the percent reduction achieved at the end of active biotreatment, and cumulative percent reduction achieved as of the most recent intrinsic bioremediation sampling event.

**Table 1. Total PAH concentrations in soil**

| Time (Yrs.) | LTU-1 | LTU-2 | LTU-3 | LTU-4 |
|---|---|---|---|---|
| | Mean ± 90% Confidence Interval (mg/kg – dry weight) | | | |
| 0.0 | 1661.6 ± 181.6 | 973.7 ± 203.8 | 2.0 ± not available | 296.6 ± 9.6 |
| 0.1 | 399.5 ± 50.6 | 259.9 ± 31.2 | 1.2 ± not available | 84.7 ± 15.7 |
| 0.2 | 642.9 ± 35.5 | 325.9 ± 137.0 | 0.9 ± not available | 60.6 ± 3.9 |
| 0.3 | Not Sampled | Not Sampled | Not Sampled | 38.2 ± 12.7 |
| 1.0 | 518.6 ± 23.0 | 207.3 ± 7.0 | 1.7 ± 0.2 | Not Sampled |
| 1.3 | 360.0 ± 19.2 | 149.8 ± 20.8 | 0.8 ± 0.3 | 20.5 ± 1.9 |
| 2.0 | 208.7 ± 44.7 | 104.0 ± 23.2 | 1.3 ± 0.4 | 15.3 ± 1.7 |
| 3.0 | 166.7 ± 25.9 | 95.9 ± 8.5 | 0.4 ± 0.4 | 18.6 ± 1.9 |
| 5.0 | Not Sampled | Not Sampled | 0.006 ± 0.002 | 8.7 ± 3.8 |
| 6.0 | 124.4 ± 82.8 | 64.6 ± 20.1 | Not Sampled | Not Applicable |

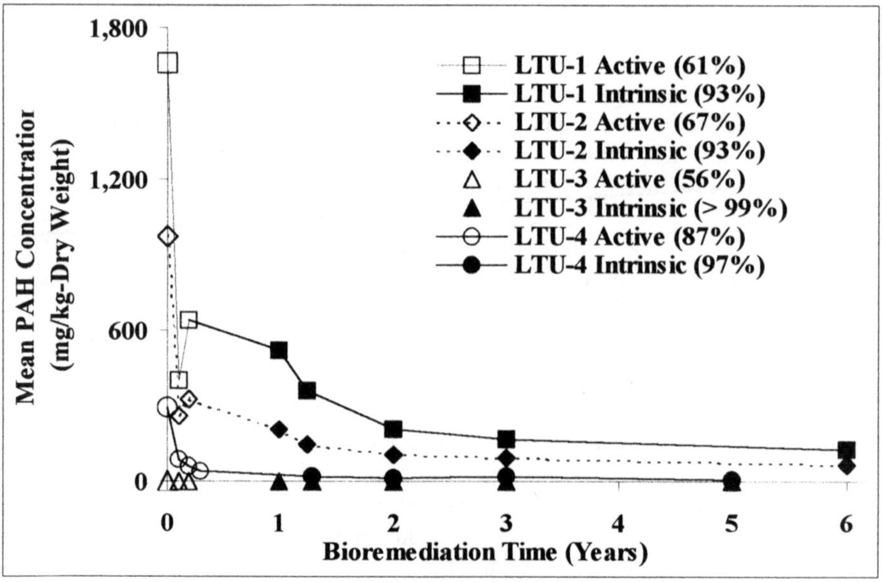


**Figure 1. Total PAH concentration in soil over time.**

During active bioremediation, the reductions noted were generally reductions of the lower molecular weight 3-ring and 4-ring compounds. Early in the intrinsic bioremediation timeframe, reduction trends focused on lower molecular weight PAHs; over time, the reduction trends were more evident for the higher molecular weight PAHs. Figure 2 shows PAH percent reductions in soil by ring size at the end of active biotreatment and as of the most current intrinsic bioremediation sampling point, 5–6 years (2-ring PAHs are not included as they were not detected in the untreated soil).

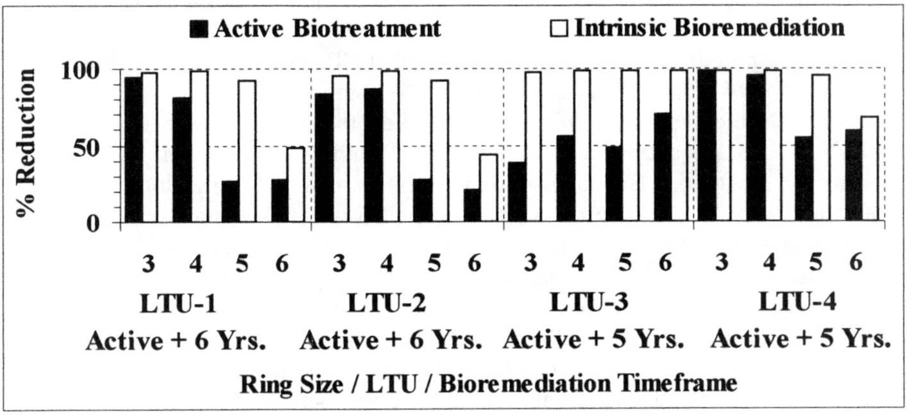


**Figure 2. PAH percent reduction in soil by ring size**

**PCB Reductions in Soil.** Table 2 and Figures 3(a) and 3(b) show total PCB concentrations in soil over time for all of the LTUs. The Figure 3 legend entries include the percent reduction achieved at the end of active biotreatment and the percent reduction achieved as of the most recent intrinsic bioremediation sampling point. Figures 4(a) and 4(b) present soil concentration reductions by homolog group at the end of active biotreatment and as of the most recent intrinsic bioremediation sampling (5–6 years); where PCB soil concentrations were reported at levels exceeding those reported in the untreated soil, percent reduction is represented with an "i" for "increase." (Monochlorinated PCBs are excluded from this figure since this homolog group was not detected in the untreated soil. Nona- and decachlorinated PCBs are also excluded from this figure due to low initial concentrations and surrogate interference during analysis.)

PCB biodegradation in the soil was greater than 26% for all LTUs during active bioremediation. Intrinsic bioremediation monitoring results for all LTUs confirm consistent, continued biodegradation of lower molecular weight PCB compounds and LTU-3 exhibited the highest biodegradation of all PCB compounds present (71%) in the soil. To further assess the impact of bioremediation on PCB concentrations in soil, further attention must be paid to PCB concentrations at the homolog level.

The biodegradation pattern for LTU-1 and LTU-2 clearly indicate that active biotreatment virtually depleted the dichlorinated PCBs, had a major impact on trichlorinated PCBs, and had little to no impact on more heavily chlorinated, less bioavailable compounds. Intrinsic bioremediation brought about continued degradation of the dichlorinated and trichlorinated compounds and introduced significant changes in the concentrations of tetrachlorinated compounds. Few of the concentration changes recorded in LTU-1 or LTU-2 for the pentachlorinated and more heavily chlorinated PCBs were statistically significant at the 90% (2-tailed Student-t) confidence level.

LTU-3 demonstrated the most notable bioremediation success in a compressed timeframe; after 5 years, all homolog groups were reduced by 31–93%, yielding a 71% reduction in total PCB concentration in soil. Over time, the degradation pattern displayed in LTU-3 was very similar to that described above: biodegradation of lightly chlorinated PCB compounds followed by biodegradation of more heavily chlorinated PCB compounds.

LTU-4 biodegradation patterns mimicked those described above in that significant biodegradation of dichlorinated and trichlorinated has been consistently recorded. But LTU-4 monitoring results for tetrachlorinated and higher chlorinated homolog groups were inconclusive with statistically significant increases and reductions recorded intermittently which resulted in the erratic concentration reported for the total PCBs in soil (Figures 3(b) and 4(b)). Soil heterogeneity may have played an import role in the variable results for LTU-4, which had by far, the largest PCB concentrations as a result of the heaviest NAPL-laden sludge loading. At this time, the reason for the variable results is not fully understood; the dissolution and desorption mechanisms occurring within the solid phase matrix are complex and introduce a myriad of possible partitioning scenarios. It is also noteworthy that LTU-4 PCBs were primarily associated with Aroclor 1260 and thus contain a higher percentage of more chlorinated PCB congeners. In comparison, LTU-1, 2, and 3 PCBs consisted primarily of Aroclor 1248, containing larger percentages of less chlorinated (3- and 4-chloro-biphenyl homologs) congeners.

**Table 2. Total PCB concentrations in soil**

| **Time** | **LTU-1** | **LTU-2** | **LTU-3** | **LTU-4** |
|---|---|---|---|---|
| **(Yrs.)** | **Mean ± 90% Confidence Interval (mg/kg – dry weight)** | | | |
| 0.0 | 11.8 ± 1.7 | 8.2 ± 0.7 | 2.8 ± not available | 113.3 ± 5.5 |
| 0.1 | 8.5 ± 1.4 | 5.0 ± 1.1 | 1.8 ± not available | 120.7 ± 4.1 |
| 0.2 | 7.2 ± 1.4 | 6.0 ± 1.1 | 1.7 ± not available | 117.3 ± 14.5 |
| 0.3 | Not Sampled | Not Sampled | Not Sampled | 73.2 ± 9.0 |
| 1.0 | 6.0 ± 0.5 | 5.1 ± 0.1 | 1.7 ± 0.2 | Not Sampled |
| 1.3 | 6.0 ± 0.3 | 4.9 ± 0.1 | 2.2 ± 0.4 | 75.4 ± 8.0 |
| 2.0 | 6.8 ± 0.6 | 4.9 ± 0.8 | 2.0 ± 0.3 | 121.7 ± 16.4 |
| 3.0 | 6.2 ± 0.7 | 4.9 ± 0.4 | 2.1 ± 0.2 | 91.1 ± 18.3 |
| 5.0 | Not Sampled | Not Sampled | 0.8 ± 0.1 | 106.3 ± 21.4 |
| 6.0 | 5.8 ± 1.8 | 4.7 ± 0.3 | Not Sampled | Not Applicable |

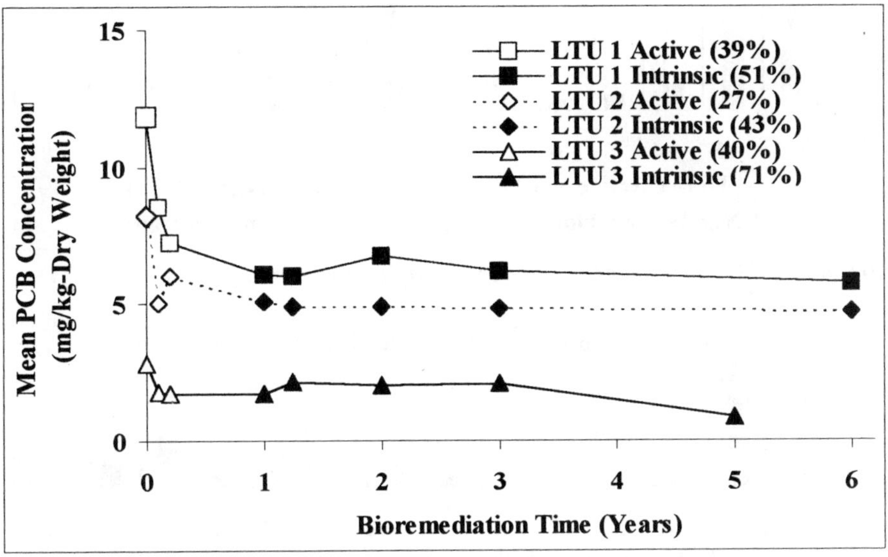


**Figure 3(a). Total PCB concentrations in soil LTU-1, LTU-2, and LTU-3.**

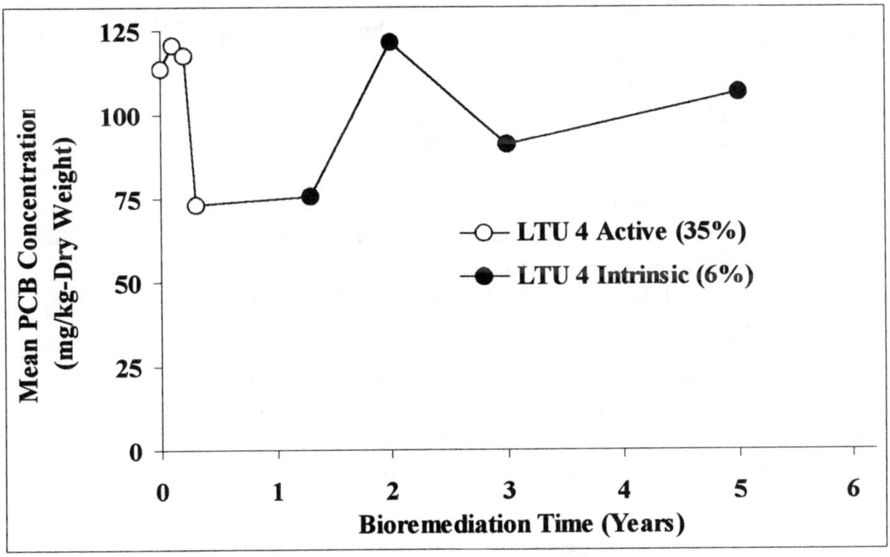


**Figure 3(b). Total PCB concentration in soil LTU-4.**

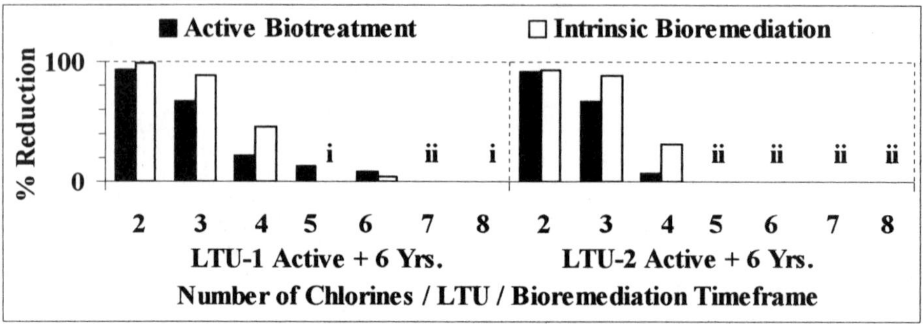


**Figure 4(a). PCB percent reductions in soil by homolog group LTU-1 & LTU-2**

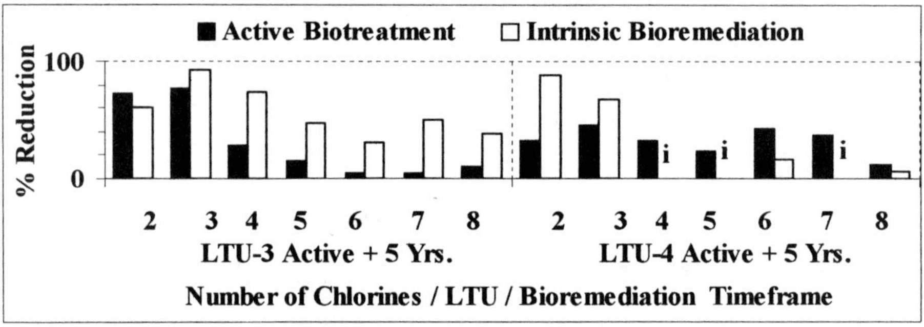


**Figure 4(b). PCB percent reductions in soil by homolog group LTU-3 & LTU-4**

**PAH and PCB Leachability.** Total contaminant concentrations in water leachate over time are shown in Table 3 and in Figure 5 for PAHs and Figure 6 for PCBs. For both figures, legend entries include the percent reduction achieved at the end of active biotreatment and the percent reduction achieved as of the most recent intrinsic bioremediation sampling event (5-6 years). As cited, active biotreatment provided swift reductions of contaminant concentrations in water leachate. PAH and PCB concentrations in water leachate were reduced 40–95%. Following active bioremediation, long-term sampling results for contaminant concentrations in water leachate have been characterized by variability.

**Table 3. Total contaminant concentrations in water leachate**

| Time | PAHs (μg/L) | | | | PCBs (μg/L) | | | |
|---|---|---|---|---|---|---|---|---|
| (Yrs.) | LTU-1 | LTU-2 | LTU-3 | LTU-4 | LTU-1 | LTU-2 | LTU-3 | LTU-4 |
| **0.0** | 91.20 | 40.00 | NS | 5.25 | 0.95 | 0.78 | NS | 3.09 |
| **0.1** | NS | NS | NS | 0.53 | NS | NS | NS | 3.57 |
| **0.2** | 6.00 | 2.00 | NS | 1.24 | 0.43 | 0.28 | NS | 1.19 |
| **0.3** | NS | NS | NS | 3.17 | NS | NS | NS | 0.46 |
| **1.0** | 0.14 | 0.46 | 1.35 | NS | 0.11 | 0.12 | 0.01 | NS |
| **1.3** | 5.83 | 3.83 | 1.36 | 0.11 | 0.05 | 0.02 | 0.01 | 3.93 |
| **2.0** | 0.30 | 0.06 | ND | 1.39 | 0.24 | 0.15 | 0.06 | 3.24 |
| **3.0** | 1.02 | 0.99 | 1.62 | ND | 0.23 | 0.22 | 0.11 | 4.59 |
| **5.0** | NS | NS | ND | 0.16 | NS | NS | 0.06 | 5.07 |
| **6.0** | 1.13 | 0.19 | NS | NA | 0.21 | 0.19 | NS | NA |

NS: Not Sampled; NA = Not Applicable; ND = Not Detected in sample

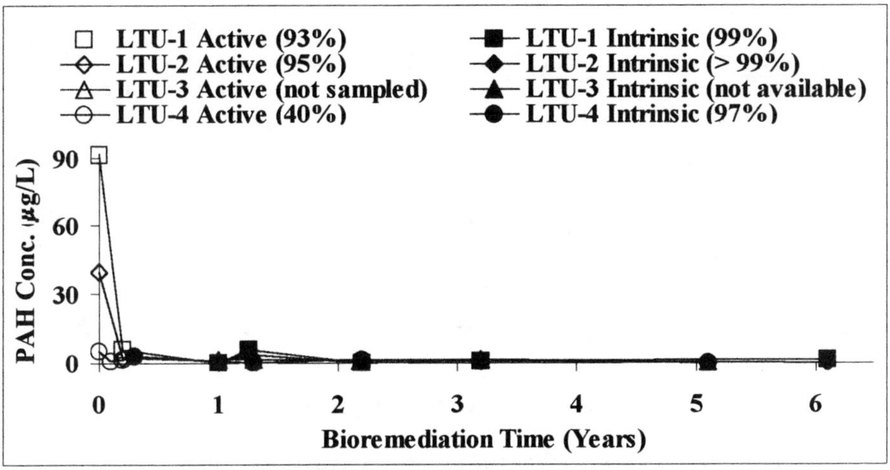


**Figure 5. Total PAH concentration in water leachate over time.**

Based on an independent study of LTU-4, the variability in water leachate PCB concentrations is explained by the presence of the much higher contaminant-laden oil and grease in LTU-4 (Ghosh et. al, 2000); we hypothesize that the presence of the heavily contaminant-laden oil and grease in LTU-4 affects contaminant partitioning behavior. Preferential dissolution and biodegradation of oil and grease constituents may result in contaminant mole fraction increases within the NAPL. Based on Raoult's Law (i.e., aqueous phase concentration of contaminant is equal to the mole fraction of the contaminant in the oil phase times the aqueous solubility of the pure contaminant), as the contaminant mole fraction of the less soluble compounds within the oil increases so do the contaminant aqueous equilibrium concentrations. This theory correlates strongly with the reported results and may explain, to a large degree, the variability of the reported water leachate concentrations. For comparison purposes, Table 4 presents oil & grease concentrations measured in LTUs 1, 2, 4 (LTU-3 O&G is not available).

**Table 4. Oil and grease concentrations and percent reductions**

| | Initial Concentration (mg/kg-dw) | % Reduction At end of Active | Most Recent Intrinsic Samples (mg/kg-dw) | % Reduction Current |
|---|---|---|---|---|
| **LTU-1** | 6,707 | 56% | 1,460 | 78% |
| **LTU-2** | 3,940 | 37% | 1,190 | 70% |
| **LTU-4** | 9,100 | Not Analyzed | 1,430 | 84% |

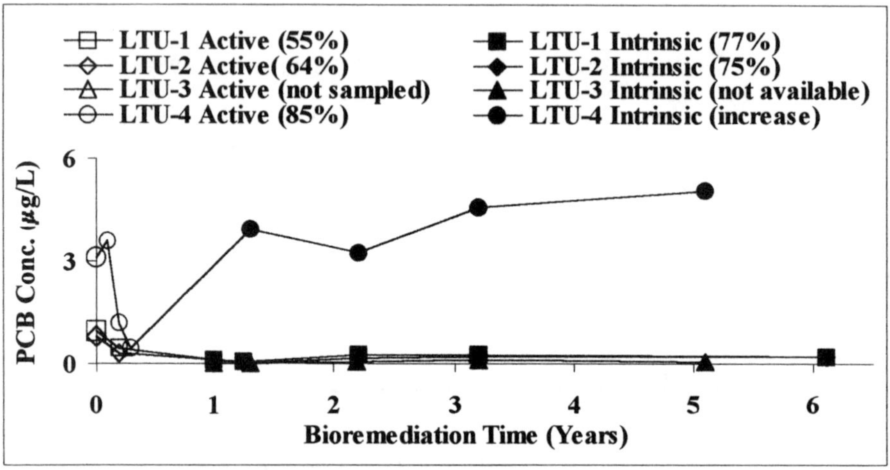


**Figure 6. Total PCB concentration in water leachate over time.**

**Toxicity Reductions.** A number of toxicity test have been executed on the LTU soils over the course of this project (Alcoa, 1998b). For untreated LTU soil, Microtox toxicity tests on luminescent bacteria showed a 50% reduction in luminescence when bacteria was exposed to test solutions containing 0.3% LTU-4 leachate and to test solutions containing greater than 50% LTU leachate for LTUs 1, 2, and 3. For all LTUs after active biotreatment, test solutions containing 100% LTU leachate did not yield a 50% reduction in luminescence. Earthworm (14-day whole soil) survival tests measured a 37% survival rate for earthworms exposed to untreated LTU-1 soil. After 2 to 3 years of bioremediation, earthworm 14-day survival rates were 97-100% for all LTU soils.

**Summary and Conclusions.** This study strongly supports that bioremediation of the PAHs in the contaminated LTU soils has reduced the threat posed to humans and the environment. As biodegradation proceeds, the more desorbable, more soluble contaminants are depleted, and the contaminants left in the soil are the strongly sorbed, less leachable components, which supports that biostabilization is being approached. This concept can also be hypothesized for the PCBs in LTUs 1, 2, and 3, but not LTU-4 which contains a more contaminant-laden NAPL and is predominantly Aroclor 1260.

From a full-scale perspective, this study confirms that implementation of land-treatment based biodegradation technologies requires adequate engineering controls to ensure no exposure of target contaminants to receptors, these controls may be lessened as determined by site-specific risk-based factors as environmentally acceptable end-points are reached.

Alcoa plans to continue operating and monitoring the LTUs over the next five years.

## References

Alcoa (1995) *Bioremediation Tests for Massena Lagoon Sludges/Sediments – Program Summary*, Volumes I and II. EHS Technology Center, Alcoa Technical Center, Alcoa Center, PA.

Alcoa (1998a) *Long Term Monitoring of Land Treatment Units in Massena, New York.* Report ETC No. 98-01-01. EHS Technology Center, Alcoa Technical Center, Alcoa Center, PA.

Alcoa (1998b) *Long Term Monitoring of Land Treatment Units in Massena, New York. Status Report 2.* Report ETC No. 98-10-01. EHS Technology Center, Alcoa Technical Center, Alcoa Center, PA.

Alcoa (2000) *PCB Destruction and Mobility Reduction.* Prepared for U.S. Department of Energy/ACQ by Alcoa, EHS Technology Center, Alcoa Technical Center, Alcoa Center, PA.

Alexander, M. (1997). "Sequestration and Bioavailability of Organic Compounds in Soil." In D. Linz and D. Nakles (Eds.), *Environmentally Acceptable Endpoints in Soils.* Annapolis, MD, American Academy of Environmental Engineers.

Ghosh, U., A.S. Weber, J.N. Jensen, J.R. Smith (2000) "Relationship Between PCB Desorption Equilibrium, Kinetics, and Availability During Land Biotreatment." *Environ. Sci. Technol*, 34, 2542-2548.

Loehr, R.C., and Webster, M. T. (1997) "Effect of Treatment on Contaminant Availability, Mobility and Toxicity." In Linz, D. and Nakles, D., (Eds.) *Environmentally Acceptable Endpoints in Soils.* Chapter 2. Annapolis, MD, American Academy of Environmental Engineers.

# BIOREMEDIATION OF CREOSOTE-CONTAMINATED SOIL BY INDIGENOUS SOIL FUNGI.

***Harrison I. Atagana*** ( Mangosuthu Technikon, P.O.Box 12363, Jacobs, Durban 4026, South Africa).
R.J.Haynes and F.M.Wallis ( University of Natal, Pietermaritzburg 3209, South Africa).

**ABSTRACT:** Fungal isolates from a creosote contaminated land and from the soil of an uncontaminated forest floor were tested for their capabilities to colonize and degrade creosote in a soil contaminated with >25,000 mg $kg^{-1}$ (w/w) creosote. The fungi included two strains of basidiomycete both of which were *Pleurotus sp.*, one from the contaminated soil and the other from an uncontaminated forest floor, and four strains of non-basidiomycete fungi, *Cladosporium sp., Fusarium sp., Penicillium sp.* and *Aspergillus sp.,* from the contaminated soil. Soil analyses was done by Soxhlet extraction and gas chromatography (Eriksson *et al.*,2000), to effect the separation of volatile and heavy components of creosote. The non-basidiomycete were found to remove between 78% and 88% creosote from the soil within the 70 days of treatment. *Pleurotus sp.* was found to remove up to 94% of the initial creosote concentration. A composite mixture of all the isolates from the contaminated soil showed better creosote removal from the soil. Barley as a bulking agent, N and P supplementation and aeration were observed to influence the rate of growth of all the fungi and improve on the removal of creosote from the soil. The results showed that fungi can be effectively used to remediate creosote contaminated soils provided that adequate conditions are provided. Studies are presently underway to find means of sustaining the growth of the fungi for longer periods for use in large scale application in bioremediation.

## INTRODUCTION

Bioremediation technologies have continued to rely on bacterial activity to decontaminate the environment. While bacterial bioremediation has continued to improve, very little attention has been given to fungal remediation of the environment.

Both bacteria and fungi are ubiquitous in soil and members of both groups contribute to the biodegradation of hydrocarbons. Although, in many soils fungi appear to be more important than bacteria in community respiration (Alexander, 1994), studies of degradation of hydrocarbons in the environment are commonly based on bacterial activity. This may, in part, be due to the difficulties presented by fungi increasing their biomass through hyphal lengthening and branching rather than binary fission, and the organisms undergoing morphological changes during their life cycle (Alexander, 1994). Such heterogeneity is related to changes in physiological activities in the course of fungal growth (Fritsche, 1992).

Fungi possess a number of characteristics that are potentially useful for application in bioremediation. They are able to grow under environmentally stressed conditions such as low nutrient availability, low water activity and at low pH values where bacterial growth might be limited (Davies and Westlake, 1979). Their hyphal growth is chemotactically directed to areas of the soil covered by the carbon substrate thus optimizing surface contact with the contaminant and increasing the bioavailability of the compound. Kirk (1969) found that bacteria was attached to the mucilaginous hyphae of *Corollospora maritima* during degradation of oil. This suggests that soil contaminants are cometalized by soil microorganisms. Under the co-metabolic conditions fungal degradation is a transformation-detoxification process rather than catabolism (Fritsche, 1992).

Although, many reports claim the success of fungi in metabolizing different hydrocarbons, only a few reports are available where direct application of fungi in bioremediation has been carried out. The potential for these organisms to degrade high molecular weight polycyclic aromatic hydrocarbons ( Wood and Wiseman, 1979; Cerniglia *et al.,* 1979; Field *et al.,* 1992; Pothuluri *et al.,* 1992; MacGillivray and Shiaris, 1993; Andersson and Henrysson, 1996) and other recalcitrant organic compounds in the laboratory by the use of their extracellular and other enzyme systems, presents an attractive area for investigation.

Researches on fungal bioremediation of hydrocarbon pollutants have been mainly on the white and brown rot fungi which are naturally not inhabitants of the soil environment. This focus is due to their ability to produce hydrogen peroxide and enzymes that have been found to degrade lignin to its final product carbon dioxide (Gold *et al.,* 1989). The low specificity of the enzymes produced by the white rot fungi is believed to enable them degrade recalcitrant, anthropogenic compounds.The white rot fungus *Phanerochaete chrysosporium* have been reported to degrade a number of xenobiotics including DDT, PCP, PCB, TCCD and PAHs (Higson, 1991; Field *et al.,* 1992; Bogan and Lamar, 1995; Novotný *et al.,* 1999). Cerniglia (1997) presents a list of PAHs that have been reported to successfully supported the growth of a large number of fungi and yeasts. This report gives a further indication that these organisms, if managed under appropriate conditions, will help address the issue of the recalcitrance of some of the PAHs in soil.

The land under investigation is in a creosote wood treatment plant. Many parts of the facility was observed to be contaminated by creosote hydrocarbons to varying degrees. Although, creosote in the soil of some parts of the facility were seen to be weathered, concentrations were observed to range between 25 000 mg $kg^{-1}$ and 5 000 mg $kg^{-1}$ . The main focus of this study is thus, the application of fungi in treating soils contaminated with high concentrations of creosote hydrocarbons many of which are resistant to bacterial degradation.

## MATERIALS AND METHODS

### Soil samples

Soil contaminated with creosote was collected from the site with a spade to a depth of about 30 cm. The soil was transported to the laboratory and stored at 4 $^{O}C$ until required.

**Soil reactors**

Eighteen polyvinyl chloride (PVC) vessels of about 3 litre capacity each, with a diameter of about 33 cm and depth of 13 cm were each filled with 2.5 kg of the heavily contaminated ( >25,000 mg C $kg^{-1}$ soil, fresh weight) soil from the experimental site. Barley grains previously infected with selected strains were used as bulking in the reactors. Polystyrene boards cut to fit into the vessels, were perforated at regular intervals to allow passage of leachate, were fitted into the vessels and supported at regular intervals underneath to prevent it from collapsing under the weight of the soil, before the soil was placed in the vessels. This was designed to prevent the soil from becoming waterlogged during the period of the experiment.

**Table 1: Treatments in soil reactors.**

| Reactor | Barley | Fungi | Nutrient |
|---|---|---|---|
| 1 | yes | *Cladosporium sp.* | no |
| 2 | yes | *Fusarium sp.* | no |
| 3 | yes | *Penicillium sp.* | no |
| 4 | yes | *Aspergillus sp.* | no |
| 5 | yes | *Pleurotus sp.* | no |
| 6 | no | no | no (control 1) |
| 7 | yes | *Cladosporium sp., Fusarium sp., Penicillium sp., Aspergillus sp., Pleurotus sp.* | no |
| 8 | no | no | yes (control 2) |
| 9 | sterile | no | no (control 3) |
| 10 | yes | *Cladosporium sp.* | yes |
| 11 | yes | *Fusarium sp.* | yes |
| 12 | yes | *Penicillium sp.* | yes |
| 13 | yes | *Aspergillus sp.* | yes |
| 14 | yes | *Pleurotus sp.* | yes |
| 15 | sterile | no | yes (control 4) |
| 16 | yes | Pleurotus sp.(from CEDARA) | yes |
| 17 | yes | Pleurotus sp.(from CEDARA) | no |
| 18 | yes | *Cladosporium sp. Fusarium sp. Penicillium sp. Aspergillus sp. Pleurotus sp.* | yes |

*KwaZulu-Natal Department of Agriculture, Cedara, South Africa.

Black polyethylene sheets were used to cover the vessels to prevent photo-oxidation of the creosote components. Distilled water was used to maintain the soil at 70 % water holding capacity. Soil pH was found to be 4.5 at the beginning of the experiment.

**Treatments**

Treatments are as shown in Table 1. Supplementary nutrients were added by means of $NH_4NO_3$ and $K_2HPO_4$ to give a C:N:P ratio of 25:5:1 as recommended for aerobic metabolism (Kennedy *et al.*, 1975). All soil boxes received the same volume of water for the duration of the experiment. Aeration was achieved by tilling the soil with a small garden trowel once every week.

**Analyses of the samples**

Soxhlet extraction as outlined in EPA Method 3540 was used to extract the hydrocarbons from the soil. The extracts were analysed by GC/FID. The GC was a Varian-3800 with argon as the carrier gas and fitted with a 30 m capillary column with

0.25 mm internal diameter and 0.25 Φm film thickness, and a flame ionization detector (FID). Two temperature programmes as described by (Eriksson *et al.,* 2000) were run in order to obtain a good separation and quantification of both the volatile and heavy components of creosote.

## RESULTS AND DISCUSSION

**Degradation of creosote in soil reactors**

All the treatments resulted in the removal of between 65% and 94% of creosote from the soils in the reactors. Removal of creosote from the control reactors ranged between 26% and 58% (Figure1). Decreases in the concentration of creosote in the first 14 days was relatively slow in all the reactors (Figures 2, 3 & 4). This was because it took some time for the fungi to establish in the soil. Reduction in hydrocarbon generally became faster in the last one month. This is attributed to the well established growths found in all the treated samples at this stage.

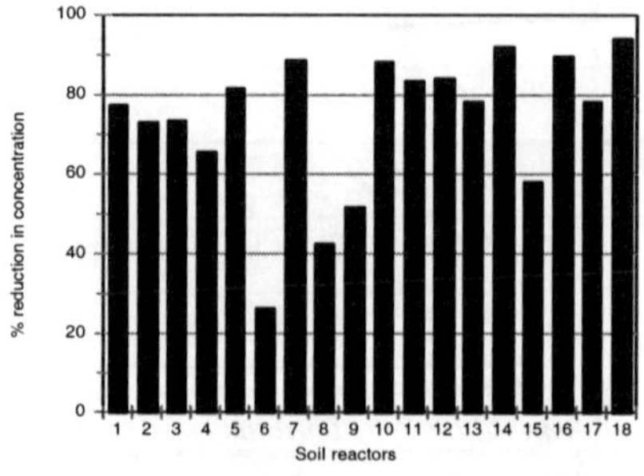


**Figure 1. Total reduction (%) in concentration of creosote in soil reactors**

The single strains of fungi in the reactors not supplemented with inorganic nutrients (N and P) (Reactors 1-5), showed good degradation of creosote (65.6%-81.7%) (Figure 1). The least reduction was observed in the reactor inoculated with *Aspergillus sp.* (Reactor 4) while the highest reduction in this group was recorded in the reactor inoculated with *Pleurotus sp.* (Reactor 5). All other strains used in this group reduced the creosote beyond 70% within the treatment period. The reactor that received a composite mixture of all the strains of fungi isolated from the contaminated soil ( Reactor 7) however, showed a higher degradation of 88.7% (Figure 1). Control 1(Reactor 6) in which there was no barley bulking, fungal inoculation nor nutrient addition, but was watered and aerated like the others, showed a significantly less reduction (26.4%) in creosote concentration (Figure 1). There was a slight improvement in creosote reduction (42.6%) in Control 2 which received nutrient supplements. Control 3 which had decontaminated barley as bulking but received no nutrient supplement showed a slightly higher reduction (51.8%) in creosote than Control 2. This result shows that nutrient (N and P) addition and barley as bulking agent, enhanced the degradation of creosote in the soil.

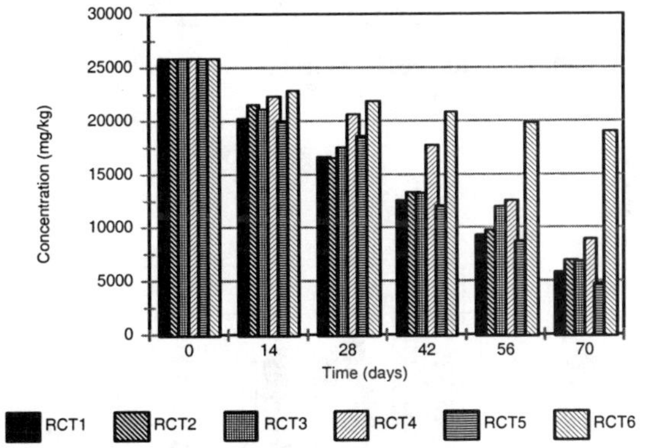


**Figure 2. Changes in concentration (mg kg$^{-1}$) of creosote in reactors 1-6**

Most fungi are not known to utilize PAHs and other hydrocarbon as their sole source of carbon and energy, and as a result the medium must be supplemented with an additional carbon source to allow fungi to metabolize such compounds (Cerniglia, 1997). Since fungi are known to metabolize their substrate by non-specific extracellular enzymes ( Martens and Zedrazil, 1992; Field *et al.*, 1992; Fritsche, 1992), the enzymes produced during growth on another substrate can readily facilitate the metabolism of another compound. Carbohydrate-rich substrates such as barley, wheat, and cotton seed are commonly used as alternative sources of carbon for cultivation of fungi for various purposes

(Rodriguez *et al.,* 1999). The growth advantage shown in barley treated reactors manifested in greater creosote degradation ( Figure 3).

Reactors 10-14 and 16 and 18 all received inorganic nutrient and fungi inoculation. In this group of reactors, *Pleurotus sp.* also showed the highest reduction (92.1%) among the reactors inoculated with single strains of fungi (Reactor 14). In

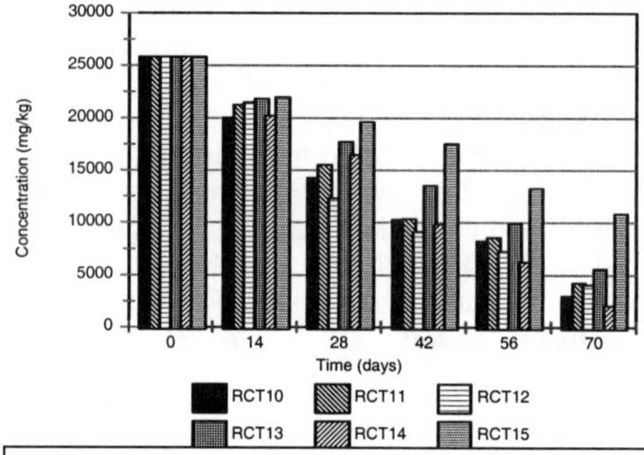


**Figure 3. Changes in concentration (mg kg$^{-1}$) of creosote in reactors 10-15**

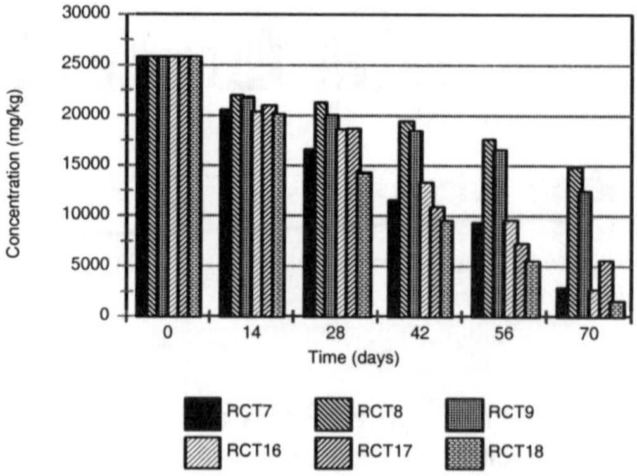


**Figure 4. Changes in concentration (mg kg$^{-1}$) of creosote in reactors 7-9 & 16-18**

all the reactors in this group, creosote degradation was observed to be significantly higher than their corresponding reactors that did not receive inorganic nutrient supplements (Figure 2 & 3). The introduction of pollutant hydrocarbon into the soil supplies a large amount of carbon which often results in an excessively high C:N and C:P ratios. This is usually unfavourable for microbial growth, hence the supplementation of the medium with N and P. Among the single strains in both groups of reactors, *Pleurotus sp* and *Cladosporium sp.* were observed to have the highest degradative capacities. Both fungi have been reported to grow on and metabolize organic substrates (Eggen *et al.*, 1999; Novotoný *et al.*, 1999).

Control 4 which was bulked with sterilized barley and received inorganic nutrient supplement but was not inoculated with any fungi showed the highest degradation among the controls (Figure 3). Since the controls did not have any biological inhibitors incorporated into them, soil bacteria, fungi and actinomycetes were observed to grow in them. This explains the relative high degradation (Reactor 15 = 58.2%) recorded in some of them. The reactor with composite mixture of fungi that also received the nutrient supplement showed the highest degradation (94%) among the entire treatment. The fact that supplementary nutrient were observed to enhance microbial degradation of creosote confirms earlier reports of Lamar *et al.* (1989).The *Pleurotus sp.* obtained from the uncontaminated forest floor was found to less effective in degrading creosote than the one isolated from the contaminated soil (Figure 4).

Of the selected creosote compounds studied, the phenolic (phenol, o-cresol, m-cresol, p-cresol) and the lower molecular weight polycyclic aromatic hydrocarbons (PAHs) (naphthalene, anthracene, phenanthrene, pyrrole, fluorene) were degraded below detectable level. The higher molecular weight PAHs ( pyrene, chrysene, flouranthene, benzo(a)pyrene) studied were well degraded but still showed some residual concentrations (Figures 5). Reduction of these compounds in the control experiments were very much lower than the other reactors.

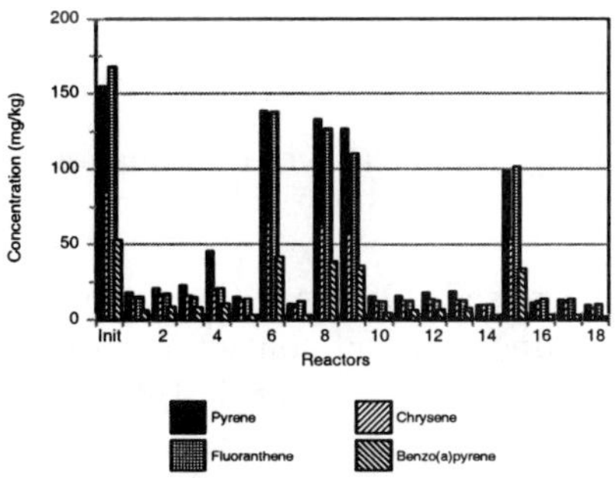


**Figure 5. Changes in concentrations (mg kg$^{-1}$) of higher molecular weight PAHs**

Through the treatment period, pH was observed to fluctuate very slightly.

These changes were however not seen to affect fungal growth nor creosote degradation. The lowest pH observed was 3.8 and the highest was 6.6. This acid range of pH favoured the growth of fungi in the soil.

## CONCLUSION

From the results obtained, it is clear that fungi have very strong potentials for the degradation of creosote compounds in soil. It was also established that the white rot fungus *Pleurotus* is a better degrading organism than the non-basidiomycete fungi studied. Mixed cultures of fungi were more effective in creosote degradation. The *Pleurotus sp.* isolated from the contaminated environment showed more degradative capabilities than the ones isolated from the uncontaminated forest soil.

Barley proved to be very effective in supporting the growth of the fungi and also in the concomitant degradation of the hydrocarbon substrate. Nutrient supplementation proved to be effective in enhancing the degradation of creosote.

The phenolic compounds and the lower molecular weight PAHs studied were to be more susceptible to fungal degradation than the higher molecular weight PAHs. The higher molecular weight PAHs were however more effectively degraded than the results obtained in another study using bacteria on the same contaminated soil..

## DRAWBACKS AND RECOMMENDATIONS

The major problems envisaged in the large-scale application of fungi for bioremediation is the maintenance of a large amount of the culture for a long period in order to facilitate a continuos spread of the fungi through the contaminated soil. It is therefore suggested that the study of bulking and continued sustenance of fungal cultures on a large scale be carried out before fungal bioremediation is embarke upon. In an outdoor system, the production of spores could also present some problems if not controlled in an enclosure. It is obvious that large spore releases in the atmosphere could constitute a health hazard.

## REFERENCES

Alexander, M. 1994. Biodegradation and bioremediation. Academic Press, San Diego. p.302.

Andersson, B.E. and T. Henrysson. 1996. Accumulation and degradation of dead-end metabolites during treatment of soil contaminated with polycyclic aromatic hydrocarbons with five strains of white rot fungi. *Appl. Microbiol. Biotechnol.* 46: 647-652.

Bogan, B.W. and R.T. Lamar. 1995. One-electron oxidation in the degradation of creosote polycyclic aromatic hydrocarbons by *Phanerochaete chrysosporium. Appl. Environ. Microbiol.* 61: (7) 2631-2635.

Cerniglia, C.E., R .L. Herbert, R. H. Dodge, P.J. Szaniszlo, and D.T. Gibson.

1979. Some Approaches to Studies on the Degradation of Aromatic Hydrocarbons by Fungi, In A.L Bourquin and H. Pritchard (eds), Microbial Degradation of Pollutants in Marine Environments. EPA Report No. EPA-600/9-79-012.EPA, Washington, DC.

Cerniglia, C.E. 1997. Fungal metabolism of polycyclic aromatic hydrocarbons: past , present and future applications in bioremediation. *J. Ind. Microbiol. Biotechnol.* 19: 324-333.

Davies, J.S. and D.W.S. Westlake. 1979. "Crude oil utilization by fungi". *Can. J. Microbiol.* 25: 146-156.

Eggen, T., E. Araneda, Ø. Vethe, and P. Sveum. 1999. Degradation of aged creosote-contaminated soil by *Pleurotus ostreatus.* In: Bioremediation technologies for polycyclic aromatic hydrocarbon compounds. (Eds) Leeson, A. and Alleman, B.C. Battelle Press, Columbus. pp 99-104.

Eriksson, M., G. Dalhammar, and A.-K. Borg-Karlson. 2000. Biological degradation of selected hydrocarbons in an old PAH/creosote contaminated soil from a gas work site. *Appl Microbiol Biotechnol.* 53: 619-626.

Field, J.A., E. De Jong,, G.F. Costa, and J.A.J. de Bont. 1992. Biodegradation of polycyclic aromatic hydrocarbons by new isolates of white rot fungi. *Appl. Environ. Microbiol.* 58: (7) 2219-2226.

Fritsche, W. 1992. Degradation of xenobiotics by fungi. In: Preprints of international symposium on soil decontamination using biological processes, Karlsruhe, Germany. 6-9 December 1992. Dechema, Frankfurt am Main. pp 31-36.

Gold, M.H., H. Wariishi, K. Valli. 1989. Extracellular peroxidases involved in lignin degradation by the white rot basidiomycete *Phanerochaete chrysosporium. ACS Sym Ser.* 389: 127-140.

Higson, F.K. 1991. Degradation of xenobiotics by white rot fungi. *Rev. Environ. Contamin. Toxicology.* 122: 111-152.

Kennedy, R.S., W.R. Finnerty, K. Sudarsanan, R.A.Young. 1975. Microbial assimilation of hydrocarbons. I. The fine structure of a hydrocarbon-utilizing *Arthrobacter sp. Arch. Microbiol.* 102: 75-83.

Kirk, P.W. 1969. Isolation and culture of lignicolous marine fungi. *Mycologia.* 61: 174-177.

Lamar, R.T., J.A. Glaser and T.K. Kirk. 1989. Use of white-rot fungi to remediate soils contaminated with wood preserving waste. Abstracts Fourth International

Conference-Biotechnology in the Pulp and Paper Industry. Raleigh, N.C. May 16-19, 1989. P. 82.

Lees, Z.M. 1996. Bioremediation of oil-contaminated soil: A South African case study. A PhD thesis submitted to the University of Natal, Pietermaritzburg, South Africa.

MacGillivray, A.R. and M.P. Shiaris. 1993. Biotransformation of polycyclic aromatic hydrocarbons by yeasts isolated from coastal sediments. *Appl. Environ. Microbiol.* 59: 1613-1618.

Martens, R. And F. Zadrazil. 1992. Screening of white rot fungi for their ability to mineralize polycyclic aromatic hydrocarbons in soil. In: Preprints of international symposium on soil decontamination using biological processes, Karlsruhe, Germany. 6-9 December 1992. Dechema, Frankfurt am Main. pp 31-36.

Novotný, C., P. Erbanová, V. Šašek, A. Kubátová, T. Cajthaml, E. Lang, J. Krahl and F. Zadra□il. 1999. Extracellular oxidative enzyme production and PAH removal in soil by exploratory mycelium of white rot fungi. *Biodegradtion.* 10: 159-168.

Pothuluri, J.V., J.P. Freeman, F.E. Evans, and C.E Cerniglia. 1992. Fungal metabolism of acenaphthene by *Cunninghamella elegans. Appl. Environ. Microbiol.* 58: 3654-3659.

Rodriguez-Vazquez, R., C. Montalvo, L. Dendooven, F. Esparza and L. Fernandez-Linaris. 1999. "Degradation of benzo(a)pyrene in soil by White Rot Fungi". In bioremediation technologies for polycyclic aromatic hydrocarbon compounds (Eds) A. Leeson and B.C. Alleman. Battelle Press, Columbus. p. 93-98.

Woods, L.F.J. and A. Wiseman. 1979. Metabolism of benzo(a)pyrene by the cytochrome P-450/P-448 of *Saccharomyces cerevisiae. Biochem. Soc. Trans.* 7: 124-127.

# BIOREMEDIATION OF DIOXIN-CONTAMINATED SOILS BY DIFFERENT TYPES OF BACTERIA

***Hiroshi Habe***, Hiroyuki Hirano, Takako Yoshida, Hideaki Nojiri, and Toshio Omori
(The University of Tokyo, Tokyo, Japan)
Kazuki Ide, Mizuyo Yotsumoto, and Hirokazu Tsuji
(Obayashi Corporation, Ltd., Tokyo, Japan)

**ABSTRACT:** Methods for bioaugmentation of polychlorinated dibenzo-*p*-dioxins and polychlorinated dibenzofurans (PCDDs/PCDFs or $Cl_XDDs/Cl_XDFs$, x=1-8) by gram-positive and gram-negative bacteria have developed. A dibenzofuran (DF)-utilizing bacterium *Terrabacter* sp. strain DBF63 and a carbazole (CAR)-utilizing bacterium *Pseudomonas* sp. strain CA10 were used as inocula. In biotransformation experiments of PCDDs and PCDFs, both the DF-grown DBF63 cells and the CAR-grown CA10 cells could degrade the tested mono- to trichlorinated DDs and DFs, and some mono- and dichlorinated DDs and DFs were converted to corresponding chlorosalicylic acid and chlorocatechols. 2,8-$Cl_2DF$ degradation by DBF63 cells and 2,3-$Cl_2DD$ degradation by CA10 cells were also investigated in model soil samples, showing that 90% of 2,8-$Cl_2DF$ was degraded by augmented strain DBF63 after 3 days and that 80% of 2,3-$Cl_2DD$ was degraded by augmented strain CA10 after 5 days. Finally, using both dioxin-degraders, we tried remediation of the actual dioxin-contaminated soil. It was revealed that both strains had potential to degrade tetra- to hexachlorinated congeners including most toxic compound, 2,3,7,8- $Cl_4DD$.

## INTRODUCTION

A considerable part of man-made, xenobiotic chemicals released to the environment strongly resists abiotic and biotic degradation. Especially, PCDDs and PCDFs are widely spread in the environment, and show extremely high persistence in soil sediments. PCDDs and PCDFs are produced as unwanted by-products in a variety of manufacturing processes such as production of herbicides, insecticides, fungicides and paper pulp bleaching as well as during combustion of solid waste in incinerator or accidental fires, and they are serious pollutants because of their high toxicity.

Recently, bioremediation has become more popular as an alternative to physical and chemical methods because of their relatively low cost and minimal impact on the environment. Hence, many researchers have focused their studies on the isolation of dioxin-degrading microorganisms that can be cultured in the laboratory, and resultantly, several aerobic bacteria able to degrade and co-oxidize PCDDs and PCDFs have been isolated. In earlier studies, we reported the isolation and characterization of some bacteria that can degrade dioxin-related compounds and their initial dioxygenase genes. One of such strains, the gram-positive bacterium *Terrabacter* sp. strain DBF63, which was isolated based on its ability to utilize DF as a sole carbon and energy source, could co-metabolize dibenzo-*p*-dioxin (DD) (Monna et al., 1993). Also, the gram-negative bacterium *Pseudomonas* sp. strain CA10, which can utilize CAR as a sole carbon, nitrogen, and energy source (Ouchiyama et al.,

1993), could co-oxidize DD and DF. Functional analyses of enzymes catalyzing the angular dioxygenation in these strains revealed that the substrate specificities and the organization of terminal oxygenase components were quite different from each other (Sato et al., 1997; Nojiri et al., 1999; Kasuga et al., 2001).

For bioremediation of dioxin-contaminated soils, it is of interest to know the possible fate of PCDFs and PCDDs in the environment by these different types of dioxin-degraders. In this report, we investigated the degradabilities of both strains for PCDDs and PCDFs, and developed application conditions for the effective degradation of dioxins in soil slurry microcosm. Finally, using both strains, we demonstrated remediation of the actual PCDDs- and PCDFs-contaminated soil.

## MATERIALS AND METHODS

**Chemicals.** 2-ClDF, 2,8-$Cl_2$DF, 2-ClDD, 2,3-$Cl_2$DD, 2,7-$Cl_2$DD, and 1,2,3-$Cl_3$DD were purchased from AccuStandard Inc. (New Haven, Conn.) All other chemicals used in this study were of the highest purity commercially available.

**Bacterial strains and culture conditions.** *Terrabacter* sp. strain DBF63 is a DF-degrading bacterium and *Pseudomonas* sp. strain CA10 is a CAR-utilizing bacterium, which were isolated from a soil sample and an activated sludge in Japan, respectively. Strains DBF63 and CA10 were cultivated at 30°C in carbon free minimal medium (CFMM) (Monna et al., 1993), and carbon and nitrogen free minimal medium (CNFMM) (Ouchiyama et al., 1993) supplemented with 0.1% (v/v) of DF and CAR, respectively. After cultivated for appropriate days, DBF63 cells and CA10 cells were collected by centrifugation, washed twice with CFMM or CNFMM, and resuspened in appropriate volume of the same medium. In strain DBF63, the growing cells were also used.

**Soils.** For preparation of dioxin-contaminated model soil, the granitic soil was used in this study. Soils were sieved (2-mm mesh), non-sterilized, and stored at room temperature. The major soil characteristics are the followings: organic carbon, 0.16%; pH 6.0; cation exchange capacity, 4.81 cmol $kg^{-1}$; maximum water-holding capacity, 24.5%; sand, 79%; clay, 10%; silt, 11%. On the other hand, the actual PCDDs- and PCDFs-contaminated soil was obtained from an incinerator site in Japan. The total dioxin concentration (PCDDs and PCDFs) was 725-ng/g-dry soil, and 11-ng Toxicity Equivalency Quantity (TEQ)/g-dry soil (the maximum acceptable concentration of dioxins in Japan is 1-ng TEQ/g-soil).

**Biotransformation analyses of PCDDs and PCDFs.** Biotransformation reactions using strain DBF63 or CA10 were performed at 30°C in glass tubes sealed with silicon caps. The tubes containing the 5 ml of washed-cell suspension and the substrate corresponding to 10 ppm (10 μg/ml) concentration were incubated on a reciprocal shaker (300 strokes/min) at 30°C for 18 h. Stock solutions of PCDFs and PCDDs (1 mg/ml) were made up in dimethyl formamide (DMF). The reaction mixtures were extracted two times with 5 ml of ethyl acetate in acidic or neutral conditions.

**2,8-$Cl_2$DF degradation by DBF63 cells and 2,3-$Cl_2$DD degradation by CA10 cells in soil slurry microcosm.** The granitic soil was treated with 2,8-$Cl_2$DF or 2,3-$Cl_2$DD at a final concentration of 1 ppm (1 μg/g-soil) from stock solutions. Resultant 2,8-$Cl_2$DF or 2,3-$Cl_2$DD-containing soils were used as dioxin-contaminated model soils in this study. One-gram aliquots were introduced into sterile Sakaguchi flasks with aluminum caps. These soil microcosms were inoculated with 5 ml of CFMM containing growing cells of strain DBF63 (3 days cultivation) or 5 ml of CNFMM containing washed cells of strain CA10 (2 days cultivation), and incubated on shaker (200 strokes $min^{-1}$) at 30°C in the dark. As a control, the inoculation of soil slurry with sterilized DBF63 cells or CA10 cells was carried out. Samples were extracted two times with 5 ml of ethyl acetate. The concentrations of 2,8-$Cl_2$DF or 2,3-$Cl_2$DD were determined after 1, 3, 5, and 7 days of incubation.

**The effect of coexistent congeners on PCDDs degradation.** 2-ClDD, 2,3-$Cl_2$DD, 1,2,3-$Cl_3$DD, and their combination (2-ClDD+2,3-$Cl_2$DD, 2-ClDD+1,2,3-$Cl_3$DD, 2-ClDD+2,3-$Cl_2$DD+1,2,3-$Cl_3$DD) were used as target substrates, and each PCDD was added to the granitic soil at final concentration of 1 ppm from each stock solution. The contaminated model soils were prepared, inoculated with washed cells of strain CA10, and incubated as described above. As a control, the inoculation of soil slurry with sterilized CA10 cells was carried out. Samples were extracted two times with 5 ml of ethyl acetate. The degradation rates of each substrate were determined with the averages of triplicate determination in control experiments being 100%.

**Monitoring of model substrate depletion.** Quantification of 2-ClDF, 2,8-$Cl_2$DF, 2-ClDD, 2,3-$Cl_2$DD and 1,2,3-$Cl_3$DD was performed, and the amounts of these substrates were determined by a model JMS-Automass 150 GC-MS system (JEOL, Ltd., Tokyo, Japan) fitted with a fused-silica chemically bonded capillary column (DB-5; 0.25 mm [inside diameter] by 15 m, 0.25 μm film thickness; J&W Scientific Inc., Folsom, Calif.). 1 ppm of 2-ClDF used as an internal standard was added to the samples from stock solution (1 mg/ml in DMF), just before the extraction with ethyl acetate. Each sample was injected into the column at 80°C in the splitless mode. After 2 min at 80°C, the column temperature was increased at 16°C $min^{-1}$ to 280°C. The head pressure of the helium carrier gas was 65 kPa. We determined the remaining rate of substrate by comparison of the peak area for molecular ions of compounds (218 for 2-ClDD, 252 for 2,3-$Cl_2$DD, and 286 for 1,2,3-$Cl_3$DD) after the appropriate incubation.

**Bioremediation of the actual PCDDs- and PCDFs-contaminated soil.** In this demonstration experiment, microcosms consisted of the soil (20 g of soil) in Sakaguchi flask (500 ml) inoculated with 100 ml of CFMM containing growing cells of DBF63 or CNFMM containing washed cells of strain CA10 as prepared above. The densities of these strains were $10^{11}$ cells/g-dry soil. The inoculation of soil slurry with sterilized DBF63 or CA10 cells was carried out as a control. The soil slurry microcosm was incubated on shaker (200 stroke/min) at 30°C in the dark for 7 days. On the other hand, PCDDs and PCDFs in the actual contaminated soils were analyzed by the EAC Corporation (Gumma, Japan) using high resolution mass spectrometry (HRGS/HRMS), based on the Environmental Protection Agency standard methods.

## RESULTS AND DISCUSSION

**Cooxidative potential of *Terrabacter* sp. strain DBF63 for PCDFs and PCDDs.** Conversion of PCDFs and PCDDs to the corresponding chlorosalicylic acid and chlorocatechols by DF-grown DBF63 cells was investigated (Fig. 1). On GC-MS analysis, it was revealed that 2-ClDF and 2,8-$Cl_2$DF are converted to the same compound whose mass spectrum exhibited fragment ions identical to those of authentic 5-chlorosalicylic acid. The yield of 5-chlorosalicylic acid from 2-ClDF and 2,8-$Cl_2$DF were 35-50% and 65-85%, respectively. When 2-ClDD was added as a substrate to washed cells of strain DBF63, we detected the single product whose mass spectrum exhibited fragment ions same as that observed for the TMS derivative of authentic 4-chlorocatechol. The yield of 4-chlorocatechol from 2-ClDD was 30-45%. Strain DBF63 also transformed 2,3-$Cl_2$DD to the product with the MS fragmentation pattern consistent with those of authentic 4,5-dichlorocatechol. The yield of 4,5-dichlorocatechol from 2,3-$Cl_2$DD was 20-35%. In addition, it was revealed that 2,7-$Cl_2$DD and 1,2,3-$Cl_3$DD were not converted to corresponding chlorocatechol.

From these findings, all the tested PCDFs and PCDDs were revealed to be degraded by DBF63 cells, with the remaining amounts of 2-ClDF, 2,8-$Cl_2$DF, 2-ClDD, 2,3-$Cl_2$DD, 2,7-$Cl_2$DD, and 1,2,3-$Cl_3$DD during transformation experiments of 18 h being 10-25%, 10-20%, 10-30%, 10-30%, 10-25%, and 70-80%, respectively (Fig. 1).

**Cooxidative potential of *Pseudomonas* sp. strain CA10 for PCDFs and PCDDs.** CAR-grown CA10 cells converted 2-ClDF to 5-chlorosalicylic acid. Based on the GC-MS analysis, it was also revealed that 4-chlorocatechol and 4,5-dichlorocatechol were formed from 2-ClDD and 2,3-$Cl_2$DD transformation experiments, respectively. These results are the same with strain DBF63, however, yield of 5-chlorosalicylic acid, 4-chlorocatechol, and 4,5-dichlorocatechol were 5-10%, 1.0-1.5%, 0.1-1.0%, respectively, that were much lower than that of strain DBF63 (Fig. 1). Moreover, the remaining amounts of 2-ClDF, 2-ClDD, and 2,3-$Cl_2$DD were 30-50%, 60-75%, and 60-70%, respectively. These results showed that strain CA10 had rather weak degradability toward PCDFs and PCDDs than strain DBF63. In addition, 4-chlorocatechol was not detected from 2,7-$Cl_2$DD transformation experiment, and 3,4,5-trichlorocatechol and catechol were not detected from 1,2,3-$Cl_3$DD, although 2,7-$Cl_2$DD and 1,2,3-$Cl_3$DD was decreased 15-35% and 10-20%, respectively. Different from strain DBF63, biotransforation experiment of 2,8-$Cl_2$DF with CA10 cells did not yield 5-chlorosalicylic acid. However, the 5-15% of 2,8-$Cl_2$DF was decreased.

**2,8-$C_2$DF degradation by strain DBF63 and 2,3-$Cl_2$DD degradation by strain CA10 in soil slurry microcosm.** As shown in Fig. 2, single inoculation of strain DBF63 into 2,8-$Cl_2$DF-contaminated model soil or strain CA10 into 2,3-$Cl_2$DD-contaminated model soil enhanced the degradation of each substrate.

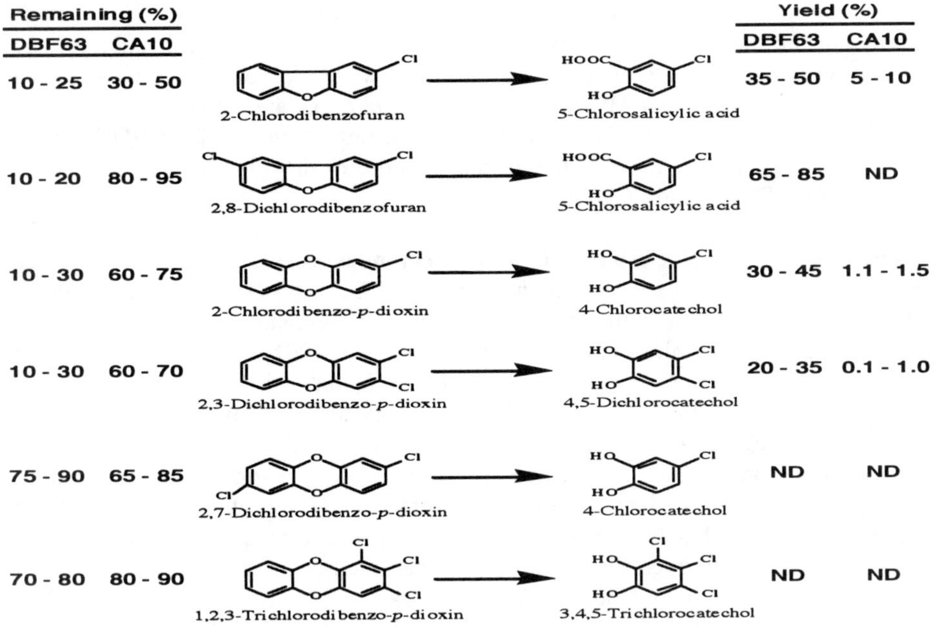


**FIGURE 1. Chlorosalicylic acid and chlorocatechols formed from PCDF or PCDD congeners by strains DBF63 and CA10. ND: Not detected.**

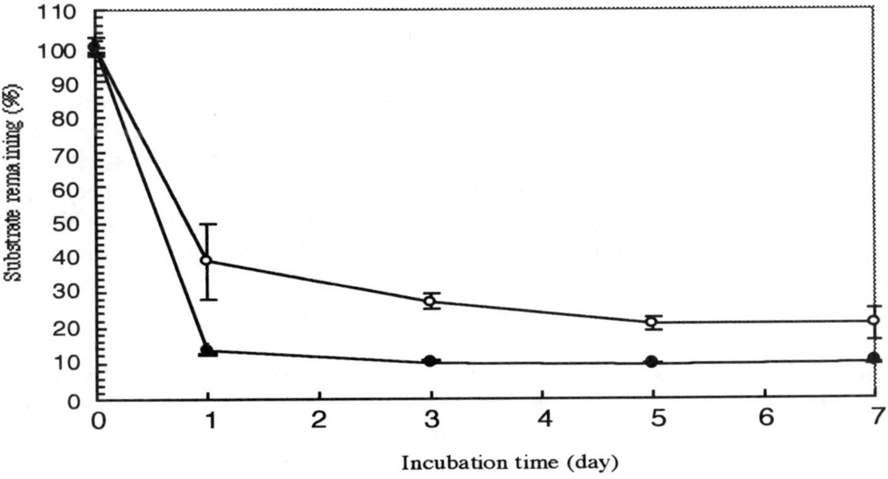


**FIGURE 2. Degradation of 2,8-$Cl_2$DF by strain DBF63 (●) and 2,3-$Cl_2$DD by strain CA10 (○) in the contaminated model soil. Control experiments did not show significant degradation (less than 20%).**

At a density of $10^9$ CFU/g-soil, 86% of 2,8-$Cl_2DF$ was removed in 1 day, and the continuing incubation after 3 days (90% degradation) did not significantly enhance the further 2,8-$Cl_2DF$ degradation. Similarly, at a density of $10^9$ CFU/g-soil, the removal of 2,3-$Cl_2DD$ was 73% after 3 days of incubation, and approximately 80% of 2,3-$Cl_2DD$ was removed by 7 days. After that, the degradation of 2,3-$Cl_2DD$ was not promoted. These results indicate that in the presence of this granitic soil, both strains could degrade PCDD and PCDF congeners.

**Degradation of PCDDs mixture in soil slurry microcosm.** Strain CA10 was also tested for the ability to degrade the mixture of PCDD substrates. To investigate the influence of the chlorine substitution pattern on the degradability of strain CA10 for PCDDs, mono- to trichlorinated DDs (2-ClDD, 2,3-$Cl_2DD$, and 1,2,3-$Cl_3DD$) were used as target substrates. In the case of strain CA10, approximately 97% of 2-ClDD, 89% of 2,3-$Cl_2DD$, and 52% of 1,2,3-$Cl_3DD$ was removed by 5 days incubation, respectively. Apparently, these results showed that the degradation rate of PCDD congeners by strain CA10 decreased with the increasing degree of chlorination. On the other hand, in the degradation experiment of mixture of two congeners, the removal rates of each substrate were similar to that in the case with the addition of a single substrate. Moreover, the degradation experiment of mixture of three congeners also exhibited the similar tendency. The degradation rates observed in strain CA10 were as follows: 2-ClDD, 95-96%; 2,3-$Cl_2DD$, 85-88%; 1,2,3-$Cl_3DD$, 46-75%. These results strongly indicate that strain CA10 is able to degrade each congener in the PCDDs mixture same as the case with a single congener.

**Bioremediation of the actual PCDDs- and PCDFs-contaminated soil by strain DBF63.** Demonstration experiment of the actual dioxin-contaminated soil remediation was performed with the soil slurry microcosm. After 7 days of treatment, the total dioxin (PCDDs and PCDFs) concentration decreased to 685 ng/g soil from 725 ng/g soil, and to 10.2 ng-TEQ/g dry soil from 11 ng-TEQ/g dry soil. The degradation rates of respective congeners were shown in Table 1. It was revealed that strain DBF63 had the potential to degrade tetra- to hexachlorinated dioxins, and that it appears to attack preferably the dioxin whose initial concentration was relatively high (Table 1). The degree of degradability of PCDDs and PCDFs seemed to be dependent on the chlorines substitution, and $Cl_4DDs$ and $Cl_4DFs$ exhibited the highest degradation rate. On the other hand, $Cl_8DD$ and $Cl_8DF$, which do not allow the attack of initial dioxygenase because no positions are free of chlorine substitution, were not degraded, and $Cl_7DD$ and $Cl_7DF$ were also hardly removed.

**Bioremediation of the actual PCDDs- and PCDFs-contaminated soil by strain CA10.** After 7 days of treatment, the total dioxin (PCDDs and PCDFs) concentration decreased to 665 ng/g soil from 725 ng/g soil, and to 9.4 ng-TEQ/g dry soil from 11 ng-TEQ/g dry soil. The degradation rates of respective congeners were shown in Table. 2. Result indicated that strain CA10 had the possibility to degrade dioxins with 4 to 7 chlorines, and the highest degradation rates were observed in $Cl_4DDs$ and $Cl_4DFs$, which were relatively low-chlorinated congeners.

**TABLE 1. Degradation of PCDD and PCDF congeners by strain DBF63 in the contaminated soil.**

| Congeners[a] | | Control | | Single inoculation of DBF63 | | Degradation amount (pg/g-dry soil) | Degradation rate (%) |
|---|---|---|---|---|---|---|---|
| | | Concentration (pg/g-dry soil) | TEQ[b] | Concentration (pg/g-dry soil) | TEQ[b] | | |
| PCDDs | 1,3,6,8-$Cl_4$DD | 180 | 0 | 155 | 0 | 25 | 13.9 |
| | 1,3,7,9-$Cl_4$DD | 103 | 0 | 90 | 0 | 13 | 12.6 |
| | 2,3,7,8-$Cl_4$DD | 44 | 44 | 40 | 40 | 4 | 9.0 |
| | $Cl_4$DDs | 1025 | - | 870 | - | 155 | 15.1 |
| | 1,2,3,7,8-$Cl_5$DD | 765 | 765 | 705 | 705 | 60 | 7.8 |
| | $Cl_5$DDs | 7000 | - | 6300 | - | 700 | 10.0 |
| | 1,2,3,4,7,8-$Cl_6$DD | 2150 | 215 | 2000 | 200 | 150 | 7.0 |
| | 1,2,3,6,7,8-$Cl_6$DD | 2800 | 280 | 2450 | 245 | 350 | 12.5 |
| | 1,2,3,7,8,9-$Cl_6$DD | 2650 | 265 | 2300 | 230 | 350 | 13.2 |
| | $Cl_6$DDs | 31500 | - | 27500 | - | 4000 | 12.7 |
| | 1,2,3,4,6,7,8-$Cl_7$DD | 28000 | 280 | 28000 | 280 | 0 | 0.0 |
| | $Cl_7$DDs | 54500 | - | 52500 | - | 2000 | 3.7 |
| | $Cl_8$DD | 45000 | 4.5 | 46000 | 4.6 | - | - |
| PCDFs | 1,2,7,8-$Cl_4$DF | 450 | 0 | 395 | 0 | 55 | 12.2 |
| | 2,3,7,8-$Cl_4$DF | 265 | 26.5 | 225 | 23 | 40 | 15.1 |
| | $Cl_4$DFs | 28500 | - | 25000 | - | 3500 | 12.3 |
| | 1,2,3,7,8-$Cl_5$DF | 1800 | 90 | 1550 | 78 | 250 | 13.9 |
| | 2,3,4,7,8-$Cl_5$DF | 4750 | 2375 | 4350 | 2175 | 400 | 8.4 |
| | $Cl_5$DFs | 71000 | - | 63500 | - | 7500 | 10.6 |
| | 1,2,3,4,7,8-$Cl_6$DF | 9900 | 990 | 9150 | 915 | 750 | 7.6 |
| | 1,2,3,6,7,8-$Cl_6$DF | 12000 | 1200 | 11000 | 1100 | 1000 | 8.3 |
| | 2,3,4,6,7,8-$Cl_6$DF | 31500 | 3150 | 29500 | 2950 | 2000 | 6.5 |
| | $Cl_6$DFs | 176250 | - | 160000 | - | 16250 | 9.2 |
| | 1,2,3,4,7,8,9-$Cl_7$DF | 21500 | 215 | 21500 | 215 | 0 | 0.0 |
| | $Cl_7$DFs | 283500 | - | 275000 | - | 8500 | 3.0 |
| | $Cl_8$DF | 110000 | 11 | 120000 | 12 | - | - |

**TABLE 2. Degradation of PCDD and PCDF congeners by strain CA10 in the contaminated soil.**

| Congeners[a] | | Control | | Single inoculation of CA10 | | Degradation amount (pg/g-dry soil) | Degradation rate (%) |
|---|---|---|---|---|---|---|---|
| | | Concentration (pg/g-dry soil) | TEQ[b] | Concentration (pg/g-dry soil) | TEQ[b] | | |
| PCDDs | 1,3,6,8-$Cl_4$DD | 180 | 0 | 150 | 0 | 30 | 16.7 |
| | 1,3,7,9-$Cl_4$DD | 103 | 0 | 87 | 0 | 16 | 15.5 |
| | 2,3,7,8-$Cl_4$DD | 44 | 44 | 36 | 36 | 8 | 18.2 |
| | $Cl_4$DDs | 1025 | - | 875 | - | 150 | 14.6 |
| | 1,2,3,7,8-$Cl_5$DD | 765 | 765 | 705 | 705 | 60 | 7.8 |
| | $Cl_5$DDs | 7000 | - | 6050 | - | 950 | 13.6 |
| | 1,2,3,4,7,8-$Cl_6$DD | 2150 | 215 | 2000 | 200 | 150 | 7.0 |
| | 1,2,3,6,7,8-$Cl_6$DD | 2800 | 280 | 2500 | 250 | 300 | 10.7 |
| | 1,2,3,7,8,9-$Cl_6$DD | 2650 | 265 | 2450 | 245 | 200 | 7.5 |
| | $Cl_6$DDs | 31500 | - | 29000 | - | 2500 | 7.9 |
| | 1,2,3,4,6,7,8-$Cl_7$DD | 28000 | 280 | 25500 | 255 | 2500 | 8.9 |
| | $Cl_7$DDs | 54500 | - | 48500 | - | 6000 | 11.0 |
| | $Cl_8$DD | 45000 | 4.5 | 43000 | 4.3 | 2000 | 4.4 |
| PCDFs | 1,2,7,8-$Cl_4$DF | 450 | 0 | 365 | 0 | 85 | 18.9 |
| | 2,3,7,8-$Cl_4$DF | 265 | 26.5 | 235 | 23.5 | 30 | 11.3 |
| | $Cl_4$DFs | 28500 | - | 25000 | - | 3500 | 12.3 |
| | 1,2,3,7,8-$Cl_5$DF | 1800 | 90 | 1600 | 80 | 200 | 11.1 |
| | 2,3,4,7,8-$Cl_5$DF | 4750 | 2375 | 4300 | 2150 | 450 | 9.5 |
| | $Cl_5$DFs | 71000 | - | 63500 | - | 7500 | 10.5 |
| | 1,2,3,4,7,8-$Cl_6$DF | 9900 | 990 | 9500 | 950 | 400 | 4.0 |
| | 1,2,3,6,7,8-$Cl_6$DF | 12000 | 1200 | 10850 | 1085 | 1150 | 9.6 |
| | 2,3,4,6,7,8-$Cl_6$DF | 31500 | 3150 | 30500 | 3050 | 1000 | 3.2 |
| | $Cl_6$DFs | 176250 | - | 166400 | - | 10000 | 5.6 |
| | 1,2,3,4,7,8,9-$Cl_7$DF | 21500 | 215 | 20000 | 200 | 1500 | 7.0 |
| | $Cl_7$DFs | 283500 | - | 247500 | - | 36000 | 10.1 |
| | $Cl_8$DF | 110000 | 11 | 109500 | 11 | 500 | 0.4 |

[a]$Cl_x$DD and $Cl_x$DF represent the amount of all other congeners except the listed congeners, whose authentic samples can be used.
[b]TEQ: Toxicity Equivalency Quantity; TEQ was calculated by multiplying TEF and each concentration. TEF: Toxicity Equivalency Factor; TEF was quoted from WHO/IPCS (1997).

During recent decades, public and scientific interest has focused mostly on environmental contamination by highly toxic 2,3,7,8-$Cl_4DDs$. In this study, it was revealed that 2,3,7,8-$Cl_4DD$ showed relatively high degradation value, being 18.2%.

In addition, strain CA10 appeared to attack PCDD and PCDF congeners having high initial concentrations because the degradation amounts of dioxins tended to increase with initial concentrations being relatively high (Table 2).

## CONCLUSION

The results showed that two different types of dioxin-degraders, *Terrabacter* sp.strain DBF63 and *Pseudomonas* sp. strain CA10, were able to degrade mono- to trichlorinated DDs and DFs at low concentration (1 ppm) in soil microcosms, and also had potential to degrade dioxins with four to more chlorines. To perform better bioremediation of the actual dioxin-contaminated soils, more improved conditions and strategies will have to be developed.

## ACKNOWLEDGMENT

This work was partly supported by the New Energy and Industrial Technology Development Organization (NEDO) (Project ID 98Z34017).

## REFERENCES

Kasuga, K., H. Habe, J. –S. Chung, T. Yoshida, H. Nojiri, H. Yamane, and T. Omori. 2001. "Isolation and characterization of the genes encoding a novel oxygenase component of angular dioxygenase from the gram-positive dibenzofuran degrader *Terrabacter* sp. strain DBF63" *Biochem. Biophys. Res. Commun. 283*(1): 195-204.

Monna, L., T. Omori, and T. Kodama. 1993. "Microbial degradation of dibenzofuran, fluorene, and dibenzo-*p*-dioxin by *Staphylococcus auriculans* DBF63." *Appl. Environ. Microbiol. 59*(1): 285-289.

Nojiri, H., J. –W. Nam, M. Kosaka, K. Morii, T. Takemura, K. Furihata, H. Yamane, and T. Omori. 1999. "Diverse oxygenations catalyzed by carbazole 1,9a-dioxygenase from *Pseudomonas* sp. strain CA10." *J. Bacteriol. 181*(10): 3105-3113.

Ouchiyama, N., Y. Zhang, T. Omori, and T. Kodama. 1993. "Biodegradation of carbazole by *Pseudomonas* spp. CA06 and CA10." *Biosci. Biotech. Biochem. 57*(3): 455-460.

Sato, S., J. –W. Nam, K. Kasuga, H. Nojiri, H. Yamane, and T. Omori. 1997. "Identification and characterization of genes encoding carbazole 1,9a-dioxygenase in *Pseudomonas* sp. strain CA10." *J. Bacteriol. 179*(15): 4850-4858.

# NITROCELLULOSE WASTE FINES: BIOREMEDIATION BY A TURNED WINDROW COMPOSTING SYSTEM

***Ingrid O'Donnell***, Dave Malcolm, Ian Swallow, Richard Ogden, Gordon Bulloch
(BAE SYSTEMS Environmental Services, Westcott, Bucks, HP18 0NP, UK.)

**ABSTRACT:** Laboratory and field-scale trials were used to assess the feasibility of using a composting technique to remediate explosive nitrocellulose (NC) waste. Subsequently, a turned windrow system was implemented to treat approximately 2000 tonnes of NC waste fines. The selected technique resulted in a rapid reduction in the molecular size of the NC polymer and a simultaneous reduction in energetic activity. Field trials demonstrated that the optimal compost composition contained NC, chicken manure, and spent coffee grounds. The maximum degradation achieved was an 88% reduction in NC content over 95 days. This corresponded with decrease in energetic content from 9.8% to <1%.

## INTRODUCTION

Dry Nitrocellulose (NC) is classified as an explosive, but when wet it is less sensitive and is therefore considered to be a flammable solid. It is insoluble in water, non-volatile and non-toxic and resembles paper mache when wet. It is used in many propellant formulations, such as those used in gun propellants, rocket motors and ejector seats etc. NC is a polymer consisting of a cellulose backbone of repeating glucose units where some of the hydroxyl groups are substituted with nitro groups. The term nitrocellulose therefore does not apply to a single molecular structure, but to a heterogeneous range of structures of varying length and degrees of nitration.

NC has been produced at the Royal Ordnance factory at Bishopton, Scotland since 1937. The manufacturing process involves the nitration of Kraft paper or cotton linters, with a mixture of concentrated nitric and sulphuric acids. Waste NC fines are a by-product of the subsequent washing, stabilisation and pulping stages. All the water effluent passes through one of two settling lagoons to remove the suspended NC waste. The liquid effluent is then further treated prior to discharge. Originally, the settling lagoons were emptied by excavation and the contents incinerated. However, this disposal route became unacceptable due to changing legislation and so bioremediation by composting was considered as an alternative treatment for approximately 2000 tonnes of accumulated NC waste fines.

**Objective.** To identify and implement a safe, cost-effective and environmentally-acceptable solution to the NC waste management problem.

## LABORATORY TRIAL

**Materials and Methods.** The ability of nine different compost amendment compositions to degrade waste NC fines was evaluated in a simple microcosm study (see Table 1). The microcosms were established in wooden ammunition boxes (30cm x 30cm x 60cm). The contents were thoroughly mixed by hand to ensure homogeneity. The moisture content was maintained at around 60% w/w and the ambient temperature at 30°C throughout the study. The contents were aerated for half an hour every four hours using an airpump (Interpet AquaAir AP4) and a 20 cm airstone. Microcosm temperatures were recorded daily. Microcosm contents were remixed by hand and sampled in triplicate on a weekly basis. The mass of the microcosm, any added water, and each sample were also recorded. Samples were subsequently submitted for chemical analysis.

**Table 1: showing the composition (in litres) of the 9 compost mixtures evaluated. Microcosms M1-M6 were based on compost amendment types previously investigated at Badger Army Ammunition Plant (Doyle *et al.*, 1986; Williams *et al.*, 1989)**

| Microcosm | NC | Sewage Sludge | Wood Shavings | NPK Solution | Horse Feed | Hay | Horse Manure | Woodchips | Sawdust | Straw | Chicken Manure | Green Waste | Spent Coffee Grounds |
|---|---|---|---|---|---|---|---|---|---|---|---|---|---|
| M1 | 3 | 4 | 12 | 1.6 | | | | | | | | | |
| M2 | 3 | | | 1 | 7.5 | 7.5 | 1 | | | | | | |
| M3 | 3 | | | | 2 | 8 | 6 | 1 | | | | | |
| M4 | 3 | | | | | | 1 | | 12 | 3 | | | |
| M5 | 3 | | | | | 6 | 10 | | | | | | |
| M6 | 3 | | | | | | 12 | | | 4 | | | |
| M7 | 3 | | | | | | | | | | 8 | | 8 |
| M8 | 3 | | | | | | | | | | 2 | 14 | |
| M9 | 3 | | | | | | | | | 4 | 6 | | 6 |

At the time of this writing, there is no one standard analytical test available for the quantification of nitrocellulose in compost samples – the presence of various other natural compounds within the compost causes significant interference. Thus, a combination of two different analytical techniques was used to demonstrate changes in NC content. Gel Permeation Chromatography (GPC) with UV detection (242 nm) was utilised to follow changes in molecular size. Thermo-Gravimetric Analysis (TGA) was utilised to determine the energetic potential of samples. Volatile matter (moisture) content and pH were also measured. A preliminary study indicated that a substantial volume/mass reduction occurred during the composting process. Therefore, changes in NC concentration did not directly indicate degradation. All results have thus been presented in terms of the total NC quantities within each microcosm.

**Results and Discussion.** The total NC remaining in each microcosm is summarised in Table 2. Analysis of the data using two-sample T-tests (99% confidence) shows that a statistically significant reduction in NC occurred between Day 0 and Day 37 in all microcosms. Apparent initial increases in NC were observed in Microcosms 1,2,3,7 & 8. This is thought to be due to more rapid degradation of other compost components compared to NC at the start of the composting process.

**Table 2: Percentage reductions in total NC according to GPC analysis. The majority of the data represent the mean of three samples and are calculated on a dry weight basis. * indicates results reported as % reduction in 'NC Equivalents' (see text).**

| Time (days) | M1 | M2 | M3 | M4 | M5 | M6 | M7* | M8* | M9* |
|---|---|---|---|---|---|---|---|---|---|
| **0** | 0 | 0 | 0 | 0 | 0 | 0 | 0 | 0 | 0 |
| **6** | -7 | -17 | -18 | 26 | 16 | 15 | -28 | -30 | 7 |
| **15** | -5 | -17 | -8 | 41 | 3 | 0 | 21 | 3 | 18 |
| **23** | 31 | 3 | 12 | 35 | 18 | 8 | 50 | 24 | 42 |
| **30** | 17 | 39 | 60 | 51 | 48 | 22 | 50 | 14 | 42 |
| **37** | 36 | 41 | 51 | 61 | 54 | 66 | 60 | 30 | 57 |

**Figure 1: an overlay of GPC traces for extracts of compost samples taken from Microcosm 7 on Days 0, 6 & 15. This indicates a progressive reduction in the molecular size of the NC polymer. Note: there is a logarithmic relationship between elution time and molecular size.**

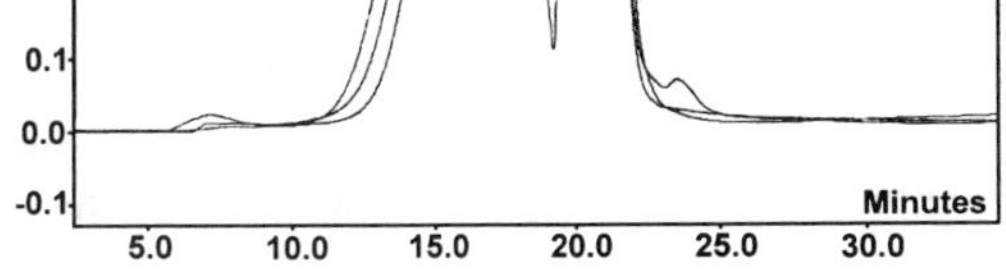


The GPC data obtained from Microcosms 7, 8 & 9 indicated a significant reduction in the size of NC molecules with time (see Figure 1). These smaller molecules are considered too small to be classified as NC and thus in these microcosms results are reported as 'NC Equivalents'. This effect was not observed in any other microcosm.

The percentage reduction in energetic materials was determined in a limited number of samples using TGA. The results show that in Microcosms 4, 7 & 9 a reduction of 30% to 40% in energetic content occurred between Day 0 and Day 30 (data not shown).

It was not possible to identify the new, smaller molecules generated in Microcosms 7, 8 and 9.

One possible candidate is nitrostarch, another explosive. However, as TGA results demonstrate a reduction in sample energetic activity this seems unlikely. It seems possible that chemical or biological processes result in both denitration and cleavage of the NC molecule, producing a variety of non-explosive products.

**Conclusions.** The results indicate that composting of NC is feasible. All microcosms showed a statistically significant reduction in NC content after 37 days with several microcosms achieving over 60% reductions. Overall, microcosms 4 & 7 were considered to have outperformed other amendment combinations

## FIELD TRIAL

**Materials and Methods.** Three small windrows (approximately 3 m wide x 6 m long x 1.5 m high) were constructed within a disused building at the site. The composition of each windrow is presented in Table 3. The NC was thoroughly blended with the various compost amendments using a ALLU SM grinding bucket (Ideachip, Oy, Hollola, Finland) attachment on a front end loader. Care was taken to ensure that the moisture content of all NC-containing materials was maintained at above 55% w/w at all times to ensure safe working conditions (Lowe *et al.*, 1995). Moisture content was measured using a portable Theta probe moisture meter (Delta-T Devices, Cambridge, UK). When required, water was added from a tank using a 12V battery-operated pump and hose.

**Table 3: showing the composition by volume of the three field trial windrows**

| Windrow A (25% NC *d.w.*) | Windrow B (22% NC *d.w.*) | Windrow C (27% NC *d.w.*) |
|---|---|---|
| • 2 parts wet NC<br>• 10 parts sawdust<br>• 2 parts straw<br>• 1 part horse manure | • 2 parts wet NC<br>• 10 parts chicken manure & spent coffee grounds 1:1 compost mix | • 3.5 parts wet NC<br>• 11.5 parts chicken manure & spent coffee grounds 1:1 compost mix |

The windrows were passively aerated via perforated plastic drainage pipe (12 cm diameter). These were laid transversely through the windrows in two rows. The pipes in each row were laid 1 m apart. The first row was located at a height of 0.5 m and the second at 1 m, arranged in a staggered configuration to optimise windrow aeration. In the interest of safety, all windrows were capped with a layer of their respective compost mixtures (minus any NC) to act as a sacrificial drying layer: surface drying of uncovered windrows was expected. Moisture content, temperature and soil gas were monitored on a fortnightly basis. In addition, triplicate samples were taken from discrete random locations on each windrow (below the sacrificial capping layer). Samples were subjected to analyses by GPC and TGA. Volatile matter (moisture) content and copper, nickel, and zinc levels were also determined.

Quantification of metal levels was undertaken as a practical alternative to weighing whole windrows in order to estimate the expected reduction in windrow mass. This is possible because the total quantity of metal within each windrow should not be changed by any biological processes and can thus be used to normalise any changes in NC concentration that might be observed.

Degradation rates appeared to tail off in Windrow B and C after Day 82. These windrows were subsequently remixed and rebuilt. In an attempt to identify the factor limiting further NC degradation, Windrow B was remixed with 1 $m^3$ of additional non-NC compost amendments (0.5 $m^3$ chicken manure : 0.5 $m^3$ spent coffee grounds), whereas Windrow C received no further additions.

**Results and Discussion.** The normalised GPC results are summarised in Table 4 and illustrated in Figure 2. A similar progressive reduction in NC molecular size to that observed in the laboratory trial was indicated for Windrows B & C. Windrow B achieved the greatest reduction in NC Equivalents (88%), whereas Windrow C only achieved a 45% reduction over the same period. This poor performance was attributed to sub-optimal aeration and other uncontrollable variations in experimental conditions. The fact that a rapid increase in degradation followed the remixing/rebuilding of Windrow B and C (Day 82) supports the hypothesis that the maintenance of aerobic conditions is necessary for effective NC degradation.

The observed reduction in NC content in Windrows B & C was accompanied by a simultaneous decrease in energetic matter content as determined by TGA (data not shown). Over the 95-day trial period the energetic content of Windrows B fell from 9.8% to <1%.

**Table 4: showing % reductions in NC[1] and NC Equivalents[2] as determined by GPC. The data has been normalised against metal concentrations to correct for changes in compost mass/volume. All analyses were conduced on a dry weight basis.**

| Days | Windrow A[1] | Windrow B[2] | Windrow C[2] |
|---|---|---|---|
| **0** | 0.0 | 0.0 | 0.0 |
| **12** | -4.7 | -27.0 | -58.9 |
| **25** | -1.5 | 31.8 | -33.3 |
| **39** | 6.6 | 77.7 | 7.3 |
| **54** | 9.1 | 70.4 | 17.9 |
| **68** | -14.9 | 69.9 | 0.0 |
| **82** | -23.0 | 45.2 | 4.5 |
| **95** | -13.0 | 87.9 | 45.2 |

**Figure 2: Mean concentration of NC and NC Equivalents in each windrow as determined by GPC.**

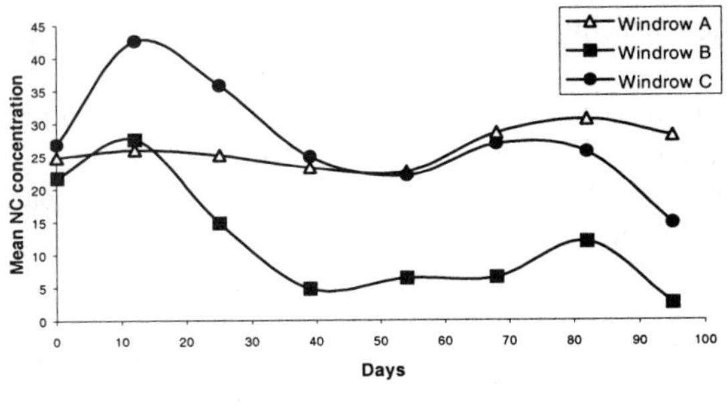

In contrast to laboratory results, Windrow A did not show any appreciable change in NC or energetic matter content during the trial. This may be due to differences in ambient temperatures during the two studies. Laboratory studies were maintained at around 30°C, but the maximum ambient temperature during the field trial was 15°C. Furthermore, Windrow A proved less able to generate its own heat under field conditions compared to Windrows B & C. This may be due to the low manure content and consequent lack of available nitrogen in Windrow A.

**Conclusions.** The field trial confirmed the laboratory study findings that compost mixtures containing chicken manure and spent coffee grounds can be utilised to effect NC degradation. In both windrows containing these amendments there appears to be a rapid decrease in the molecular size of the NC fraction soon after mixing. At the same time, TGA analysis indicates a reduction in energetic activity.

## FULL-SCALE BIOREMEDIATION

**Materials and Methods.** Based on the above evidence, it was recommended that the full-scale remediation of 2000 metric tonnes of waste NC fines be undertaken using a Windrow B/C-type amendment composition in a turned windrow composting system. This would ensure effective aeration and homogenisation. A similar system has previously been implemented at Umatilla Army Depot (Hermiston, Oregon) for the successful bioremediation of TNT and RDX-contaminated soil (Chong & Cubbage, 1996).

The project was managed in accordance with detailed Risk Assessments and Method Statements as well as stipulations laid down by the Scottish Environmental Protection Agency (SEPA) and UK Health and Safety Executive (HSE). An isolated area of hardstanding 180 m × 40 m with adequate road access and water supply was used and only minimal preparation of the "Treatment Area" proved necessary. A portable office/wash facility was installed and vehicle access routes were delineated in order to minimise cross-contamination of 'clean' and 'dirty' areas. In addition, a wooden "Mixing Area" was constructed to prevent sparking from metal machinery with concrete or stones during mixing and to control the escape of unmixed NC. After treatment the NC-contaminated Mixing Bay was dismantled and burned at a pre-existing incineration facility on site.

The NC settling lagoons were drained and excavated in 4 batches to comply with restrictions on the quantities treated at any one time. The NC waste fines were mixed in the volume ratios: 2 parts waste NC fines : 3 parts chicken litter : 3 parts spent coffee grounds. Mixing was carried out using firstly a 2m$^3$ excavator bucket and then specialist ALLU SM grinding bucket to achieve a homogenous mix. The moisture contents of all NC and NC mixtures were monitored throughout the project using a portable Theta moisture meter to ensure that moisture content was always above 55%.

Each batch of compost was then laid out in 4 windrows (3 m wide x 1.5 m high x 150 m long) on the Treatment Area. These were covered with breathable/semi-permeable fleeces, which prevent water ingress but allow diffusion of gases. Fleeces were removed when necessary for windrow turning, sampling or monitoring operations by means of a fleece winding kit (Morawetz, Austria). Windrows were turned regularly using a tractor-towed TSIVB windrow turner (Morawetz, Austria). Windrows were monitored on a daily basis for temperature, moisture content and soil gas content. Every 3 weeks immediately after turning, six composite samples were taken from each windrow. Samples were analysed for pH, moisture content and NC by GPC, TGA and Rotter Impact methods. Once active treatment was complete (approximately 8 weeks), the compost was removed from the Treatment Area and allowed to mature for several months in a stockpile. Stockpiles were not turned but monitoring and sample analysis continued. These samples were also tested for compliance with existing compost quality standards.

**Preliminary Results.** Further results from GPC testing, TGA, Rotter Impact and Compost Quality Testing (including phytotoxicity studies) are still awaited.

**Table 5: Concentration of NC Equivalents according to GPC in compost samples. The results represent a mean ± standard deviation, but have not been corrected for the mass/volume loss during composting. Entries marked # represent samples taken from maturation stockpiles rather than windrows. For windrow samples the mean represents 24 replicates, but samples taken from stockpile represent a mean of only 6 replicates.**

| BATCH 1 | | BATCH 2 | | BATCH 3 | | BATCH 4 | |
|---|---|---|---|---|---|---|---|
| **Days** | **NC (g/kg)** | **Days** | **NC (g/kg)** | **Days** | **NC (g/kg)** | **Days** | **NC (g/kg)** |
| 6 | 421 ± 49 | 1 | 530 ± 99 | 18 | 383 ± 72 | 0 | 322 ± 53 |
| 32 | 388 ± 65 | 22 | 399 ± 63 | 54 | 372 ± 73 | 26 | 204 ± 60 |
| 70# | 122 ± 26 | 51 | 237 ± 34 | 74 | 206 ± 59 | 40 | 114 ± 17 |
| 105# | 97 ± 58 | 85# | 183 ± 24 | | | | |
| 139# | 135 ± 61 | 121# | 128 ± 116 | | | | |
| 175# | 200 ± 99 | 141# | 161 ± 102 | | | | |
| 195# | 77 ± 70 | | | | | | |

The GPC data to date (see Table 5) again indicate that a time-dependent reduction in NC molecular size occurred during composting. Thus, the data has again been reported as 'NC Equivalents'. Significant sample variation was found in several cases, this is likely to be due to inherent heterogeneity within the compost and the small number of replicates used in some cases. Overall, however, the GPC results demonstrate a substantial and reproducible reduction in NC content over time. The maximum reduction achieved so far being 81% over 195 days. The final concentration in the finished compost was approximately 77 g/kg, which is below the "non-reactive" limit of 100 g/kg (10%) proposed by Lowe *et al.* (1996).

## OVERALL CONCLUSIONS

The results demonstrate that NC can be rapidly and economically degraded under appropriate composting conditions. Further work may be required to determine the exact nature of the degradation mechanism and whether the mechanism is predominantly chemical, biological or a combination of both. Further work is also needed to identify metabolites generated by the process.

Work is on-going in improving current analytical techniques; in particular, a highly effective methodology employing Differential Scanning Calorimetry has recently been developed. However, due to the difficulties inherent in analysing NC in a compost environment, final assessment of all results obtained and use of the finished compost will be based on a detailed risk assessment approach, rather than on simply defining "safe" target limits. In the UK, a similar approach is required by law for dealing with land contamination. An acceptable use for the finished compost, or the need for further treatment, will be determined based on a critical risk assessment using information gathered from the range of biological, chemical and physical tests detailed above.

The objectives of this work were to identify and implement an effective and commercially-viable remediation strategy to treat accumulated NC wastes. The results presented demonstrate that a turned windrow composting system is capable of effectively degrading this hazardous material and produces a non-hazardous product, which is potentially suitable for reuse.

## REFERENCES

Doyle,RC, JD Isbister, GL Anspach and JF Kitchens. 1986. *Composting Explosives/Organics Contaminated Soils*. US Army Toxic and Hazardous Materials Agency, Aberdeen Proving Grounds, MD. Report Number: AMXTH-TE-CR-86077

Lowe,WL, LH Myers and W Brinton. 1995. *Composting Of Nitrocellulose Fines - Hazards Analysis*. US Army Environment Center, Aberdeen Proving Grounds, MD. Report Number: SFIM-AEC-ET-CR-95083

Lowe,WL, LH Myers and JM Savage. 1996. *Composting Of Nitrocellulose Fines: Regulatory And Logistical Feasibility RAAP Installation - Position Paper*. US Army Environmental Center, Aberdeen Proving Ground, MD. Report No. SFIM-AEC-ET-CR-96152

Chong,MH and CH Cubbage. 1996. *Cost Report: Windrow Composting To Treat Explosives Contaminated Soils At Umatilla Army Depot Activity (UMDA)*. US Army Environmental Center, Aberdeen Proving Ground, MD. Report No. SFIM-AEC-ET-CR-961854

Williams,RT, PS Ziegenfuss and PJ Marks. 1989. *Field Demonstration - Composting Of Propellants-Contaminated Sediments At The Badger Army Ammunition Plant (BAAP)*. US Army Toxic and Hazardous Materials Agency, Aberdeen Proving Grounds, MD. Report Number: CETHA-TE-CR-89061

# TOSS TREATMENT OF 2,4-DNT CONTAMINATED SOIL AT AN EXPLOSIVES MANUFACTURING PLANT IN SWEDEN

***Tommy Lundgren***, SYDKRAFT SAKAB AB, Solna, SWEDEN
Roger B. Green, Waste Management, Inc., Cincinnati, Ohio

**ABSTRACT:** SAKAB has performed a pilot study of the TOSS (Two Step Static) bioremediation process for treating approximately 200 tons of 2,4-dinitrotoluene (2,4-DNT) contaminated soil. The TOSS process combines sequential anaerobic and aerobic stages to achieve treatment. In the first stage 2,4-DNT contaminated soil was combined with a carbon source (starch and glucose), an inoculum, and water to achieve anaerobic conditions. The resulting mixture was placed in a bermed construction and maintained under anaerobic conditions for 44 weeks. Over this period the concentration of 2,4-DNT was reduced from a starting concentration of 4,000 mg/kg to 19 mg/kg meeting the cleanup standard of 20 mg/kg. Similarly, the total concentration of other nitrotoluene explosives was reduced from 5,300 mg/kg to 22 mg/kg. The production of amino-substituted degradation intermediates was detected by GC/MS during anaerobic treatment, but these intermediates were not detected at the end of treatment. After anaerobic treatment, the soil was mixed with yard waste and formed into a 2-meter high pile to begin the aerobic treatment. The pile was located on top of a synthetic liner system with leachate control. Results of leachate quality monitoring demonstrate very low concentrations (< 2 µg/L) of leachable 2,4-DNT.

## INTRODUCTION

Dyno Nobel's Gyttorp facility traces its history back to 1864 when Alfred Nobel founded his first company Nitroglycerin Aktiebolag. The Swedish explosives manufacturer produced dynamite and other explosives at the site from 1864 to 1995. Over this long period of production, significant amounts explosives were released to the environment. SAKAB was contracted by Dyno Nobel to perform a pilot test on degradation of 2,4-DNT, its isomers, and nitrotoluenes. The treatment goal was to get below the Swedish guideline value for 2,4-DNT of 20 mg/kg.

Based on a successful laboratory treatability study, the TOSS process, developed by Waste Management, Inc., was selected as the treatment technology for a pilot demonstration (Eckman, et al, 1999). The TOSS process is designed to degrade and immobilize explosive compounds through a sequence of reducing (anaerobic) and oxidizing (aerobic) treatment stages (Green et al., 1998). In the first stage of treatment, explosives contaminated soil is combined with an electron donating substrate (hydrocarbon), a microbial inoculum (cow manure and/or anaerobic digester sludge), and water to achieve anaerobic conditions. The resulting mixture is placed in a treatment cell and is maintained under anaerobic conditions. This first stage of treatment facilitates the chemical reduction of nitroaromatic and nitramines in a solid phase system. Nitroaromatics are reduced to corresponding amines in multistep process involving nitroso and hydroxylamine intermediates. The second or aerobic stage consists of combining the anaerobically treated soil with composted yard waste and constructing the mixture into a biopile.

The TOSS process utilizes a significantly lower ratio of organic amendments than windrow composting which results in lower bulking of soil to be treated. However, this means that the TOSS process, unlike windrow composting, is not self-heating, which may limit its application in low temperature climates (Jerger and Woodhull, 2000). A unique soil heating system, comprised of electrically heated water circulated through plastic pipes, was designed and implemented. The heating system allowed successful operation through the winter months.

## MATERIALS AND METHODS

**Soil Preparation.** 200 metric tons (130 $m^3$) of soil contaminated with nitrotoluenes, including 2,4-DNT, had been previously excavated at the Dyno Nobel explosives production site near Örebro, Sweden. The soil was prepared for treatment by combining the soil with manure and a carbon source in a rotary drum-screening unit (REMU Somin metalli Oy, Finland). Material greater than 100 mm in diameter, including lumps or chunks of explosives, was separated from the soil during the screening process.

**Anaerobic Treatment.** The resulting 250 tons (170 $m^3$) of the amended soil mixture was spread out and further homogenized when transferred to the TOSS treatment cell by a backhoe. Water was added to the mixture to near field capacity. The TOSS treatment cell consisted of an existing concrete basin that was leak proofed. Utilizing existing infrastructure was done to reduce project costs. The treatment cell was 1.2 m deep, 23 m long, and 6 m wide. A removable roof was built on top of the concrete construction to provide protection from inclement weather.

Anaerobic treatment lasted for a period of 308 days (44 weeks). Although the treatment period spanned the winter months with temperatures as low as -5 °C, the TOSS cell was maintained at a temperature of approximately 25 °C, variation between 22 °C (winter) and 34 °C (summer). The cell was left undisturbed during the anaerobic treatment except for the addition of water to maintain a moisture content near saturation, and periodic soil sampling. Redox potential was measured on a weekly basis and was quite constant at approximately – 400 mV with a minimum of – 386 mV and maximum of –459 mV. No temperature or pH correction has been performed on the redox value.

**Aerobic Treatment.** When the anaerobic stage of treatment was complete the soil was excavated and combined with 50 $m^3$ of compost (matured yard waste). The approximately 300 $m^3$ of resulting soil-compost mixture was placed on a synthetic liner that sloped towards a water collection pond. Percolation through the material was estimated at 300 mm/year based on rainfall of 600 mm/year.

**Sampling and Analysis.** Soil samples were collected at frequency of one sample per 4 $m^3$ of anaerobic material. Discrete samples were collected at 39 equally spaced locations in the cell at 0, 63, 91, 133, 203, 238, and 308 days of anaerobic treatment. Temperature, moisture, and redox potential were monitored and controlled during the test.

Laboratory analysis of soil samples was performed at the Dyno Nobel laboratory. Dyno Nobel developed the analytical method used. The soil sample is dried, crushed and sieved trough a 2 mm mesh screen, 2 grams of sample is directly extracted with toluene, and directly injected onto non-polar capillary column (DB-1, 0.32 mm diameter, 30 m length). The instrument used is a Varian 3300.

A calibration standard comprised of a mixture of 11 isomers of dinitrotoluene, dinitroethylbenzene (DNEB), and dinitroxylene (DNX) was used. The transient reduction products of DNT were monitored for using GG-MS by Akzo Nobel.

## RESULTS

The result of 308 days of anaerobic treatment on 2,4-DNT and total nitrotoluenes (excluding 2,4-DNT) are summarized in Figure 1 and Figure 2, respectively. The concentration of 2,4-DNT was reduced from an initial mean concentration of 4,000 mg/kg (dry weight basis) to 19 mg/kg. Similarly, the total concentration of other nitrotoluene explosives was reduced from 5,300 mg/kg to 22 mg/kg. The following compounds were identified by GC-MS during anaerobic treatment: 5-amino-6-nitrotoluene, 2,6-dinitrotoluene, 2-nitro-6-aminotoluene, 2,4-dinitro-1,3-dimethylbenzene, 2,4-dinitrotoluene, and 2-nitro-6-aminotoluene. These compounds were transient and not detected after the second round of sampling at 63 days.The Swedish guideline value for 2,4-DNT is 20 mg/kg with 95 percent confidence. This means that a maximum of 5 percent of the samples (0.05 times 39 samples = 2 samples) are permitted to have higher concentration than the guideline value. Although the mean 2,4-DNT concentration at 308 days was less than 20 mg/kg, 4 samples exceeded the 20 mg/kg limit. The local environmental authority required that a leachability study be performed to demonstrate that the anaerobic treated material would not impact ground water.

A leachability test, recommend by the Naturvårdsverket (Swedish EPA), was used to determine site-specific cleanup values. The leachability test is a shaker test method, utilizing an "end over end" apparatus with deionized water, sulphuric acid pH adjustment to 4, and 1.22 kg of soil sample. The Swedish Geotechnical Institute (SIG) performed the leachability test. Results of the test indicated that only a small amount of 2,4-DNT (0.6 g) would be released to the groundwater from the treated soil during the first 14 years.

The leachate collected from liner containing the compost-amended soil was collected and analyzed monthly. Results for 2,4-DNT and nitrotoluene analysis of these leachate samples are presented in Table 1. These results confirmed the leachability test results, which predicted that only a small amount of DNT should leach from the TOSS treated soil. The higher 2,4-DNT and

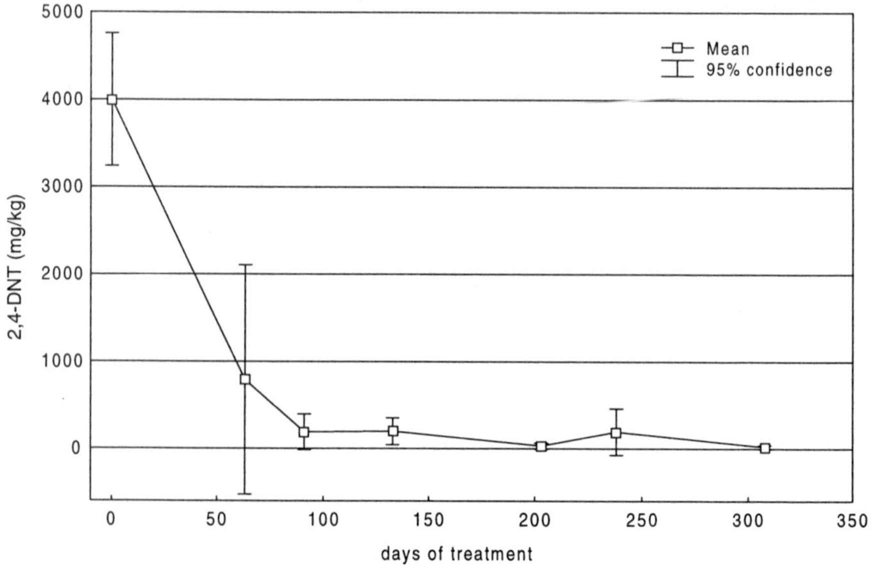


**Figure 2. 2,4-DNT versus time during anaerobic treatment.**

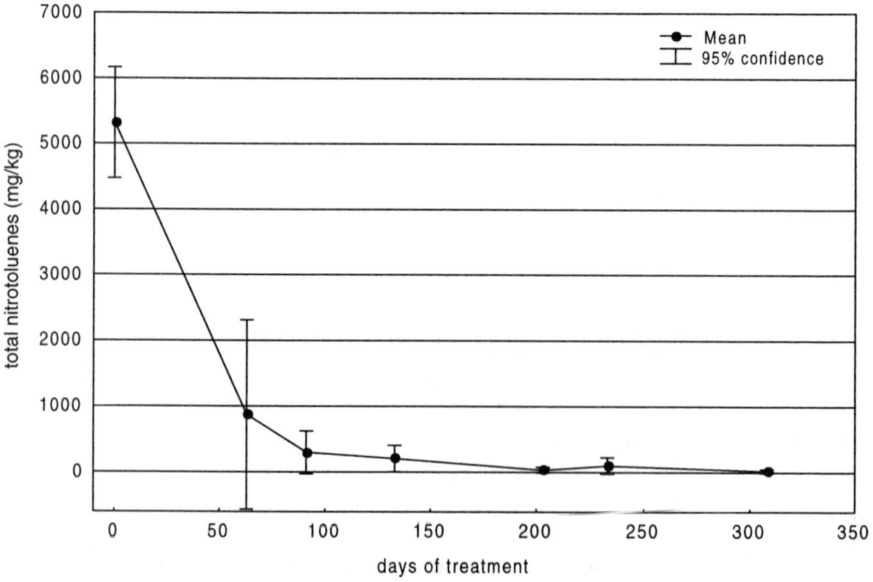


**Figure 1. Total nitrotoluenes, excluding 2,4-DNT, versus time during anaerobic treatment.**

nitrotoluene concentrations detected at July 11 sampling were determined to be attributable to explosives bound to particulate contamination. This observation was based on the fact that when the samples were filtered through a 0.45 μm cellulose acetate filter no 2,4-DNT or nitrotoluenes were detected.

**TABLE 1. Concentration of nitrotoluenes and 2,4-DNT in leachate samples.**

| Sampling date | Nitrotoluenes total (μg/L) | | 2,4-DNT (μg/L) | |
|---|---|---|---|---|
| | Unfiltered | Filtered | Unfiltered | Filtered |
| 07/11/00 | 21 | <2 | 8 | <2 |
| 07/11/00 | 14 | <2 | 6 | <2 |
| 08/02/00 | < 2 | — | < 2 | — |
| 09/04/00 | < 2 | — | < 2 | — |
| 10/02/00 | < 2 | — | < 2 | — |
| 12/13/00 | < 2 | — | < 2 | — |

## CONCLUSIONS

In summary, results demonstrate that the TOSS process could effectively degrade 2,4-DNT to levels below the Swedish guideline value of 20 mg/kg allowing the soil to be reused as cover material. Substantial removal of other nitroaromatics was also achieved, with a combined removal efficiency of 99.5 percent. While reduction products were identified during the anaerobic stage, these compounds were transient and were not detected after approximately two months of treatment. This suggests that the aerobic stage of the TOSS process, which is used to facilitate the binding of reduced intermediates to soil organic matter, may not be necessary in all cases.

The test also demonstrates that given adequate provisions, the TOSS process could be effectively used in low temperature conditions.

## REFERENCES

Eckman, G, T. Lundgren, A. Handel, and R. Green. 1999. Sequential Anaerobic-Aerobic Treatment of Explosives-Contaminated Soil. Presented at: XIII SAFEX International Congress, May 26-29. Dublin, Ireland.

Green, R.B., P.E. Flathman, G.R. Hater, D.E. Jerger, and. P.M. Woodhull. 1998. Sequential Anaerobic-Aerobic Treatment of TNT-Contaminated Soil. In: G.B. Wickramanayake and R.E Hinchee, (eds), *Designing and Applying Treatment Technologies: Remediation of Chlorinated and Recalcitrant Compounds*, pp. 271-276. Battelle Press, Columbus, OH.

Jerger, D.E. and P. Woodhull. 2000. Applications and Costs for Treatment of Contaminated Soil. In: J.C. Spain, J.B. Hughes, and H.-J. Knackmuss, (eds), *Biodegradation of Nitroaromatic Compounds and Explosives,* pp. 407-408. Lewis Publishers, Boca Raton, FL.

# ANAEROBIC/AEROBIC BIODEGRADATION OF ENERGETIC COMPOUNDS

***Jan L. Davel***, Makram T. Suidan, Moustafa A. Moteleb, Jeff R. Macomber and Jay Kim (University of Cincinnati, Cincinnati, Ohio)
Neal R. Adrian (U.S. Army Construction Engineering Research Laboratory, Champaign, Illinois)

**ABSTRACT:** Two laboratory scale anaerobic fluidized-bed reactors (AFBRs) were operated under identical mass loading, temperature and pH conditions. The first was charged with granular activated carbon (GAC) and the second with sand as attachment medium. The reactors were fed synthetically prepared solutions of TNT and ethanol, and the degradation phenomena observed. The GAC was preloaded with TNT before start-up of the reactors, and on this AFBR it was observed that feed ethanol was converted to gas- and aqueous phase carbon dioxide until the start of volatile fatty acid (VFA) production. Once the oxidation potential of the pre-loaded TNT had been depleted, excess ethanol was available and VFA production could commence, followed shortly afterwards by methane production. From this point onwards AFBR effluent was observed to contain numerous TNT degradation intermediates, most of which were removed in the following activated sludge reactors. During the second phase of the study the TNT in the feed solutions was replaced with RDX and HMX, and only one RDX intermediate was observed in the AFBR effluent. More complete removal of energetic compounds was observed with GAC as attachment medium than with sand.

## INTRODUCTION

2,4,6-Trinitrotoluene (TNT), hexahydro-1,3,5-trinitro-1,3,5-triazine (RDX) and octahydro-1,3,5,7-tetranitro-1,3,5,7-tetrazocine (HMX) are organic energetic compounds classified by the US EPA as priority pollutants. The number of contaminated sites where these compounds are found is increasing due to demilitarization activities and disposal of obsolete munitions. The current methods of waste disposal include discharge to deep sea, landfills and incineration for solid wastes, and GAC adsorption followed by incineration of the spent carbon for liquid wastes.

Biological transformation of these compounds as a treatment alternative has been studied extensively, and the most commonly noted transformations involve reductive pathways (McCormick et al., 1976; Stahl and Aust, 1993). Commonly noted intermediates include aminodinitrotoluenes, diaminonitrotoluenes, 2,4,6-triaminotoluene (TAT), azoxy dimers, as well as unknown products. Transformation of TNT under sulfate reducing and methanogenic conditions has been noted (Boopathy et al., 1993), and 3% mineralization to $^{14}C\text{-}CO_2$ under aerobic conditions has been reported (Boopathy et al., 1994).

**Objective**. This research project investigates two-stage anaerobic/aerobic bioremediation as a treatment alternative. AFBRs were employed in the first treatment stage to remove the energetic compounds TNT, RDX and HMX in the presence of ethanol as co-substrate. The AFBRs were charged with different types of attachment media to observe how this influences the removal of these compounds. The AFBR effluent was subjected to a second aerobic activated sludge treatment stage to remove any anaerobic degradation intermediates.

## MATERIALS AND METHODS

Two 9.1 L AFBR units were operated side by side under identical mass loading conditions. Their temperatures were maintained at 35°C and pH at 7.2 approximately. At the start of the study, the first AFBR was charged with 1 kg of GAC, pre-loaded with 200 grams of TNT, and the second charged with an equal volume of sand. The AFBRs were seeded with a culture from a municipal anaerobic digester. Synthetically produced wastewater containing 95 mg/L TNT and 560 mg/L ethanol as co-substrate, along with a solution containing essential nutrients and a pH buffer, were continuously fed to both AFBRs at a flow rate of 6.9 L/day.

The effluent from each AFBR was treated in separate aerobic activated sludge reactors. Each of these reactors consisted of a 6.0 L polyethylene porous pot placed within a larger 18 L HDPE container. Air was continuously bubbled through the mixed liquor, keeping the biomass suspended and at a dissolved oxygen (DO) concentration of approximately 7 mg/L. The pH was maintained at approximately 7.6.

During the second phase of the study the TNT in the feed solutions was replaced with RDX and HMX, at concentrations of 16 mg/L and 1.1 mg/L respectively.

**Chemicals**. TNT, RDX, HMX and all intermediates were provided to the University of Cincinnati by Construction Engineering Research Laboratory (CERL) in Urbana, IL. Salts, nutrients, and carbonate were purchased from Fisher Scientific, Pittsburgh, PA, and were 95 % purity or greater.

**Gas Composition**. Effluent gas samples from the AFBRs were analyzed for nitrogen, oxygen, carbon dioxide and methane with a Hewlett Packard 5890 Series II gas chromatograph (GC) equipped with a thermal conductivity detector (TCD). Helium was used as the carrier gas through a 3.2 mm ID, 3 m steel column packed with 45/60 molecular sieve (Hewlett-Packard Company, San Fernando, CA).

**Chemical Oxygen Demand**. Samples were analyzed for COD in accordance with Method 508C (Standard Methods for the Evaluation of Water and Wastewater (1985) 16th ed.). Filtered samples were acidified with 85% *o*-phosphoric acid to a pH of 2 and purged with pre-purified nitrogen for 10 minutes to strip out sulfide. Prepared Hach COD test vials (range 0 – 150 mg/L) and a Bausch and Lomb Spectronic 70 spectrophotometer (Bausch & Lomb, U.S.A.)

were used for the analysis.

**VFAs and Alcohols**. The VFAs, acetate and propionate, as well as the alcohols ethanol and methanol, were analyzed by aqueous injection into a Hewlett Packard 5890 Series II GC equipped with flame ionization detector (FID) and 60/80 Carbopack B5% Carbowax 20 M glass column (Supelco, Bellefonte, PA).

**Dissolved Organic Carbon and Inorganic Carbon**. Inorganic carbon (IC) was measured by injection into a Shimadzu TOC Model 5000 analyzer (Shimadzu Corp., Tokyo, Japan). For measurement of the dissolved organic carbon (DOC), the pH of the aqueous sample was first reduced to below 2 with hydrochloric acid. This was followed by purging with ultra pure air, containing less than 0.1 ppm hydrocarbons, to strip inorganic carbon as $CO_2$. The DOC was then measured using the same instrument.

**TNT and Intermediates**. TNT, 2-A-4,6-DNT, 4-A-2,6-DNT, 2,4-DA-6-NT, and 2,6-DA-4-NT were analyzed using a Hewlett Packard (HP) High Pressure Liquid Chromatograph (HPLC) 1050 series equipped with a diode array detector. The column used was a Supelcosil LC-18-DB (150 x 4.6 mm, Supelco, Bellefonte, PA). An isocratic mobile phase of 50% dibasic sodium phosphate buffer (2.5 mM) and 50% acetonitrile was used to elute the amino-nitro compounds. The flow rate was kept at 0.7 ml/min and detection limit of TNT was at 17 ppb.

**TAT**. TAT was analyzed separately on the HPLC, HP 1050 series using a Supelcosil LC-CN column (250 x 4.6 mm, Supelco, Bellefonte, PA). The mobile phase consisted of 1 % acetonitrile and 99 % of a 2.5 mM dibasic sodium phosphate buffer. The flow rate was kept at 1.0 ml/min.

**HMX, RDX and Intermediates**. The compounds HMX and RDX, as well as the nitroso-RDX intermediates hexahydro-1-nitroso-3,5-dinitro-1,3,5-triazine (MNX), hexahydro-1,3-dinitroso-5-nitro-1,3,5-triazine (DNX) and hexahydro-1,3,5-trinitroso-1,3,5-triazine (TNX), were analyzed on the HPLC, HP 1050 series using an Alltech LiChrospher RP-18 column (250 x 4.6 mm, Alltech Associates, Inc., Deerfield, IL). The mobile phase consisted of water and methanol in a 60:40 ratio, at a flow rate of 1.0 ml/min.

## RESULTS AND DISCUSSION

**Ethanol Utilization**. The attachment medium of the GAC-AFBR was preloaded with 200 g of TNT. This allowed a very clear observation during the initial start-up of this AFBR, of the probable degradation mechanism under these treatment conditions. The effluent from this reactor was monitored daily over the first few weeks for TNT and TNT intermediates, ethanol, VFAs, DOC, IC, methane and $CO_2$ production.

The carbon balance over this period is indicated in Figure 1. It shows the total influent carbon to the reactor and the individual accounted for carbonaceous components, as well as the combined accounted for total of these components. From Day 50 onwards it was observed that these concentrations did not vary much, and operational steady-state was assumed.

The period preceding that, however, provided some insight into the degradation mechanisms which may be occurring. Over the first 9 days, the initial increase and decrease in DOC concentration corresponded to the ethanol concentrations observed in the effluent. These could be ascribed to an initial acclimation period of the microorganisms, during which time the available ethanol was not utilized effectively. From Day 10 up until Day 22 of operation, the effluent did not contain any ethanol, VFAs, TNT or TNT degradation intermediates. No methane was observed either, only the equivalent conversion of the feed ethanol to gas- and aqueous phase carbon dioxide.

From Day 22 the effluent DOC corresponded to the VFAs observed in the effluent. From Day 28 methane production started, and gradually increased as the VFAs were decreasing. From these data it was concluded that ethanol was oxidized while TNT was being reduced according to the following reaction:

$$\underset{\text{(ethanol)}}{1.5\ C_2H_6O} + \underset{\text{(TNT)}}{C_7H_5N_3O_6} \xrightarrow{\text{microorganisms}} \underset{\text{(TAT)}}{C_7H_{11}N_3} + 3\ CO_2 + 1.5\ H_2O$$

From this equation it can be concluded that the combined oxidation capacity of the pre-loaded TNT and all TNT fed to the reactor from start-up would have to be satisfied before any excess ethanol would be available to the microorganisms. From Figure 1, the point at which excess ethanol does become available, would appear to coincide with the commencement of acetogenesis, namely Day 23.

This was strongly supported by comparing the cumulative influent TNT and ethanol on a molar basis, as shown on Figure 2. The pre-loaded TNT is shown to represent the majority of TNT introduced to the reactor, slowly increasing on a cumulative basis due to the daily TNT feed. Similarly, the cumulative ethanol to the reactor is shown to increase steadily. These cumulative influent ethanol data have been corrected for the ethanol observed in the effluent and VFAs observed from Day 23 onwards. These corrections do not impact significantly on the cumulative influent ethanol plot since they are very small compared to the large influent ethanol loading.

Also indicated on this figure is the theoretical ethanol requirement for the complete reduction of TNT, according to the above Equation 1. The cumulative effective influent ethanol is shown to intersect this theoretical ethanol requirement line at Day 20. The observed data from Figure 1 shows acetogenesis to start at Day 23. This difference is assumed to account for the ethanol required for microbial growth, which is not incorporated into the above oxidation-reduction equation.

**Degradation Performance**. No TNT or TNT degradation intermediates were observed during the first 50 days of operation, which is ascribed to adsorption to the GAC. From Day 50 onwards, the main degradation intermediate observed was the reduced form of TNT, namely TAT. The only other incomplete reduction intermediate which was observed was the compound 2,4-diamino-6-nitrotoluene (DANT).

These results are reflected in Table 1. It is clear that the effluent concentrations obtained for these compounds is significantly lower in the GAC-AFBR than in the sand medium AFBR. This observation is attributed to the superior microbial attachment properties of the GAC, and possibly also the catalytic properties of the GAC surface.

**TABLE 1. Degradation intermediates and DOC observed in the two AFBRs.**

| Energetic Compound | Sand Attachment Medium AFBR (mg/L) | GAC Attachment Medium AFBR (mg/L) |
|---|---|---|
| TAT | 36.7 ± 1.6 | 22.6 ± 3.2 |
| DANT | 0.5 ± 0.2 | n.d. |
| TNT | n.d. | n.d. |
| DOC | 54.0 ± 7.4 | 21.3 ± 5.4 |

**Secondary Aerobic Activated Sludge Treatment**. Secondary aerobic activated sludge treatment units oxidized the reduced/partially reduced TNT intermediates contained in the AFBR effluent streams. From the removal results indicated in Table 2, it can be seen that these concentrations entering the secondary treatment stage were lowered but not completely removed.

**TABLE 2. Degradation intermediates and DOC observed in the two aerobic activated sludge reactors.**

| Energetic Compound | A/S reactor for sand AFBR effluent (mg/L) | A/S reactor for GAC AFBR effluent (mg/L) |
|---|---|---|
| TAT | 11.6 ± 3.2 | 3.1 ± 1.1 |
| DANT | n.d. | n.d. |
| TNT | n.d. | n.d. |
| DOC | 27.0 ± 5.2 | 14.2 ± 9.7 |

**RDX and HMX Degradation**. As a second phase of the study, TNT in the feed solution was replaced with 16 mg/L RDX and 1.1 mg/L HMX. No RDX, HMX or degradation intermediates were observed on the GAC-AFBR. This observation is not attributed to adsorption, though, since no break-through of these compounds was observed throughout the seven month operation period. In a follow-up study (not reported in this manuscript) the feed composition to this AFBR was altered to include 95 mg/L TNT and operated for an additional nine month period. Even under these conditions no break-through of RDX, HMX or degradation intermediates were observed.

Similar to the TNT-feed phase of the study, the sand AFBR showed less complete degradation during the RDX/HMX-feed phase of the study than observed on the GAC-AFBR. This was evident not only from the difference in DOC concentrations, but also from the regular observation of both RDX and HMX in the effluent of the sand AFBR. In addition, the effluent from this reactor regularly contained the RDX degradation intermediate MNX, although only a qualitative analysis of this intermediate was possible. No residual RDX, HMX or intermediates were observed in the effluent of the activated sludge unit.

**Applications of Research Findings**. The close correlation between the theoretical ethanol requirement and the observed ethanol utilization for the complete reduction of TNT to TAT, is an important step toward establishing practical design criteria for bioremediation efforts. Although the operation of anaerobic fluidized bed reactors at these ethanol loadings would appear to be possible, further research in this field may reveal other considerations affecting operational performance.

Comparison of the two types of attachment media clearly indicates that GAC provides superior removal efficiencies compared to sand. This finding provides valuable support to the use of fluidized-bed technology, specifically using GAC as attachment medium. Upon investigation of the performance of fluidized-beds for the removal of RDX and HMX, it is clear that water contaminated with these compounds is treatable using an anaerobic/aerobic sequential process.

The research results show definite benefits for incorporating aerobic activated sludge as secondary treatment step, even though complete removal of these compounds was not achieved. Moreover, the intermediates resulting from this treatment approach were not identified. In spite of these uncertainties, it is clear the two-stage anaerobic/aerobic treatment procedure holds promising possibilities for the treatment of organic energetic compounds.

## REFERENCES

Boopathy, R., J. Manning, C. Montemango, and C. Kulpa. 1994. "Metabolism of 2,4,6-Trinitrotoluene by a *Pseudomonas* Consortium under Aerobic Conditions." *Current Microbiol. 28*: 131-137.

Boopathy, R., M. Wilson, and C. F. Kulpa. 1993. "Anaerobic Removal of 2,4,6-Trinitrotoluene (TNT) Under Different Electron Accepting Conditions: Laboratory Study." *Water Env. Rev. 65*(3): 271-275.

McCormick, N. G., F. E. Feeherry, and H. S. Levinson. 1976. "Microbial Transformation of 2,4,6-Trinitrotoluene and Other Nitroaromatic Compounds." *Appl. Environ. Microbiol. 31*: 949-958.

Stahl, J. D. and S. D. Aust. 1993. "Metabolism and Detoxification of TNT by *Phanerochaete Chrysosporium*." *Biochemical and Biophysical Research Communications. 192*(2): 477-482.

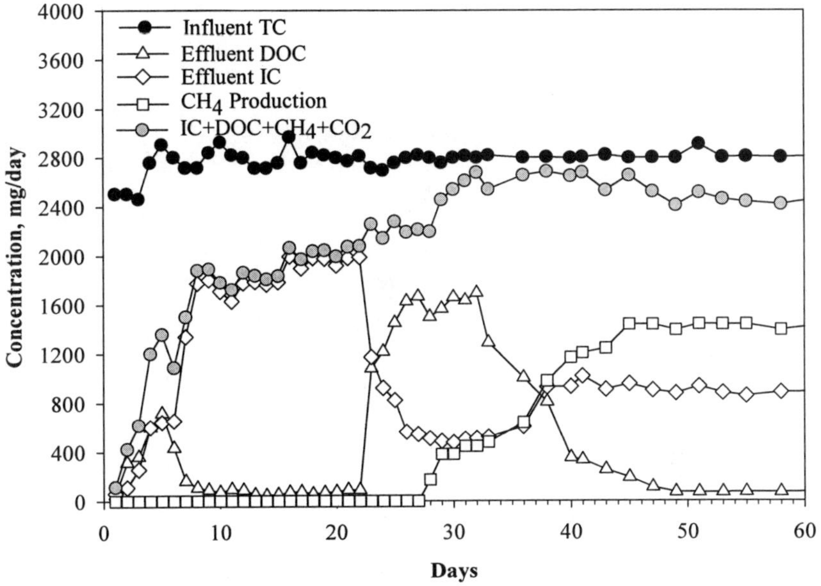


**FIGURE 1. Carbon balance of the TNT feed GAC-AFBR during start-up period. TC = total carbon, DOC = dissolved organic carbon, IC = inorganic carbon.**

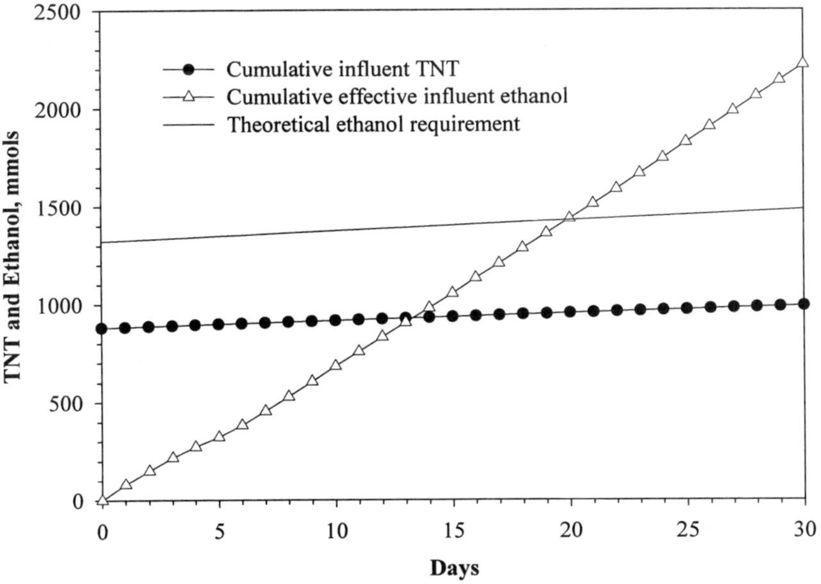


**FIGURE 2. Theoretical ethanol requirement for the complete reduction of cumulative influent TNT, compared with cumulative effective influent ethanol.**

# BIOLOGICAL TREATMENT OF AMINE WASTES FROM THE GAS INDUSTRY

***J.R. Gallagher*** and J.A. Sorensen
University of North Dakota Energy & Environmental Research Center
Grand Forks, ND

**ABSTRACT:** At a decommissioned sour gas-processing plant, the burial of amine reclaimer unit sludges has resulted in subsurface contamination with amines, amine byproducts, and salts. Data generated by prior investigations of the natural attenuation processes controlling the environmental fate of these contaminants were used to develop a remediation strategy for the amine-contaminated soils. The biodegradability of the amines, coupled with their relatively low mobility, indicated that bioremediation was an appropriate technology. The presence of salts required leaching of the soil. Based on these conditions, a 500 $m^3$ biopile was designed and constructed in 1998. The biopile, which included aeration, irrigation, and leachate collection systems, was operated in 1998, 1999, and 2000. The results from the biopile operation show that all identified amines, significant amounts of organic and inorganic nitrogen species, salts, and organic carbon were removed.

## INTRODUCTION

Alkanolamines are commonly used by the natural gas industry to remove hydrogen sulfide, carbon dioxide, and other acid gases from natural gas ("sour" gas if hydrogen sulfide is present). Sour gas makes up a significant portion of the natural gas produced in both the United States and Canada. About 13% of the natural gas produced in the United States is sour enough to require treatment (Prouty 1994). In Canada about 40% of the annual net production is sour. Alberta produces approximately 83% of the total Canadian gas and is the largest producer of sour gas in North America. There are at least 394 gas-sweetening facilities in the United States that utilize amine-based acid gas removal (AGR) units (Tannehill and Galving 1993). The fate and remediation options for treating amines, and amine contamination have not previously been examined.

## BACKGROUND

Alkanolamines are organic derivatives of ammonia. Monoethanolamine (MEA) is considered to be the most basic of the alkanolamines, and is therefore the most reactive toward acid gases. In gas-sweetening systems, alkanolamines can be involved in a variety of reactions that produce chemical species that may have a role in alkanolamine-related subsurface contamination. The products formed by these reactions range from simple breakdown products (for example ammonia, water, and hydrogen gas) to complex nitrogenous organic compounds, referred to here as thermal/oxidative (T/O) reaction products, about which little is known concerning their chemical identity and properties. Some of the T/O reaction products commonly associated with MEA include *N*-(hydroxyethyl) imidazolidone (HEI); *N*-(hydroxyethyl)-1,2-ethylenediamine (HEED); *N,N* -bis-(hydroxyethyl) ethylenediamine (BHEED); and 2-oxazolidinone (OX). These are produced by MEA reactions with $CO_2$ and/or carbonylsulfide (Skinner et al. 1995). These T/O reaction products typically occur in sludges that accumulate in the reboiler tanks and filters of amine-based AGR units.

The operation of AGR units creates a variety of wastes that may be introduced to the environment. The managed and unmanaged waste streams may be composed of spent alkanolamines, sludges from process unit tank bottoms, and process system filters. The T/O reaction products discussed above are typically present in the waste streams in varying concentrations, depending on the process conditions. Additional components of the wastes include additives such as corrosion inhibitors and anti-foaming agents, benzene and trace metals (Sorensen et al. 1998).

The release of these waste streams to the subsurface environment is generally controlled by waste management techniques. Most gas sweetening facilities today employ techniques that minimize or eliminate the release of amine-related waste to the environment. However, past management practices may have resulted in significant subsurface impacts at some sites. For instance, in Canada leaking process waste pits and storage tanks containing spent alkanolamine solutions from changeover operations have been reported to be the most common source of groundwater contamination at gas processing and conditioning plants. Many of the pits are excavated into surface material and are either lined with compacted natural clay or unlined, a situation more often found at older plants. Onsite and offsite landfills have also been used to dispose of spent alkanolamines and thus have also been a source of contaminants in groundwater (Hardisty et al. 1990).

## OKOTOKS SITE DESCRIPTION

The Okotoks Gas Plant is located approximately 25 km south of Calgary in south central Alberta. The climate of the Okotoks area is semiarid to moist, with an average annual precipitation of 488 mm/year. The frost-free period for the area is between 75 and 90 days. Average temperatures in the region are typically below 10°C 8 to 9 months of the year. The near-surface geology of the site is dominated by unstratified glacial sediments; comprised of clay-rich glacial till with intercalated gravels, sands and silts.

The results of an environmental assessment conducted in 1992 as part of the plant's decommissioning process revealed the presence of a layer of sludge-contaminated soil at a depth of 2 to 3 m in the area of the former amine sludge disposal pits. A strong ammoniacal odor in the sediments, and elevated organic nitrogen concentrations suggested that the contamination was MEA-related.

## PREVIOUS RESEARCH

Preliminary biodegradation assessments were performed on parent amines and T/O products using an uncontaminated soil from the site. These studies showed that MEA was readily biodegraded under aerobic conditions. Anaerobic biodegradation was also observed, but at rates an order of magnitude less than aerobic conditions. All T/O products tested were also readily biodegraded under aerobic conditions, although anaerobic testing was not performed (Gallagher et al. 1996). Tests with the Okotoks soil for mobility showed that MEA has a distribution coefficient ($K_d$) ranging from 2.2 to 4.9, which all fall within the low mobility class ($2< K_d <10$), as defined by Dragun (1988). These results indicate that soil washing would not be an effective method for removing MEA-related contamination from the Okotoks soil. As anticipated, a bulk of the removal would have to come as a result of biodegradation.

Biodegradation testing of the contaminated soil presented a different case. This soil contained, in addition to the parent amine, MEA, a number of the T/O products, and a variety of unknown amines and byproducts. Biodegradation in soil slurry reactors showed rapid reduction in organic matter in the first 15 days, but metabolism was slower thereafter. Operation of the bioreactors for 24 days reduced the COD and organic nitrogen concentration 79% and 59%, respectively, at 8% soil concentration. Toxicity of the amine-related components resulted in reduced removals at higher slurry concentrations. The complete data set of these experiments is provided in Gallagher et al. (1996).

Laboratory-scale landfarming experiments were conducted with the amine contaminated soil to examine four variables: 1) phosphorus dose, 2) tillage frequency, 3) loading rate of contaminated soil, and 4) pH adjustment. The values of the variables were selected to bracket the practical range that may be used or encountered in a full-scale landfarming operation. The soil samples were analyzed for moisture, pH, ammonia, nitrate plus nitrite, and microbial counts. Samples from the beginning and end of the study period were analyzed for toxicity, COD, and TKN.

The results of the landfarm experiments showed that the most important variable affecting bioremediation was soil loading. Phosphorus addition resulted in small increases in bioactivity. The most effective landfarm removed 51% of the initial COD, 68% of the initial TKN, and 66% of the organic nitrogen. The data showed that toxicity of some components in the contaminated soil resulted in reduced bioactivity, longer lag times, and reduced removals. Based on Microtox®, toxicity in all land farms at the lower soil loadings (10 and 17.5 wt%) was reduced to zero. However, the results of additional assessments via seed emergence, root elongation, and earthworm survival indicated that significant toxicity remained even in the lower soil-loading conditions. Further testing showed that the toxic fraction of the treated soil is extractable in both water and methanol, which suggests that those components may be leachable from the soil. The identity of these toxic components is not known. The complete data set of these experiments are provided in Gallagher et al. (1997).

These experiments suggested that an engineered biopile would be the most cost effective method of treating these soils. The biopile would ensure that the soil is well aerated, can be leached to reduce initial toxicity, and reduce final toxicity, and allow for addition of additives to improve remediation.

## BIOPILE DESIGN AND OPERATION

The biopile containment cell measures 40 m long by 10 m wide by 1.5 m deep. Above a 25-mil reinforced polyethylene (RPE) liner is a thin layer of crushed gravel. Filter fabric caps the gravel layer and lies directly beneath the treatment soils. The soil layer is gently mounded and enclosed by a 25-mil RPE cover. Approximately 450 $m^3$ of treatment soils and 50 $m^3$ of straw are housed within the constructed cell. Soil additives include 2.58 $m^3$ calcium chloride ($CaCl_2$) as well as 10–34–00 (% nitrogen–phosphorus–potassium) liquid fertilizer (≈2036 kg). Straw was added to act as a bulking agent, thereby increasing the porosity and permeability of the biopile. The addition of $CaCl_2$ effectively increases the permeability of the soil, thus allowing water to move more easily through the biopile. The chloride salt of calcium was selected as an inexpensive salt that had no nitrogen and was water soluble. Fertilizer was added as a nutrient source for the microbial population to improve the biodegradation rate.

Aeration, irrigation, and a leachate collection system define the dynamic components of the structure. The aeration system is one large air vent (100-mm perforated PVC pipe) and four equally spaced flow ducts (50-mm perforated pipe) that run the entire 40-m length. The flow ducts are powered by an external blower unit. Adequate soil moisture is also necessary to obtain appropriate microbial activity, although if the soil is saturated or nearly so, aeration will be inhibited. Therefore, water is periodically added to the biopile using an irrigation system. The timing and amount of water application is determined by soil moisture measurements taken on a weekly basis. The irrigation system comprises five semipermeable hoses, equally spaced, which also run the full length of the structure. The hoses are fed by an external freshwater supply, powered by a water-pumping system. The results of the toxicity evaluations conducted as part of the earlier activities indicated that the toxic fraction of the soil contamination is water soluble, and therefore collection of the leachate generated during biopile operation is necessary. Leachate is collected in a sump located directly below the crushed gravel layer and temporarily stored in a reinforced external tank. For the Okotoks site, leachate is disposed in an injection well located on-site.

Construction of the biopile was initiated on July 13 and completed on July 20, 1998. The aeration system was activated and the first 2000 imperial gallons (9100 liters) of water were added to the biopile on August 5, 1998. Another 2000 imperial gallons of water were incorporated into the biopile on September 25, 1998. Aeration of the biopile was stopped on November 16, 1998, as required by Alberta regulations. The biopile was remixed on November 27 to break up any channels that may have formed over the summer and fall. No sampling or monitoring activities were conducted over the winter months. Reactivation of the aeration system and rewetting of the biopile was initially scheduled to begin in March 1999 and continue until June of 1999, but changes in the CanOxy corporate structure in early 1999 caused the reactivation to be delayed until July 1999. The biopile was then operated in the aeration mode through July and August 1999. In September, the data indicated that biodegradation of the amine-related materials was likely complete and that the leachable fraction was all that remained. Thus, in September 1999 the biopile began to be operated in the leaching mode, which meant that water application and leachate collection were conducted on an essentially continuous basis. The leaching operation continued through October 1999 and was halted for the winter of 1999-2000, and was conducted again from May through September of 2000. Over the course of the leaching operations in 1999 and 2000 approximately 85,000 imperial gallons (386,000 liters) of water were applied to the biopile, which amounts to roughly 3 pore volumes.

## SOIL SAMPLING AND ANALYSIS

Soil sampling was conducted at the start, and bimonthly throughout the active treatment periods. For sampling, the biopile was artificially divided into eight equal-sized grids. At each sampling period, three grids were randomly selected and sampled with a coring device. Within each grid, a core was collected representing the 0 to 60 cm zone, and one for the 60 to 120 cm zone. To generate baseline data, a composite sample was submitted to a commercial laboratory in Calgary for the following analyses: hydrometer sediment size analysis, pH, conductivity, sodium adsorption ratio, bulk density, moisture, soluble salts, soil organics, and metals. The results of these analyses are given in Table 1. Routine analytical parameters included TKN, ammonia, nitrate+nitrite-N ($NO_x$-N), and total organic carbon (TOC), which were considered the key soil character parameters for evaluating the general activity of the biopile over the course of the study period.

**TABLE 1. Physical and chemical characteristics of a composite soil samples from the biopile.**

| Parameter | July 17, 1998 | Nov. 11, 1998 |
|---|---|---|
| % Sand | 34.3 | |
| % Silt | 28.6 | |
| % Clay | 37.1 | |
| pH | 7.3 | |
| Electrical Conductivity (dS/m) | 7.37 | 13.9 |
| Sodium Absorption Ratio | 2.9 | 7.9 |
| Bulk Density (g/cc) | 1.44 | |
| Moisture (%) | 16.9 | |
| Chloride | 587 | 1150 |
| Sulfate | 110 | 297 |
| Calcium | 398 | 105 |
| Magnesium | 111 | 85.4 |
| Potassium | 214 | 367 |
| Sodium | 183 | 307 |
| Total Purgabable Hydrocarbon | 2.1 | |
| Total Extractable Hydrocarbon | 124 | |

**RESULTS AND DISCUSSION**

Nitrogen-related data were generated from 20 sampling events over the course of the 3 partial seasons of operation. The TKN, ammonia-N ($NH_3$-N), NOx-N, TOC, and total organic nitrogen (TON, which is calculated by subtracting ammonia-N from TKN) concentrations of samples taken from the beginning and end of the each operating season are presented in Tables 2. These data show that from the start of 1998 operations through 1999, the TKN and TON generally decreased while ammonia and NOx-N concentrations increased with time. This general pattern is expected during biodegradation of the nitrogenous organic compounds (i.e. amines and T/O reaction products), since ammonia is produced as a result of the biodegradation of alkanolamines and other organic nitrogenous compounds and eventually may be oxidized to nitrate (Gallagher et al. 1996). In 2000, TKN and TOC values reached their lowest levels and remained relatively steady over the course of the summer. From September 1999 to June 2000 there were significant drops in ammonia and NOx-N, which suggests that biodegradation of the alkanolamines and the T/O reaction products was complete. This conclusion is supported by the analytical results from soil samples collected in October 1999 that were analyzed for specific amines and their associated thermal/oxidative (T/O) reaction products. Those results (Table 3) show that all amines and T/O reaction products were brought to levels below detection limit by the end of the 1999 season.

**TABLE 2. Mean concentrations of nitrogen components and organic matter in the biopile.**

| Date | TKN (mg/kg) | $NH_3$-N (mg/kg) | NOx-N (mg/kg) | TOC (mg/kg) | TON (mg/kg) |
|---|---|---|---|---|---|
| 07/17/1998 | 6250 | 480 | 130 | 28550 | 5770 |
| 11/26/1998 | 4870 | 1350 | 840 | 18300 | 3520 |
| 07/23/1999 | 6420 | 2000 | 1670 | 20220 | 4420 |
| 10/25/1999 | 4880 | 550 | 2420 | 17350 | 4330 |
| 06/02/2000 | 5200 | 560 | 1400 | 20400 | 4650 |
| 09/01/2000 | 3020 | 80 | 130 | 26680 | 2940 |

Although the concentrations of amines and T/O reaction products were below the detection limit in the biopile at the end of the 1999 season, the concentration of salts (as measured by electrical conductivity, or EC) and inorganic nitrogen compounds (as measured by ammonia and nitrate-nitrite) were above the regulatory standards. The results of the analyses of soil samples collected over the course of the summer of 2000, during which time water was being applied to the biopile on an essentially constant basis, show steady decreases in EC, SAR, ammonia, and nitrate+nitrite. The analytical results show that by the beginning of September 2000, all soil parameters for which Alberta has regulatory standards were below those standards. The September 2000 results indicate that remediation of the soil was complete.

## CONCLUSIONS

The engineered biopile resulted in rapid remediation of the amine contamination of the soil. Though spread out over the course of three years, the actual operation time of the biopile was approximately 11 months. This suggests that similarly contaminated soils in a warmer climate could be remediated in less than a year.

The cost of treating the 450 $m^3$ of contaminated soil amounted to \$46 per $m^3$, not including engineering and analytical. Costs would be much reduced on a larger scale and in locations not requiring the RPE liner.

**TABLE 3. Amine and associated T/O product concentrations through the 1998 and 1999 biopile operating seasons.**

| Analyte | July 17, 1998 | Nov 11, 1998 | Sept 27, 1999 | Oct 22, 1999 |
|---|---|---|---|---|
| MEA | 800 | 6700 | 190 | BDL[a] |
| HEI | 10,000 | BDL | BDL | BDL |
| OX | 4300 | <100 to 21,000 | 100 to 200 | BDL |
| HEED | trace | BDL | BDL | BDL |
| BHEED | trace | BDL | BDL | BDL |
| TEHEED | trace | BDL | BDL | BDL |

**a. BDL = below detection limit.**

## ACKNOWLEGEMENTS

Funding was provided by the Canadian Association of Petroleum Producers (CAPP), Canadian Occidental Petroleum Ltd. (CanOxy), the U.S. Department of Energy (DOE), Gas Research Institute (GRI), Environment Canada, and the National Energy Board of Canada.

## REFERENCES

Dragun, J., 1988. *The soil chemistry of hazardous materials*, Hazardous Materials Control Research Institute, Silver Spring, Maryland.

Gallagher, J.R., Thompson, J.S., Sorensen, J.A., and C.R. Schmit. 1996. *An Assessment of the Subsurface Fate of Monoethanolamine at Sour Gas Processing Plant Sites – Phase I Report*. Canadian Association of Petroleum Producers, Calgary, Alberta.

Gallagher, J.R., and J.A. Sorensen. 1997. *An Assessment of the Subsurface Fate of Monoethanolamine at Sour Gas Processing Plant Sites – Phase II Report*, Canadian Association of Petroleum Producers, Calgary, Alberta.

Hardisty, P.E., Dabrowski, L.S., Lyness, L.S., Scroggins, R., and P. Weeks. 1990. "Nature, occurrence and remediation of groundwater contamination at Alberta sour gas plants." Proceedings of the *International Symposium on Oil and Gas Exploration and Production Waste Management Practices*, New Orleans, Louisiana.

Prouty, J.L., 1994. "Sprucing up natural gas." *Gas Research Institute Digest*. 17(2):18-21.

Skinner, F.D., McIntush, K.E., and M.C. Murff. 1995. *Amine-Based Gas Sweetening and Claus Recovery Process Chemistry and Waste Stream Survey*. Gas Research Institute Topical Report. GRI 95/0187.

Sorensen, J.A., Fraley, R.H., Gallagher, J.R., and C.R. Schmit. 1996. *Background Report on Subsurface Environmental Issues Relating to Natural Gas Sweetening and Dehydration Operations*. Gas Research Institute Topical Report, GRI 95/0143.

Sorensen, J.A., Aulich, T.R., Hawthorne, S.B., and J.R. Gallagher. 1998. *Amine-Based Gas Sweetening Fluids: Waste Stream Characterization and Subsurface Transport and Fate*, Gas Research Institute Topical Report, GRI 98/0388.

Tannehill, C.C., and C. Galvin. 1993. *Business Characteristics of the Natural Gas Conditioning Industry*, Gas Research Institute Topical Report, GRI 93/0342.

# ENHANCING BIODEGRADABILITY OF REFRACTORY AROMATICS: PRETREATMENT WITH ELEMENTAL IRON

***Jennie R. Perey***, Seok-Young Oh, Brian Lubenow, Daniel K. Cha, Pei Chiu and Chin-Pao Huang
University of Delaware, Newark, Delaware

**ABSTRACT:** Nitro, azo, and chlorinated aromatic compounds are commonly present in industrial wastewaters. Aerobic biological treatment is often inefficient to remove these chemicals from wastewater because of their resistance to aerobic biodegradation. The goal of this study is to investigate the feasibility of using elemental iron to pretreat wastewater through chemical reduction of these aromatics before the activated sludge process. Our hypotheses are that these compounds will become more biodegradable upon reduction with iron and that the combined treatment is more effective and/or less expensive than the activated sludge treatment alone. Preliminary data from batch experiments indicate that nitro and azo compounds are rapidly removed from water through both adsorption and chemical reaction. The reaction rate was found to depend on iron type, pH, and other solutes in the wastewater. These aromatics were reduced to their amino counterparts, such as aniline and sulfanilic acid, which were readily biodegraded. Our results suggest that the sequential abiotic-biological treatment may be a cost-effective approach to degrade refractory aromatics in industrial wastewaters.

## INTRODUCTION

Nitro, azo, and chlorinated aromatics are common industrial chemicals. These compounds are found in industrial wastewaters from various manufacturing processes. Aerobic biological treatment is often ineffective to treat wastewaters containing these aromatic pollutants because of their relatively oxidized nature and resistance to aerobic degradation. A large body of environmental literature indicates that elemental iron in an aqueous environment can chemically reduce nitro, azo, and halogenated compounds. Elemental iron has been used since the mid-1990s to degrade pollutants in groundwater. To date, however, iron has not been applied to wastewater treatment. We hypothesize that using iron to pretreat wastewaters containing these aromatics prior to biological oxidation can enhance degradation efficiency and reduce treatment cost. The two primary objectives of this study are (1) determination of the feasibility of iron pretreatment and (2) investigation of the effects of parameters that affect the design and performance of the combined iron-biological treatment system.

## MATERIALS AND METHODS

**Chemicals.** 2,4-Dinitrotoluene (DNT) and standards for its reduction products and intermediates (diaminotoluene, 2-amino-4-nitrotoluene, and 4-amino-2-nitrotoluene) were obtained from Aldrich (Milwaukee, WI). Nitrobenzene and its reduction product, aniline, were purchased from Fisher Scientific (Pittsburgh, PA). Nitrosobenzene, an intermediate of nitrobenzene reduction, was purchased from Sigma (St. Louis, MO). Orange G dye was purchased from Sigma. Orange IV dye was purchased from Spectrum Quality Products (New Brunswick, NJ). Sulfanilic acid was purchased from Sigma. Trichlorobenzene (TCB), three dichlorobenzene isomers, and chlorobenzene were purchased from Aldrich. Pentachlorophenol (PCP) was purchased from Fluka (Buchs, Switzerland).

Three types of iron were screened for effectiveness as a reducing agent. High-purity iron granules (1-2 mm diameter, 99.98% pure, metals basis) and powder (<10 micron, 99.5%+ pure, total) were purchased from Alfa Aesar (Ward Hill, MA). Scrap iron used in this study was obtained from Master Builders, Inc. (Aurora, OH). Master Builders scrap iron is known to contain 3.2% carbon, 0.65% manganese, and 0.6% sulfur (Reardon, 1995). The specific surface areas of high-purity iron granules, Master Builders scrap iron, and high-purity iron powder were determined by nitrogen sorption with a Quantachrome Nova High Speed Gas Sorption Analyzer. Surface area was determined with pre-set five point BET methods. The five point method measures the volume of $N_2$ adsorbed to the iron surface at five values of $p/p_0$, calculates the corresponding surface area, and determines how well the experimental data fit the BET equation. Samples were vacuum degassed for about 20 hours at 200°C before surface area analysis.

**Batch Experiments for Nitro and Azo Aromatics.** Batch reduction experiments were conducted in an anaerobic glove box using 8-mL borosilicate vials containing 5 mL of aqueous solution and 1 or 2.5 g of iron. The vials were horizontally placed in a rack and rotated on an orbital mixer at 100 rpm. All batch reduction experiments were conducted at room temperature (21 ± 1 °C). Replicate vials were set up for each experiment and, at each sampling time, one of the vials was sacrificed and supernatant was vacuum filtered through a 0.2 µm glass fiber filter for analysis by a high performance liquid chromatograph (HPLC) or a UV-VIS spectrophotometer. The changes of concentrations of the parent compounds were monitored as a function of time, under various conditions such as initial pH, ionic strength, type and size of iron particles. In order to investigate the pH effect, several buffers were used to maintain pH during experiments. HEPES [*N*-(2-hydroxyethyl)-piperazine-*N'*-(2-ethanesulfonic acid)], acetate, and CAPSO [3-(cyclohexylamino)-2-hydroxy-1-propanesulfonic acid] buffers were used at 0.2 M to give initial pH of 7.7, 4.3, and 9.9, respectively.

To quantify adsorbed parent compounds and reduction products, the supernatant was decanted and sorbed compounds were extracted by adding acetonitrile to the iron remaining in the vial. Vials were shaken for approximately 1 min and the content was filtered for HPLC analysis. A Varian HPLC equipped with a guard column and reverse-phase Supelcosil C-18 (Supelco) column and Alltima C-18 column was used for the analysis of dinitrotoluene, aminonitrotoluenes and diaminotoluene, nitrobenzene and its reduction products (aniline and nitrosobenzene), orange G and orange IV dyes and dye reduction products (aniline and sulfanilic acid).

In addition to HPLC, quantification of dyes was performed with a Hewlett Packard 8452A Diode Array Spectrophotometer. Orange G and orange IV were found to have maximum absorption wavelengths ($\lambda_{max}$) at 475 nm and 460 nm, respectively. The spectrophotometer software allowed for analysis of standards and construction of calibration curves.

**Batch Experiments for Chlorinated Aromatics.** Batch experiments to study the reduction of chlorinated aromatics by iron were also carried out in a glove box. 1,2,4-TCB was used as the representative chlorinated benzene because of its high water solubility (~40 mg/L) relative to other TCB isomers. Two experimental methods were employed: sampling from a larger reactor (60-mL bottle) and sacrificing smaller replicate reactors (9-mL vials) at different times. pH control during the reaction was achieved via manual addition of concentrated sulfuric acid in the 60-mL reactor and by using a pH buffer (HEPES, acetate, or phosphate) in 9-mL reactors. Preliminary data show that TCB reduction is relatively slow and thus may be hindered by its rapid and strong adsorption to the scrap iron. Therefore, experiments were conducted using high-purity iron powder (<10 μm, >99.5%) to provide high surface areas and to minimize TCB adsorption.

Hexane extraction was used to quantify aqueous TCB. Hexane extracts were analyzed by splitless injection into a HP5890 gas chromatograph (GC) equipped with a flame ionization detector (FID) and a 30-m HP-5 capillary column. The oven temperature was maintained at 40°C for 1 min and increased at 20°C per minute to 200°C. The injector and detector temperatures were 150°C and 200°C, respectively. Nitrogen was the carrier gas. Quantification of TCB and products (1,2-, 1,3-, and 1,4-dichlorobenzenes, chlorobenzene, and benzene) was based on external calibration curves prepared using analytical standards.

## RESULTS AND DISCUSSION

**Batch Reduction Experiments.** Aqueous 2,4-DNT was completely removed in 60 min in the presence of scrap iron and about 80% of the DNT disappeared was recovered as 2,4-DAT (Figure 1). Similarly, over 90% of nitrobenzene was removed from water with scrap iron in 30 min and aniline was the predominant product. Orange G and orange IV dyes were also rapidly removed from solution, with the decrease in concentrations of these dyes coinciding with increase in aqueous concentrations of the reduction products, aniline and sulfanilic acid, respectively. Table 1 summarizes the results from reduction of each parent compound by scrap iron as well as product recovery. The data indicate that iron is capable of rapidly reducing azo and nitro compounds to their amino counterparts.

In order to obtain mass balance, iron granules were collected after each experiment, extracted with acetonitrile, and analyzed with HPLC. Analysis of iron granules showed that the rapid removal of DNT from solution was partially due to adsorption to the iron surface, but the adsorption was reversible and the adsorbed DNT was available for reaction. DNT was completely reduced to DAT via 2-amino-4-nitrotoluene and 4-amino-2-nitrotoluene as intermediates (data not shown). The total mass recovery ranged from 94.8 to 108.8%, indicating that all of the added DNT was accounted for in the reduction experiment. Similar results were obtained with nitrobenzene. Analysis of aqueous and sorbed nitrobenzene, nitrosobenzene, and aniline gave mass recoveries of 83.7% to 108.2%.

Compared to nitroaromatic compounds, the adsorption of azo dyes to the scrap iron was minimal. The majority of the orange IV reduction product was recovered in the aqueous phase. The mass recovery ranged from 96.9 to 112.5%

for orange IV, but the mass recovery for orange G was slightly lower (about 80%). The orange G dye used in these experiments had a purity of approximately 86%, which may have contributed to the low mass balance.

Measurement of batch TCB and PCP reduction rates was less successful due to the low reactivity (and therefore long reaction time) of these chlorinated aromatics toward iron. Analysis of TCB was also complicated by volatilization losses caused by hydrogen evolution resulting from iron oxidation by water.

**Effects of Parameters.** Two types of iron were compared in batch reduction experiments: Alfa Aesar high-purity iron granules and Master Builders scrap iron. For all azo and nitro compounds tested, scrap iron was a more effective reductant than pure iron. Results of orange G reduction by these two irons are shown in Figure 2. The faster and more complete reduction by scrap iron may be in part due to difference in surface area, which was measured to be 1.28 $m^2/g$ for scrap iron and 0.0417 $m^2/g$ for high-purity iron granules. Because of its lower cost and higher reactivity towards nitro and azo aromatics, it is expected that scrap iron will be a better choice for wastewater applications.

In addition to iron type, pH appeared to be another critical factor. While scrap iron was generally more effective than high-purity iron, at low pH the high-purity iron was also able to completely reduce azo and nitro compounds within 30 min. On the other hand, although high-purity iron was completely non-reactive at pH 10, scrap iron at the same pH was still effective in reducing DNT. Similar pH effects were also observed for the azo dyes (data not shown).

We also examined the effect of salts on nitro and azo reduction kinetics. $CaCl_2$ at 0.001 and 0.01 M increased the orange G reduction rate over deionized water (Figures 3). $CaCl_2$ also enhanced DNT reduction rate by scrap iron over DI water. Similar enhancement was observed with $MgCl_2$ and NaCl, suggesting that the rate enhancement was probably caused by chloride, whereas the cation was less important. The role of chloride in iron corrosion is well-documented (Uhlig and Revie, 1985; Reardon, 1995) and it is likely that increased corrosion by chloride may have resulted in greater reduction of the azo and nitro compounds.

**Effect of Iron Pretreatment on Biodegradability.** Biodegradation experiments were performed with acclimated mixed cultures maintained in laboratory SBRs initially seeded with mixed liquors from an industrial wastewater treatment plant in Seaford, DE. Figure 4 shows the biodegradation data of nitrobenzene and its reduction product, aniline. Clearly, iron pretreatment enhanced the biodegradation of nitrobenzene by transforming it to the more biodegradable form of aniline. Sulfanilic acid, a reduction product of orange IV, was almost completely biodegraded within 24 hours. Results of batch biodegradability assessments are summarized in Table 2. These results demonstrate that iron pretreatment enhances biodegradability of refractory aromatics in subsequent conventional biological treatment processes.

## ACKNOWLEDGMENT

The authors gratefully acknowledge the financial support of the Water Environment Research Foundation (WERF) for this study.

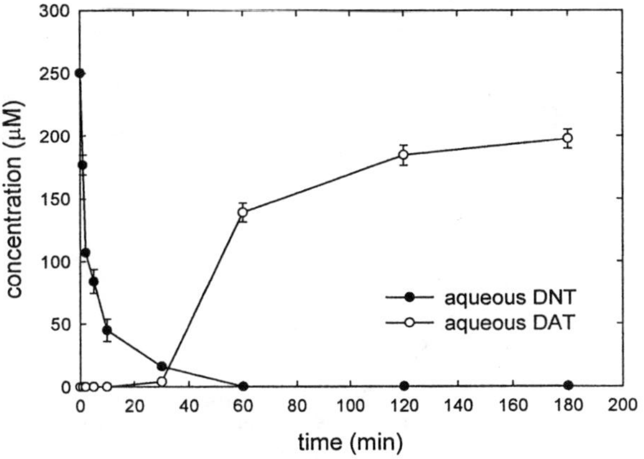


**Figure 1. Aqueous concentrations of DNT and DAT during DNT reduction. Conditions: 1g iron/5mL solution; 0.1M HEPES buffer; initial pH = 7.4.**

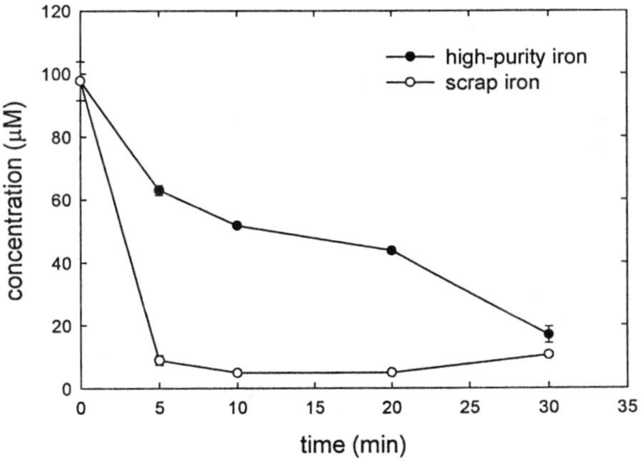


**Figure 2. Effect of iron type on orange G reduction rate. Conditions: 2.5g iron/5mL solution; 0.2M HEPES buffer; initial pH = 7.7; 100 rpm mixing.**

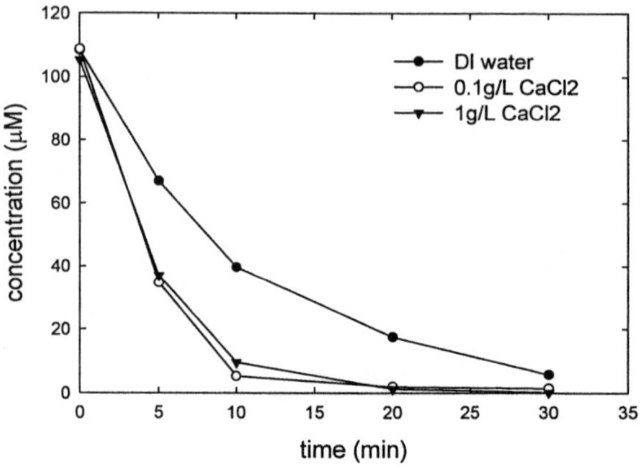


**Figure 3. Effect of $CaCl_2$ on orange G reduction by scrap iron.**

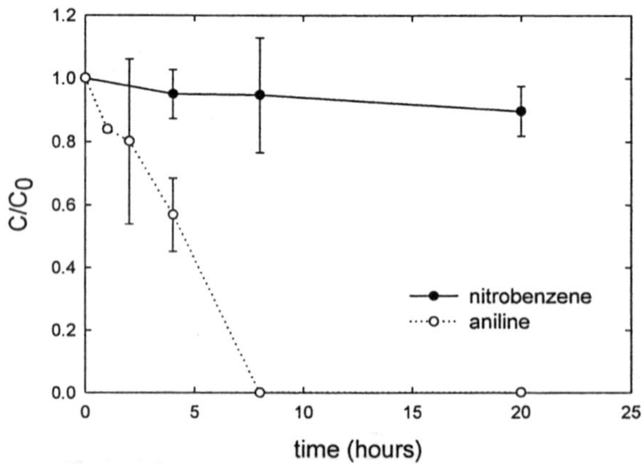


**Figure 4. Biodegradation of nitrobenzene and its reduction product, aniline.**

**Table 1. Summary of batch anaerobic reduction experiments with 2,4-dinitrotoluene, nitrobenzene, orange G, and orange IV. Percentages indicate amount of compound remaining in solution at the end of the experimental period.**

| | | | % remaining in solution | | | |
|---|---|---|---|---|---|---|
| | | | Aqueous | | Sorbed | |
| Parent compound | Major product | Iron amount/type/time | Parent | Product | Parent | Product |
| 2,4-dinitrotoluene | 2,4-diaminotoluene | 1g scrap, 180min | 0.0% | 79.1% | 0.0% | 29.6% |
| nitrobenzene | aniline | 1g scrap, 120min | 0.0% | 70.8% | 0.4% | 23.9% |
| orange G | aniline | 1g scrap, 30min | 9.9% | 40.4% | 3.4% | 19.2% |
| orange IV | sulfanilic acid | 2.5g scrap, 30min | 2.3% | 93.6% | 4.6% | 12.0% |

**Table 2. Summary of batch biodegradation experiments with acclimated biomass. Percentages indicate amount of compound remaining in bioreactor flask after a 24-hour period.**

| | | % remaining in solution at t=24h | |
|---|---|---|---|
| Parent compound | Major product | Parent | Product |
| nitrobenzene | aniline | 89.6% | 0.0% |
| orange G | aniline | 100.0% | 0.0% |
| orange IV | sulfanilic acid | 100.0% | 5.9% |

**REFERENCES**

Reardon, E. J. (1995) Anaerobic corrosion of granular iron: measurement and interpretation of hydrogen evolution rates. *Environ. Sci. Technol.*, **29**, 2936.

Uhlig, H. H. and Revie, R. W. *Corrosion and Corrosion Control.* Wiley, 1985.

# DDT DEGRADATION BY CHEMICAL OXIDATION AND WHITE ROT FUNGI

***Trine Eggen*** *(Jordforsk Centre for Soil and Environmental Research, Norway)*
Per Sveum (Deconterra AS, Norway).

**ABSTRACT:** The sand fraction (2.0-0.06 mm) of DDT [1,1,1-trichloro-2,2-*bis*(p-chlorophenyl)ethane] contaminated soil from a forest nursery garden was treated with chemical oxidation and white rot fungi. The experiments demonstrated that the white rot fungi, *Pleurotus ostreatus*, is able to degrade DDT in contaminated soil. Chemical pretreatment with Fenton's reagent reduced the effect of white rot fungi treatment. However, Fenton's reagent alone showed to be an effective degrader of DDT.

## INTRODUCTION

Since the 1940s, DDT [1,1,1-trichloro-2,2-*bis*(p-chlorophenyl)ethane] has been used worldwide to control agricultural pests and insect-born diseases like malaria. In Norway, use of DDT was banned in 1970; however, forest nursery garden were exempted from the ban until 1989. Following a survey of forestry nursery gardens, it was decided to remove approx. 2000-4000 tons of DDT contaminated soil for further treatment.

DDD [1,1-dichlor-2,2-*bis*(p-chlorophenyl)ethane] and DDE [1,1-dichloro-2,2-*bis*(p-chlorophenyl)ethylene] are transformation products of DDT, but are also byproducts from the production of DDT. DDT and its metabolites are found to be degraded in sediments. Recently, it has been demonstrated that even DDE can be degraded under anaerobic conditions in sediments (Quensen et al., 1998), and some of the factors controlling this process have been elucidated (Quensen et al., 2001). However, the half-life for some of the compounds is very high and even several decades after being deposited, their concentrations in the sediments are found to be at above-toxic levels.

Both white rot fungi (Barr and Aust, 1994; Higson, 1991) and Fenton's reagent are treatment methods that have been shown to degrade a range of recalictrant organic pollutants. The major part of the work connected to white rot fungi and DDT is connected to liquid culture systems (Bumpus and Aust, 1987; Bumpus et al., 1985; Fernando et al., 1989; Katayama et al., 1992; Köhler et al., 1988; Osano et al., 1999) or artificially contaminated soil (Fernando et al., 1989).

Fenton's reagent, $Fe^{2+}$ and $H_2O_2$, has been examined as an oxidizing agent for refractory organic chemicals, but also for xenobiotic environmental pollutants such as Polychlorinated biphenyl (Dercova et al., 1999), chlorophenols (Barbeni et al., 1987), trinitrotoluene (Arienzo et al. 1998; Li et al., 1997), and methyl tert-butyl ether (Yen and Novak, 1995). The $Fe^{2+}$ ion reacts with peroxide, forming hydroxyl radicals that are reactive and strong oxidizers. Chemical treatment has been thought of as a possible pre-treatment step for enhanced bioremediation by degrading the target compounds to by-products that are more readily biodegradable.

The aim of this work has been to study the effect of white rot fungi, alone or after a chemical oxidation pre-treatment, on DDT degradation in highly contaminated soil.

## MATERIAL AND METHODS

**Soil.** DDT (DDX, a collective term used for DDT, DDD and DDE) contaminated soil from a nursery garden (Prestebakke forest nursery garden, Kornsjø) in southern Norway was used in the experiments. Soil property was: pH 4.5; organic Carbon 4.2%, total Nitrogen 0.11%, $NH_4$-N 6.6 mg/kg, $NO_2$+$NO_3$-N 20.7 mg/kg and total Phosphate 565 mg/kg. The soil was fractionated and the sand fraction (2.0-0.06 mm) was used in the degradation experiment.

**Pre-Treatment with Chemical Oxidation (Fenton's reagent).** A preliminary experiment for screening of process parameters like slurry-, iron-, and $H_2O_2$ concentration was performed (not shown).

Based on this screening a 20% slurry concentration, 200 mg $Fe^{2+}$/L ($FeSO_4$ $7H_2O$, AppliChem, Darmstadt) and 5% $H_2O_2$ (50% $H_2O_2$, Solvay Interox Ltd. Warrigton) were selected for the further experiments. The degradation experiment was performed in 3Lglass beakers (0.6 kg soil and 2.4 liquid) and soil samples (n=1) were taken after 5 min, 30 min, 1 h, 2 h and 12 h. Soil was kept in suspension with a mechanical stirrer (Heidolph RZR2041). In the initial slurry solution, the pH was 6.6-6.8, and deceased to 5.8-6.1 and 2.7-2.8 after adding iron solution and hydrogenperoxyde solution, respectively. Fenton's reaction was quenched by adding concentrated $H_2SO_4$.

The chemical oxidation reaction was verified in bench scale (5 kg soil and 20 L liquid). Soil samples (n=2) were taken after 5 min and 2 h.

**White Rot Fungi Treatment.** Spent fungal substrate, *Pleuoruts ostreaus*, was obtained from a commercial Oyster mushroom producer. Fungal substrate and soil were homogenized (1:4 w/w), poured in glass beakers (3 L), covered with perforated parafilm and incubated for 4 weeks at room temperature. The soil and fungal mass was frequently aerated from the bottom of the beaker. Soil moisture contents was adjusted during the incubation period. Soil samples (n=3) were sieved (< 2 mm) for removal of fungal substrate before being analyzed for DDX.

***Chemical analysis.*** Analysis of DDT, DDD and DDE included both para-para and ortho-para isomers. Reference standards of all compounds were purchased from Dr. Ehrenstorfer, Germany and used as external standards for calibration. Soil samples (5 g dried material) were extracted for 2 h with dichloromethane (20 ml) before precipitation of samples overnight at 4 °C. The supernatant was diluted with acetone: iso-octane (3:7) and analyzed by gas chromatography with electron capture detector (ECD) (Hewlett Packard 5890 series II). The column was HP-5, 30 m x 0.32 mm i.d., 0.25 μm film thickness and the carrier gas was helium at a constant flow of 1.7 ml/min. Injections (2 μl) were made with the injector (240

°C) in the splitless mode. The oven was programmed from 80 °C (1.5 min) at a rate of 20 °C/min to 160 °C (0 min) followed by 4 °C/min to 270 °C (0 min) and finally at 10 °C/min to 300 °C (5 min). The temperature of the ECD was 300 °C.

## RESULTS AND DISCUSSION

The best obtained degradation result in the preliminary screening test was chosen as test conditions for further experiments: 20% slurry concentration, 200 mg $Fe^{2+}$/L and 5 % $H_2O_2$. Since the optimization of Fenton's reagent process was not the focus of this project, further modification of the process conditions was not conducted.

The laboratory scale degradation experiment demonstrated that 50 % degradation was obtained within 5 minutes treatment (Figure 1). Extending the treatment to twelve hours resulted in an increase to approx. 80 % reduction. Yeh and Novak (1995) showed a rapid degradation rate of MTBE in soil, where the reaction was completed within a few minutes. Li et al. (1997) obtained completed reaction within 300 minutes for TNT degradation in soil, while Arienzo et al. (1998) obtained the highest degradation rate within the first 4 h and continued degradation for the next 20 h.

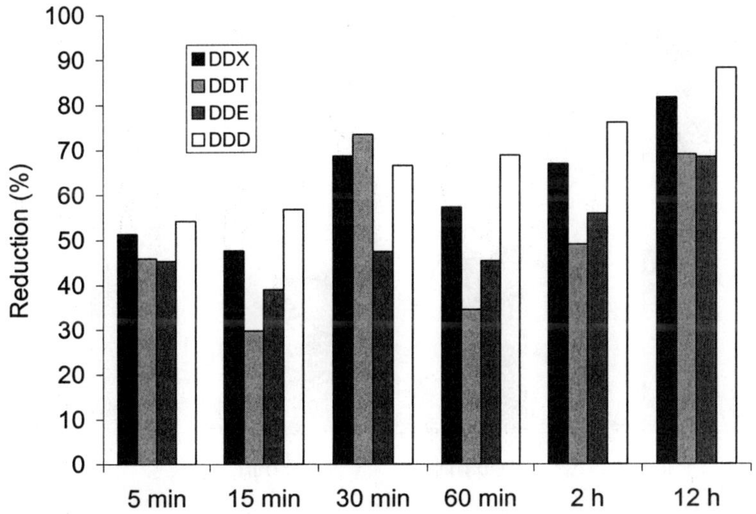


**FIGURE 1. DDX reduction, given as percent, over time during Fenton oxidation. Initial concentration of DDX in the soil sample was 880 mg/kg.**

Bench scale experiments verified the results obtained in the laboratory scale experiment, 50 percent degradation within 5 minutes, even though the initial DDX concentration was 380 mg/kg and 880 mg/kg, respectively (Figure 2).

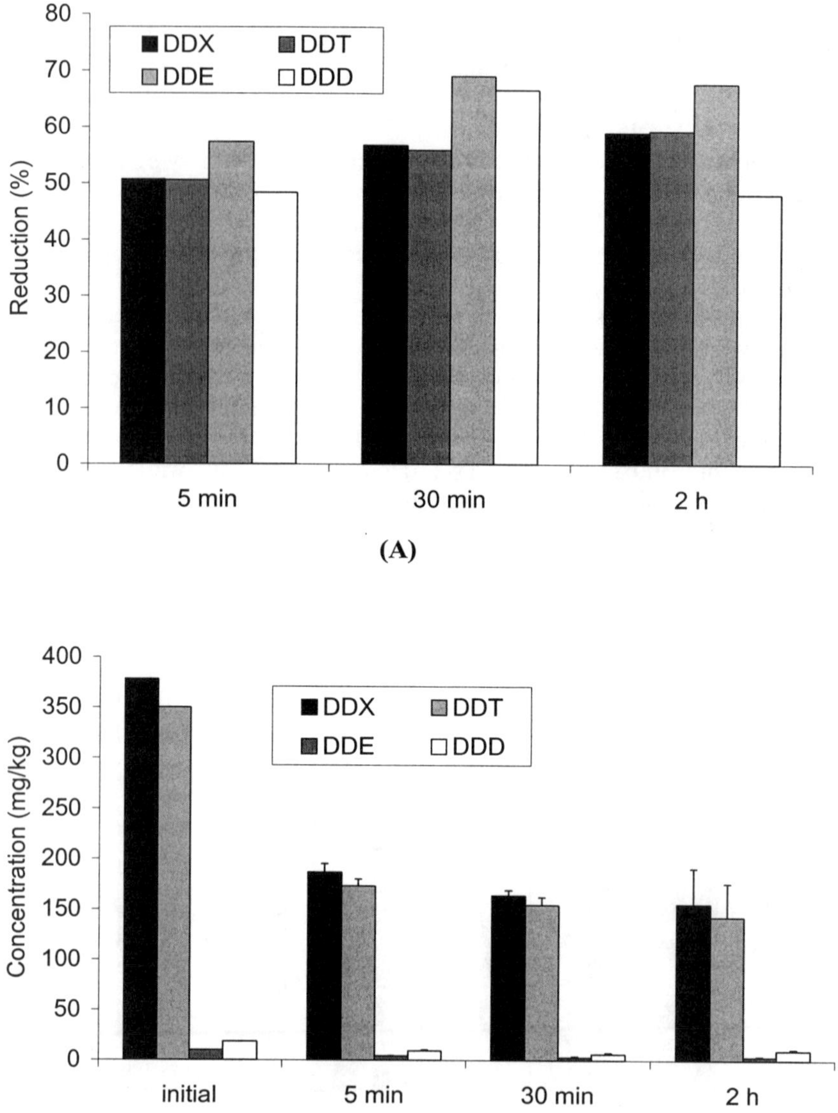


**(B)**

**FIGURE 2. DDX reduction in sand fraction performed in bench scale, given in percent (A) and concentration (B).**

In sand without chemical pre-treatment the DDX concentration decreased from 560 mg/kg to 317 mg/kg after four weeks incubation with *Pleurotus ostreatus* (Figure 3). In pretreated sand the DDX concentration decreased from 378 mg/kg to approximately155 mg/kg during the Fenton's reagent treatment, and after a four week incubation with *Pleurotus ostreatus* the final DDX concentration was 117 mg/kg (Figure 4). This corresponds to 43% and 69% degradation for untreated and pretreated soil, respectively (Figure 5). Since as much as 59 percent of the measured DDX degradation was achieved during the chemical treatment, the white rot fungi inoculation had less effect in pretreated soil (25 %) compared to untreated soil. This may be due to an inhibitory effect of chemical treatment on microbial activity. Büyüksönmez et al. (1998) showed a decline in cellular survival with increasing hydrogen peroxide and $Fe^{2+}$ concentration on the hazardous-compound-degrading bacterium, *Xanthobacter flavus* FB71. In order to reduce the toxic effect of hydrogen peroxide, which itself is harmful to microorganisms, these bacteria were acclimatized to high hydrogen peroxide concentrations prior to the addition of Fenton's reagent. This result is therefore an effect connected to the free radicals produced during the Fenton's reaction. Further investigation of the optimization of the Fenton's reaction process and a better understanding of the reduced effect of white rot fungi in the following fungal treatment is needed.

## CONCLUSIONS

The experiments demonstrated that the white rot fungi, *Pleurotus ostreatus*, is able to degrade DDT in contaminated soil. Chemical pretreatment with Fenton's reagent reduced the effect of a subsequent white rot fungi treatment. However, Fenton's reagent treatment alone effectively degraded DDT.

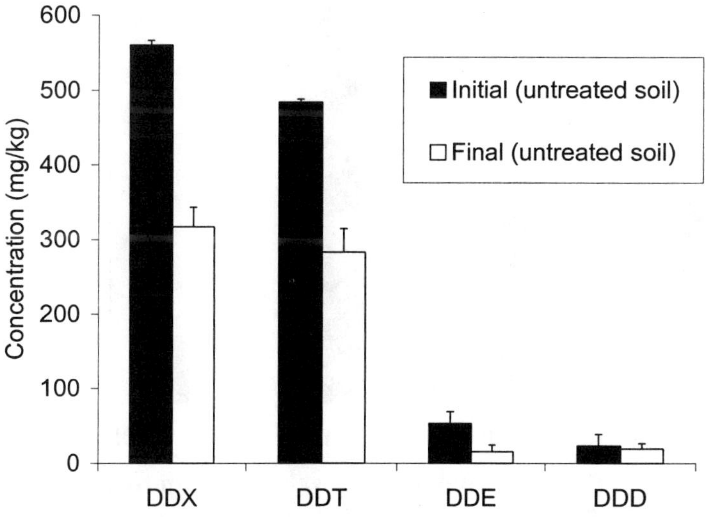


**FIGURE 3. DDX degradation after 4 weeks incubation with *Pleurotus ostratus* in soil without chemical oxidation pretreatment.**

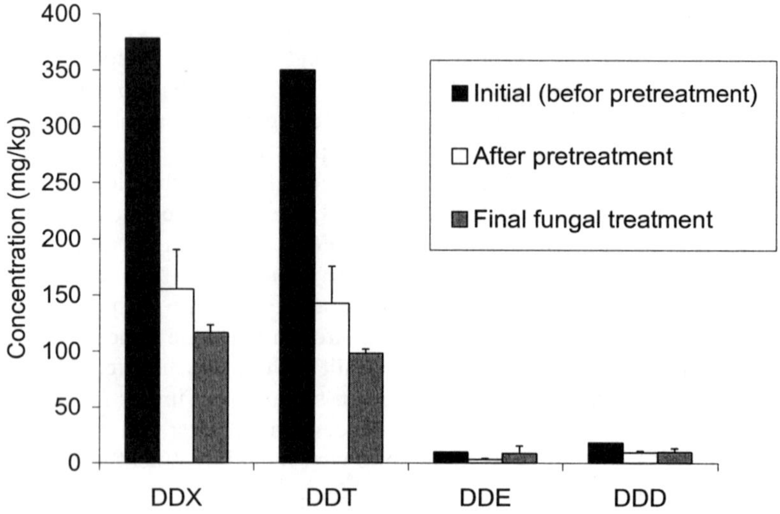


**FIGURE 4. DDX degradation after 4 weeks incubation with *Pleurotus ostratus* in soil with chemical oxidation pretreatment.**

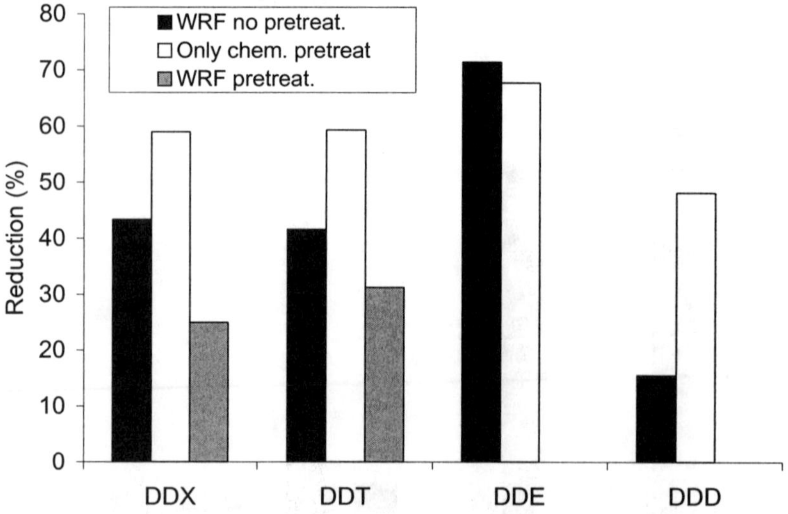


**FIGURE 5. Percent DDX degradation after white rot fungi treatment in soil without pretreatment (WRF no pretreat), after chemical oxidation alone (only chem. pretreat), and after white rot fungi treatment in chemical pretreated soil (WRF pretreat).**

## ACKNOWLEDGMENTS

The project has been funded by the Ministry of Agricultural Department. The DDT analysis have been performed at the Norwegian Crop Research Institute, Pesticide Laboratory.

## REFERENCES

Arienzo, M., S. D. Comfort, M. Zerkoune, Z. M. Li, and P. J. Shea. 1998. "Pilot-scale Devices for Remediation of Munitions Contaminated Soil". *J. Environ. Sci. Health, A33*(8):1515-1531.

Barbeni, M., C. Minero, E. Pelizzetti, E. Borgarello, and N. Serpone. 1987. "Chemical Degradation of Chlorophenols with Fenton's Reagent". *Chemosphere 16*(10-12):2225-2237.

Barr, D. P., and S. D. Aust. 1994. "Mechanisms White Rot Fungi use to Degrade Pollutants". *Crit. Rev. Environ. Sci. Technol. 28*(2): 79A-87A.

Bumpus, J. A. and S. D. Aust. 1987. "Biodegradation of DDT [1,1,1-Trichloro-2,2-bis(4-chlorophenyl)ethane] by the White Rot Fungus *Phanerochaete chrysosporium*. *Appl. Environ. Microbiol. 53*(9):2001-2008.

Bumpus, J. A., M. Tien, D. Wright, and S. D. Aust. 1985. " Oxidation of Persistent Environmental Pollutants by a White Rot Fungus". *Science 228*:1434-1436.

Büyüksönmez, F., T. F. Hess, R. L. Crawford, and R. J. Watts. 1998. "Toxic Effects of Modified Fenton Reactions on *Xanthobacter flavus* FB71". *Appl. Environ. Microbiol. 64*(10):3759-3764.

Dercova, K., B. Vrana, R. Tandlich, L. Subova. 1999. Fenton's Type Reaction and Chemical Pretreatment of PCBs. *Chemosphere 39*(15):2621-2628.

Fernando, T., S. D. Aust, and J. A. Bumpus. 1989. "Effects of Culture Parameters on DDT [1,1,1-Trichloro-2,2-bis(4-chlorophenyl)ethane) Biodegradation by *Phanerochaete chrysosporium*". *Chemosphere 19*(8/9):1387-1398.

Higson, F. K. 1991. "Degradation of Xenobiotics by White Rot Fungi". *Rev. Environ. Cont. Toxicol. 122*:111-152.

Katayama, A., S. Uchida, and S. Kuwatsuka. 1992. "Degradation of White-Rot Fungi under Nutrient-Rich Conditions". *J. Pesticide Sci. 17*(4):279-281.

Köhler, A., A. Jäger, H. Willershausen, and H. Graf. 1988. "Extracellular Ligninase of *Phanerochaete chrysosporium* Burdsall has no Role in the Degradation of DDT". *Appl. Microbiol. Biotechnol. 29*:618-620.

Li, Z. M., S. D. Comfort, and P. J. Shea. 1997. "Destruction of 2,4,6-Trinitrotoluene by Fenton Oxidation". *J. Environ. Qual. 26*:480-487.

Osano, A., G. M. Siboe, J. O. Ochanda, and J. O. Kokwaro. 1999. "Biodegradation of DDT: The Role of Pleurotus sp., a Lignicolous Fungus". In: Alleman, B. C., and A. Leeson (eds). Bioremediation of Nitroaromatic and Haloaromatic Compounds, pp. 69-74. Proceedings from the Fifth International In Situ and On-Site Bioremediation Symposium, April 19-22. San Diego, USA.

Quensen, III, J. F., S. A. Mueller, M. K. Jain, and J. M.. Tiedje. 1998. "Reductive Dechlorination of DDE to DDMU in Marine Sediment Microcosms". *Science 280*:720-724.

Quensen, III, J. F., J. M.. Tiedje, M. K. Jain, and S. A. Mueller. 2001. "Factors Controlling the Rate of DDE Dechlorination to DDMU in Palos Verdens Margin Sediments under Anaerobic Conditions." *Environ. Sci. Technol. 35*(2):286-291.

Yeh, C. K., and J. T. Novak. 1995. "The Effect of Hydrogen Peroxide on the Degradation of Methyl and Ethyl Tert-butyl Ether in Soil". *Wat. Environ. Res. 67*(5):828-834.

# EFFECT OF SOIL CONTENT ON COMPOSTING OF DIESEL FUEL CONTAMINATED SOILS

***Rafik M. Hesnawi*** and Daryl M. McCartney
University of Manitoba, Winnipeg, Manitoba, Canada

**Abstract**: The purpose of this study was to investigate the effect of soil content on the thermophilic stage of composting. A total of thirty-six bench-scale biocell reactors were prepared. Each biocell started with 6.5 kg (wet basis) of biosolids and woodchips. The ratio of woodchips to biosolids used was 0.4:1 (wet basis) to achieve a target moisture content of 55%. Sand or silt was added to the biocells in increasing amounts. The amount of sand added in the compost mixture ranged between 0% to 64% on a wet weight basis, while the amount of silt ranged from 0% to 48% on a wet weight basis. Soils were contaminated with 5000 mg of diesel fuel per kg of dry soil before addition to biocells. Diesel-$^{14}C$ octadecane mixture was also added to the reactors to trace the mineralization of hydrocarbon as $^{14}CO_2$. The biocell aeration and temperature were monitored and controlled to simulate the thermophilic stage of composting. Each run lasted for 14-days. As a measure of composting microbial activity, the relative heat generation and volatile solids removal were measured. The results of this study indicated that thermophilic microbial activity was not affected by the sand or silt content tested. No significant release of $^{14}C$ as $^{14}CO_2$ from reactors was observed during the fourteen days of composting. It is expected that the production of $^{14}C$ would increase with increased composting time. While the thermophilic microbial activity observed was not affected with sand or silt content tested, further investigations are suggested to confirm our work with clay soil and to compare those results with other researchers.

## INTRODUCTION

Composting has been shown to be effective for the degradation of soil contaminated with explosives (Williams and Keehan, 1992), pesticides (Liu and Cole, 1996), and chlorophenol (Laine and Jorgensen, 1997; Benoit and Barriuso, 1995). Composting of petroleum hydrocarbon contaminated soils has been conducted from bench scale to large pilot studies. In bench scale studies, the degradation of 3 to 4 ring polycyclic aromatic hydrocarbons (PAHs) was almost completed (70 to 90%) during 25 weeks of composting (Wischmann and Steinhart, 1997). A similar study yielded a 90% degradation, in five days, for easily degraded naphthalene, as well as 80% degradation for slowly degraded chrysene and pyrene, in 15 days (Civilini, 1996). In another study, a 50 % degradation rate was achieved in 105 days during composting of highly weathered petroleum hydrocarbon contaminated soil (Beaudin et al., 1996). In pilot scale studies, composting of petroleum contaminated soil resulted in 58% to 80% reduction of alkanes and 60% reduction of PAH content after 6 to 8 months of composting (Al-Daher et al., 1998; and O'Reilly et al., 1997). In short, composting technology has shown considerable potential to treat soil contaminated with petroleum hydrocarbons.

In practice, higher degradation rates may be achieved by optimizing the environmental conditions (moisture, oxygen, pH, nutrients and carbon rich environment) in the composting matrix and/or by increasing the bioavailability of contaminants. The amount of soil incorporated in the mixture to be composted may also affect microbial activity and degradation rates. One study found that 40% contaminated soil by volume (soil type was not discussed) in the compost mix resulted in lower microbial activity and significantly reduced the degradation of explosives (William and Keehan, 1993). Soil content of more than 80% by weight significantly effected thermophilic microbial activity and the pesticides degradation. The soil used consisted of 27% sand, 32% silt, 19 clay, and 22% gravel (Liu and Cole, 1995). Another study indicated that increasing of clay soil content above 24% by wet weight resulted in sharp decrease in thermophilic microbial activity during composting of diesel fuel contaminated soil (Wytrykush, 2000). These studies indirectly imply that the soil fraction required would vary depending on the contaminant characteristics presents, and soil type as well. As such, there is a need to estimate the amount of soil to be included in compost for each case or project.

This study was conducted to investigate the effect of soil content on the thermophilic stage of composting. It is expected that this phase of study would provide information on the maximum amount of soil that should be added to the compost in order to design an effective composting system for petroleum hydrocarbon contaminated soil.

## MATERIALS AND METHODS

**Biocell Apparatus.** A total of four biocells were constructed. Each biocell was air tight with a working volume of 28-L. The schematic diagram of a biocell apparatus is shown in Figure 3.1. Each biocell was insulated with 5cm thick pink fiberglass and aluminum-reflecting blanket in order to minimize the heat loss. The insulting biocells were then placed in a temperature-controlled chamber set at 45°C. The biocells were designed as described in Larsen and McCartney (2000) so that the compressive loading could be attached to the biocells, oxygen could be supplied to the biocells, and temperature could be controlled and measured. Exhaust gas was led through three $CO_2$ sorption NaOH vessels which were connected in series.

**Soil Material Preparation and Analysis.** Two soils were used in the study. Commercially available white silica sand and pure washed silt were chosen to represent coarse-grained and fine-grained soils, respectively. The soil moisture contents of the soils were adjusted with distilled water to 50 % of the maximum moisture holding capacity. Diesel fuel, spiked with labeled octadecane1-$^{14}$C (supplied by the Sigma Chemical Co., St. Louis, MO) at activity of 8 μci/g of diesel fuel, was added to the soil to yield 5000-mg kg$^{-1}$ dry soil.

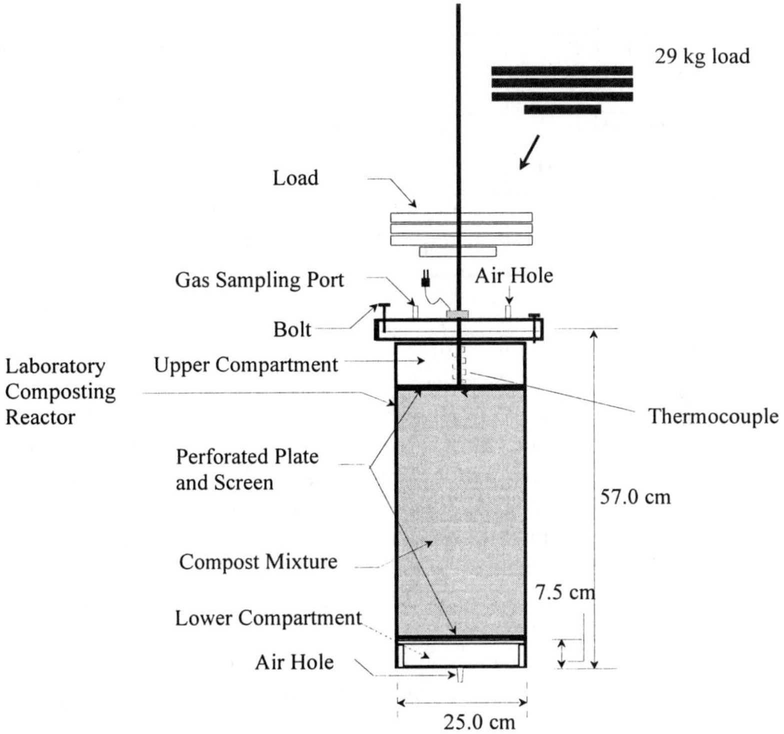


**FIGURE 4.1. Schematic diagram of a biocell.**

**Compost Feedstock Material Preparation and Analysis.** Dewatered municipal biosolids and woodchips (4.75 mm to 12.5 mm) were used as the compost feedstock. The biosolid and woodchip materials were analyzed for moisture content (MC), total solids (TS) and volatile solids (VS) using the APHA standard method 2540-G. Organic carbon contents and C:N ratios were calculated using methods presented in Rynk et al. (1992). Wet ($D_{wb}$) and dry ($D_{db}$) bulk densities were determined using the Core method (Carter, 1993). The Total Kjeldahl nitrogen (TKN-N) was determined using semi-Micro Kjeldahl method, excluding nitrite and nitrate (Carter, 1993). Except where specified otherwise, all measurements were performed in five replicate analyses. A summary of material characteristics is presented in Table 1. To achieve a starting moisture content of 55 % (wet basis) the woodchips were mixed with biosolids at ratio of 0.4:1 (wet basis). The ratio of biosolids to woodchips resulted in C:N ratio of about 20 and free air space of 0.5 ($cm^3/cm^3$).

**TABLE 1. Summary of the individual feedstock characteristics.**

| Parameter | Biosolids | Woodchips |
|---|---|---|
| MC (% wet basis) | 70.20 ± 0.002 | 17.02 ± 1.20 |
| OC (% dry basis) | 24.82 ± 0.002 | 53.51 ± 0.004 |
| TKN-N (% dry basis) | 3.49 ± 1.10 | 0.80 ± 0.65 |
| C: N | 7.1 | 66.9 |
| $D_{wb}$ (Kg/m$^3$)[1] | 934.8 ± 6.92 | 197.4 ± 5.81 |

[1] mean of triplicates

**Microbial ActivityTesting.** To investigate the effect of sand and silt content on the microbial activity, a total of thirty-six bench-scale biocell reactors were run. Twenty-three biocells used sand and thirteen biocells used silt. The amount of sand added ranged from 0% to 64% on a wet weight basis, while the amount of silt ranged from 0% to 48% on a wet weight basis. The amount of compost feedstock (biosolids and woodchips) was kept constant at 6.5 kg wet weight at the start of each run. Each experimental run included four biocells with each biocell representing one soil content. The insulated biocells were placed in a temperature-controlled chamber and a compressive load of 29 kg (5.7 kPa) was applied to each biocell. Each biocell ran for two weeks.

Air was supplied to each biocell to ensure oxygen was not limiting the reactions. The biocells were maintained under aerobic conditions by pumping air through the biocell for 15 min in each hour at rate of 1100 ml/min using a Chron Trol XT timer and Masterflex speed controlled 6-600 rpm pump (Cole Parmer Instrument Company). Aeration was conducted from the bottom to the top of the biocells. The feed air was passed through a 6 L Erlenmeyer flask filled with distilled water to ensure the air was hydrated before oxygen flowing into the biocells.

Each biocell was equipped with a thermocouple. The temperature was measured using a Fluke 52 K/J thermometer. At the start-up (time zero) the temperature control chamber and reactors were measured. The chamber temperature was then set at 45°C. During the experimental run, the biocells and chamber temperatures were measured daily for a 14-day test period. The experimental run was considered complete when reactor temperatures dropped below the chamber temperature for 24 hours.

The exhaust gas of each biocell was collected in a series of three $CO_2$ traps. The series of $CO_2$ traps consisted of sorption vessels filled with 4 N NaOH solution. The $CO_2$ traps were removed and replaced on Day 8.

**Microbial Activity Analysis.** Microbial activity was measured using the relative heat generation and volatile solid removal. $^{14}CO_2$ release from biocells was monitored in some of the biocells. The temperature monitoring results for each biocell were used to construct relative heat generation values (RHG). To obtain these values, each reactor temperature was subtracted from the chamber temperature. Cumulative RHG values for each biocell were plotted against soil content. For each biocell VS mass balances were calculated. Measurement of $^{14}CO_2$ resulting from the mineralization of $^{14}C$-labeled octadecane was conducted on a daily basis throughout the test period.

## RESULTS AND DISCUSSION

**Relative Heat Generation.** To provide valuable information on the effect of soil content on degree of microbial activity, the biocell temperatures were used to construct relative heat generation values. The relative heat generation values for sand and silt soils are presented in Figure 1 and Figure 2, respectively. For compost mixed with sand, the results showed no significant linear relationships ($r^2 = 0.047$) of data between sand content of 0 to 40% (17 data points), while data between sand content of 40 to 64% (7 data points) showed a strong linear relationship ($r^2 = 0.941$). This suggests that a change in microbial activity may have occurred when contaminated sand content exceeded 40% by wet weight.

For compost mixed with silt, the results showed no significant linear relationships ($r^2 = 0.076$) of data points between silt content of 0 to 20% (6 points). The linear regressive fit of data points between silt content of 20 to 48% (8 points) yield an $r^2$ value of 0.512. In general, however, the reduction in heat generation did not correlate to silt content.

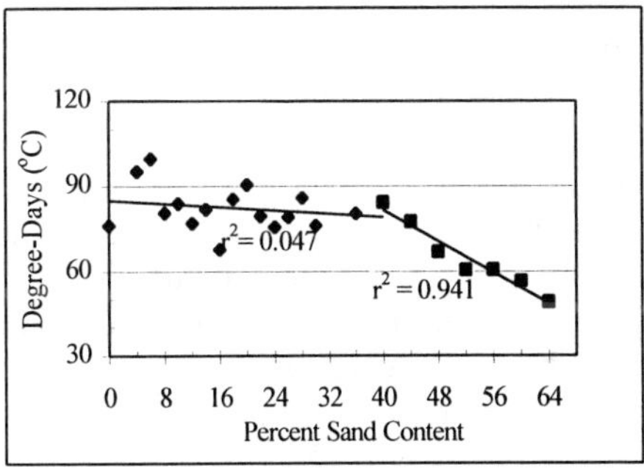


**FIGURE 1. Effect of sand content on relative heat generation.**

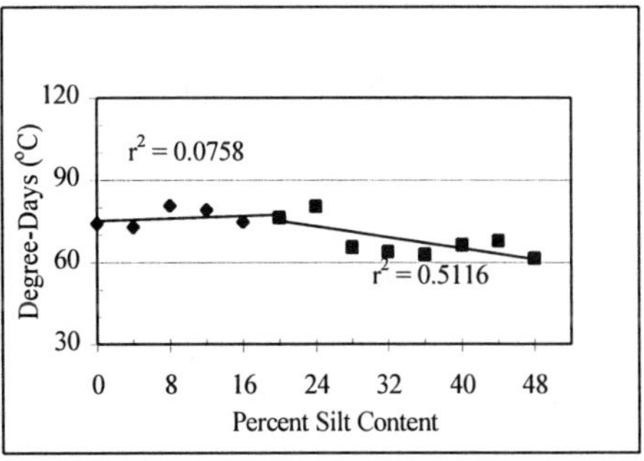


**FIGURE 2. Effect of silt content on relative heat generation.**

**Volatile Solid Removal.** The percentage of VS removal as a function of sand and silt content in the compost mixture is presented in Figure 3 and Figure 4, respectively. The maximum VS removal observed for compost mixed with sand and silt was 18% and 16%, respectively. The VS removal, however, was poorly correlated to soil content. Also, the VS removal observed did not correspond with the relative heat generation values observed. The variability in volatile solids removal values may have resulted from sampling variability. To overcome the problem, there is a need to improve the sampling and analysis methods. Generally

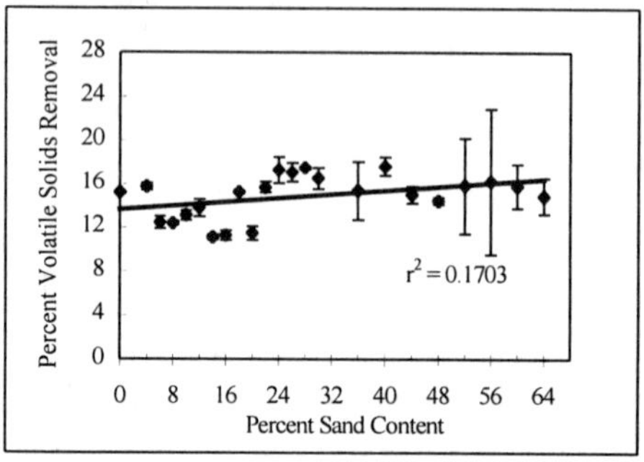


**FIGURE 3. Effect of sand content on volatile solids removal.**

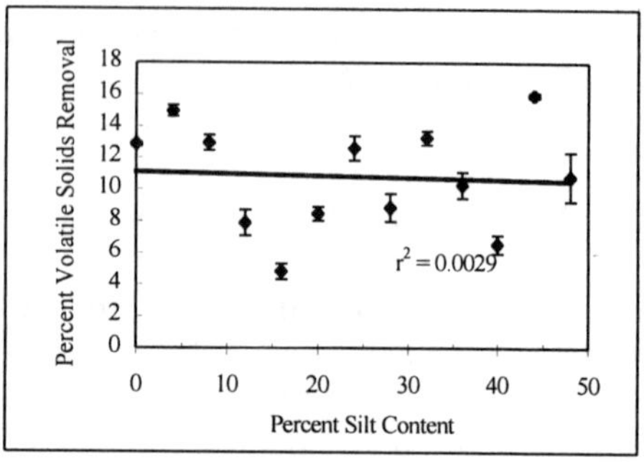


**FIGURE 4. Effect of silt content on volatile solids removal.**

the lack of trends observed between soil content and amount volatile solids removed suggest that the degree of microbial activity was not affected with soil content tested.

**Mineralization of Labeled Octadecane.** The production of $^{14}CO_2$ from both composting mixtures was minimal. Most of the $^{14}CO_2$ were detected in the first trap, with lesser amounts in the second or third trap. The total $^{14}CO_2$ released from composting mixtures was 0.01% and 0.003% of that initially in the composting mixture for both sand and silt, respectively. The findings of our study were supported by the findings obtained in the literature. Wytrykush (2000) in lab scale composting of diesel fuel contaminated clay soil found that no release of anthracene during the first 12 days of co-composting of diesel fuel contaminated sandy soil. However, after 103 days of composting, 23% of the $^{14}C$ anthracene transformed into $^{14}CO_2$. These studies indicated co-composting environment promote the formation of bound residues and that could make it difficult for microorganism to degrade diesel hydrocarbons in a short time. Since little or no carbon mineralization was observed during the two weeks of composting, no further investigation of $^{14}C$ degradation was considered at this stage of investigation.

## CONCLUSIONS

In general there was a lack of correlation between the VS removal, heat generation, and soil content data. Sample variability may have played an important role in these observations. Further work will: (1) determine VS variability in the feedstock and finished compost; (2) investigate effect of soil texture, including clay, and (3) conduct longer term studies.

## REFERENCES

Al-Daher, R., N. Al-Awadhi, and A. El-Nawawy. 1997. "Bioremediation of Damaged Desert Environment using the Windrow Soil Pile System in Kuwait". *Environmental International*, 24 (1): 175–180.

Beaudin, N., R. F. Caron, R. Legros, J. Ramsay, L. Lawlor, and B. Ramsay. 1996. "Co-composting of Weathered Hydrocarbon Contaminated Soil". *Compost Science & Utilization,* 4(2): 37-45.

Benoit, P. and E. Barriuso. 1995. "Effect of Straw Composting on the Degradation and Stabilization of Chlorophenols in Soil". *Composting Science and Utilization*, 3 (3): 31-37.

Carter, M. R. 1993. *Soil Sampling and Methods of Analysis*. Lewis Publishers, Boca Raton, Fl. U.S.A.

Civilini, D. C., M., De Bertoldi, and N. Sebastianutto. 1995. "Composting and Selected Microorganisms for Bioremediation of Contaminated Materials". In M. de Bertoldi, P. Bert, and P. Tiziano (Eds.), *In the Science of Composting*, pp. 884-891. Blackie Academic and Professional, London.

Laine, M.M. and, K.s. Jørgensen. 1997. "Effective and Safe Composting of Chlorophenol-Contaminated Soil in Pilot Scale". *Environmental Science and Technology*, 31: 371-378.

Larsen, K. L., and D. McCartney. 2000. "Effect of C:N Ratio on Microbial Activity and N Retention: Bench-Scale Study Using Pulp and Paper Biosolids". *Compost Science & Utilization*, 8(2): 147-159.

Liu, X., and M. A. Cole. 1996. "Minimum Effective Compost Addition for Remediation of Pesticide Contaminated Soil". In M. de Bertoldi, P. Bert, and P. Tiziano (Eds.), *In the Science of Composting*, pp. 903-912. Blackie Academic and Professional, London.

O'Reilly, K., T. Simpkin, K. Sobczak, and, T. Moreno. 1997. "Efficacy and Economics of Composting Aged Materials at a Refinery". *In Situ and On-Site Bioremediation*, Fourth International Symposium, **2**: 73 – 78.

Rynk, R. 1992. *On Farm Composting Handbook.* Northeast Regional Agricultural Engineering Service, Ithaca, New York.

Willimas, R. T., and K. R. Keehan. 1992. "Hazardous and Industrial Waste Composting". In Hoitink and Keener (Eds.), *In Science and Engineering of Composting*, pp. 363-381. Renaissance Publishers, Worthington, Ohio.

Wischmann, H., and H. Steinhart. 1997. "The formation of PAH Oxidation Products in Soils and Soil/Compost Mixtures". *Chemosphere*, 35 (8): 1681-1698.

Wytrykush L. 2000. "Effect of Soil Loading Rate on Microbial Activity During Co-Composting of Diesel Contaminated Clay Soil". M.S. Thesis, Civil and Geological Engineering, Manitoba University, Winnipeg, Manitoba, Canada.

# AN INTEGRATED PHYSICAL-BIOLOGICAL TREATMENT FOR ALKYLAROMATIC CONTAMINATED SOILS

***F. Fabiani****, C. Di Leo*, A. Robertiello*, A. Bittoni*, E. Sordini*, M. Molinari**
* EniTecnologie, Monterotondo, Italy
** AgipPetroli, Roma, Italy

**ABSTRACT**: Diffused old contamination was found in sandy/silty soils coming from an industrial site where lubricating oils additives are now produced. The oil contained predominantly branched alkyl benzene and naphthalene derivatives, with hot spots having concentrations up to 7,000 mg/kg.

The site is critical from an environmental viewpoint, because it is near a river and the groundwater level is very close to ground surface (-3 m).

On the basis of laboratory and pilot scale experimental results the following scheme of an integrated treatment of heavily contaminated soils has been designed:

- physical (multi gravity) separation of a suitable soil/water mixture, to remove the coarse fraction from a slurry, leaving the finest part of the soil (about 20 % of the total matter), containing more than 80 % of the contaminants
- forwarding the slurry to the wastewater treatment plant of the factory where the active sludge microbial population is able to degrade the contaminants
- bioremediation of the coarse fraction of the soil that continues to exceed regulatory limits, by landfarming.

## INTRODUCTION

The site assessment of an industrial site in which lubricating-oil additives are now produced, showed diffused contamination, predominantly by branched alkyl benzene and naphthalene derivatives, with hot spots having concentrations up to 7000 mg/kg of TPH (Total Petroleum Hydrocarbons).

The site is critical from an environmental viewpoint, because it is near a river and the groundwater level is very close to ground surface (-3 m).

The hydrogeological characterization showed the presence of a gravelly soil in a sandy/silty matrix, with big stones making preferential paths for water circulation. The permeability was $10^{-5}$ m/s.

## MATERIALS AND METHODS

**Soil characterization.** Soil samples were analyzed for the presence of TPH. The soil, previously dewatered by adding anhydrous sodium sulphate, was extracted with methylene chloride by 15 minutes of sonication. The centrifuged supernatants were analyzed by gas chromatography with a DANI 1000 equipped with a flame ionization detector (FID) at 300°C. The GC was operated as follows:

40°C for 2 minutes; 150°C 3°C/min, 150°C for 5 minutes, 280°C 5°C/min; 280°C 5 minutes.

The granular separation was carried out using 500 g of soil and a Octagon 200 sieve (4 sieves of 2 mm and 600, 200, and 63 μm respectively) using 4 L of tap water. The fines (below 63 μm) were collected in a settler from which water was removed by siphoning. The sieves, containing the soil and the fines were air dried, weighed and analyzed for the presence of TPH.

**Soil treatment.** Trials were carried out by the *Istituto per il Trattamento dei Minerali (CNR, Consiglio Nazionale delle Ricerche)* using an equipment suitable for the soil fractions separation. As shown in Figure 1, the soil fraction smaller than 5 mm was treated in an attrition cell and then wet sieved, obtaining the > 0.5 mm fraction. The remaining stream was subjected to a multi gravity separation for the removing the fine fraction.

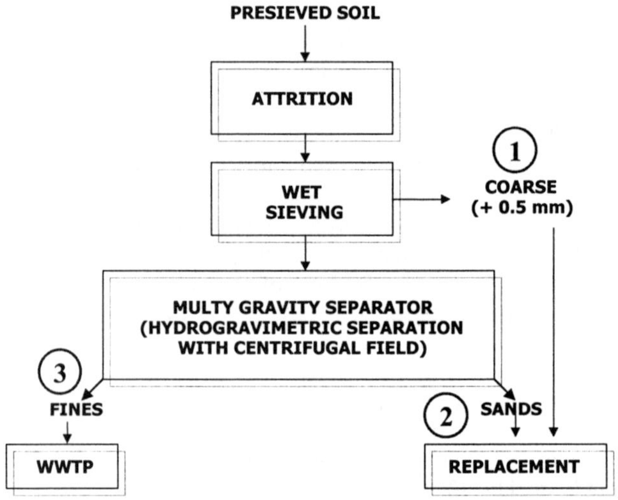


**FIGURE 1. Scheme of physical soil treatment**

Slurry samples were extracted with methylene chloride (30 minutes agitation) and the centrifuged supernatants were analyzed by gas chromatography as previously described.

Biodegradation studies of the contaminants in the slurries were performed using laboratory equipment (VITTADINI MARK-2 mini pilot plant) that simulates an activated sludges wastewater treatment plant (WWTP) (Figure 2). The bench scale pilot plant was started up by using activated sludges and wastewater generated by the industrial plant working on the site. Continuous tests with an Hydraulic Retention Time (HRT) of 37.5 hours were carried out at room temperature, adding a 5% slurry to the wastewater.

**FIGURE 2. WWTP simulator**

Landfarming laboratory trials were carried out on 2.7 kg of coarse soils, amended with 300 g of municipal solid waste compost and fertilized with a solution of ammonium nitrate and sodium phosphate (C:N:P ratio equal to 100:5:1). Periodically, the soil was watered and mixed to assure the right moisture content (75% of field capacity) and oxygen transfer.

## RESULTS

Figure 3 shows an example of a gas chromatogram of the contaminants extracted from the soil. Because these compounds are highly branched, they are typically scarcely biodegradable.

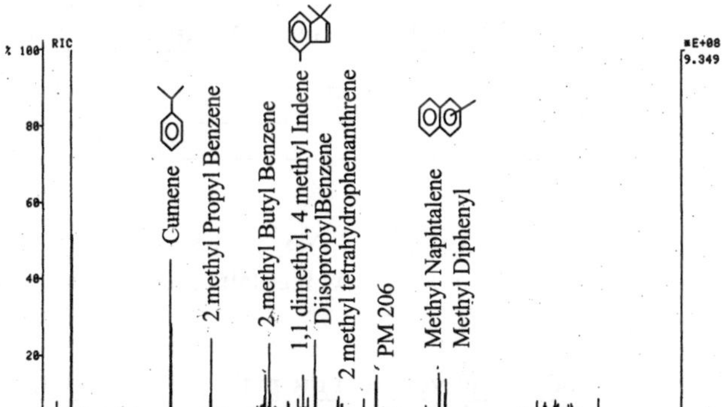


**FIGURE 3. Gas chromatography of the contaminants with the chemical structures**

Results of granulometric separation showed that more than 80% of the contaminants were associated with the finest fraction (< 63 μm), which in turn represented about 20% of the soil by weight (Figure 4).

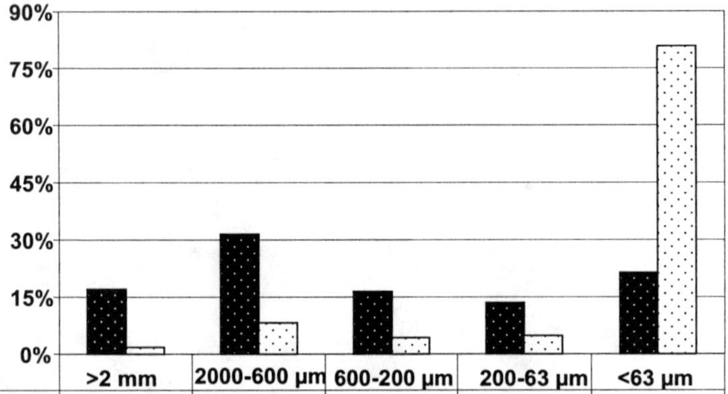


| | >2 mm | 2000-600 μm | 600-200 μm | 200-63 μm | <63 μm |
|---|---|---|---|---|---|
| ■ %in soil | 17.1% | 31.5% | 16.4% | 13.5% | 21.3% |
| □ %TPH on total | 1.8% | 8.3% | 4.3% | 4.8% | 80.8% |

**FIGURE 4. Distribution of contaminants on a soil sample having 7,000 mg/kg of TPH**

Physical separation tests were performed on two soil samples, having TPH concentrations of 2,774 and 6,970 mg/kg respectively. In the first case (Table 1) the treatment led to contaminant concentrations below the Italian regulatory limits (750 mg/kg of TPH having C atoms number > 12). The streams named 1 and 2 in Table 1 and Figure 1, can be considered non hazardous. The stream named 3 was forwarded to the bench scale pilot WWTP for the biodegradation of the contaminants.

**TABLE 1. Characterization of process streams**

| Process stream | 1 (coarse) | 2 (sands) | 3 (fines) |
|---|---|---|---|
| dry on total (%) | 50.7 | 40.7 | 8.6 |
| water (%) | 21.2 | 58.2 | 95.9 |
| TPH (mg/kg) | 40 | 143 | 31,230 |

Table 2 shows the influent and effluent COD, TPH and suspended solids values: all the parameters in the effluent are below the Italian regulatory levels for treated wastewaters.

**TABLE 2. Input and output values of the main parameters of the WWTP**

| Parameter (mg/l) | Influent | Effluent |
|---|---|---|
| COD | 1,294 | 66 |
| TPH | 65 | 0.8 |
| Suspended solids | 2,075 | < 80 |

In the second case, using the soil having had a contaminant concentration of 6,970 mg/kg, streams 1 + 2 had 1,100 mg/kg of TPH and were treated by landfarming, which led to a reduction of the contaminant concentrations (613 mg/kg) to below the Italian law target, after 120 days of treatment.

## CONCLUSIONS

Owing to the critical hydrogeological conditions and the nature of the contaminants, conventional approaches of reclamation could not be applied to the site under investigation. The results of this work point out that an innovative physical-biological process may be suitable.

To consolidate the attained results and to evaluate the cost of the whole intervention, field tests are planned for the second part of 2001.

## ACKNOWLEDGEMENTS

We wish to thank G. Belardi of the *Istituto per il Trattamento dei Minerali (CNR, Consiglio Nazionale delle Ricerche)* for the ideas and the aid supplied to address the removal of the fines from the soil.

# DEEP LANDFARMING IN SILTY or CLAYEY SOILS: EFFECTIVE AND INEXPENSIVE

***A. Prandi***, C. Xella, G. Dondi, F. Mutti, B. Grandi: Water & Soil Remediation S.r.l., Levata di Curtatone (MN), Italy.

**ABSTRACT:** Some polluted sites are made up of silt and clay with pollution which reaches quite a considerable depth: here, traditional "in situ" remediation has notable limits and difficulties because of low permeability and depth of intervention.

Efforts to overcome this problem have not been documented, if at all. We have, in fact, found no previous documentation on approaches to tackling this problem. In the light of our own recent field experiences, this problem can be overcome by an innovative application of In Situ Deep Landfarming (up to 7 meters below ground surface) in combination with "Pump & Treat".

## INTRODUCTION

The technology described in this paper applies to remediation of soils polluted by biodegradable substances, such as hydrocarbons, solvents, petrochemicals, organic compounds in general, where their migration reaches considerable depth (up to 7 meters) both in fine lands or very fine. For example, in the silty, silt-clayey and clayey layers, diffused and episodic saturation take place along with major anisotropy of both physical properties and pollution grade.

These conditions make "in situ" application difficult, often impractical, not very effective, and expensive: the main limiting factor is the transport of fluid and in situ mixing, represented by the heterogeneous fluid transfer in the subsurface environments. In biological or oxidation methods, these fluids are the carriers of electron donors, electron acceptors and nutrients.

Three field scale experiences are reported; all of which took place in Northern Italy.

## SITE CONDITIONS

Case Study 1 refers to a loss of virgin naphtha from a pipeline polluting a rural area of 6,000 $m^2$ to a depth of 4 meters (Figure 1a). Site lithology: silt, clay and episodic sandy layers.

Case Study 2 is an ex service station, where gasoline and diesel polluted an area of 200 $m^2$ to a depth of 7 meters. Site lithology: silt from 0 to -5 meters a the layer of sand from –5 to –6 meters (Figure 1b).

Case Study 3 refers to an area where a hydrocarbon terminal operated for 50 years (Figure 1c), and where gasoline, diesel and fuel oil contaminated an area of 4,000 $m^2$ to a depth of 4 meters. Site lithology: clayey silt and episodic sandy layers.

Sites 2 and 3 are located in urban areas.

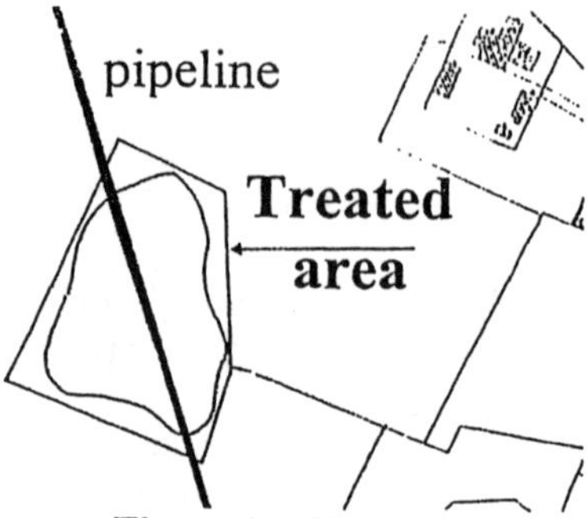


**Figure 1a: Site 1**

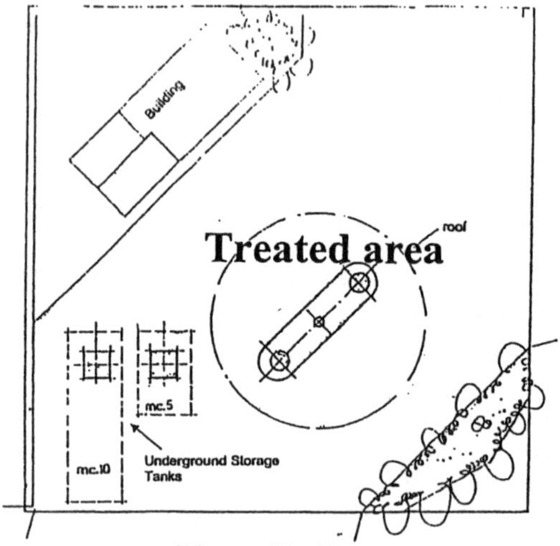


**Figure 1b: Site 2**

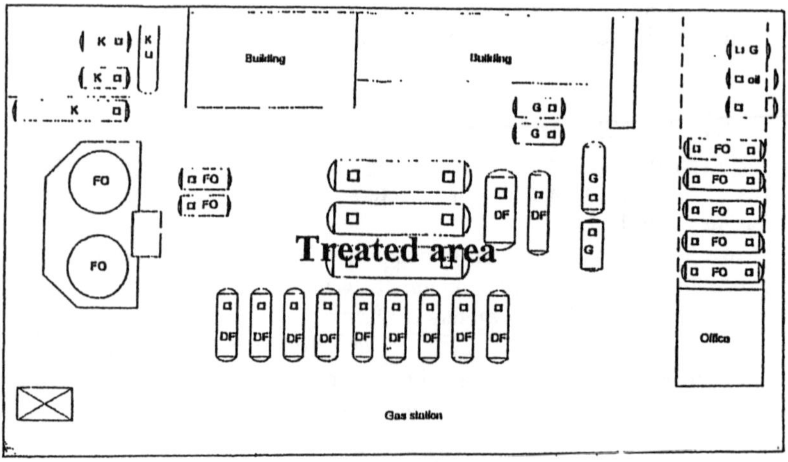


**Figure 1c: Site 3**

**FIGURE 1. SITE MAPS**

## PRELIMINARY TREATMENTS

In all cases the free phase (NAPL) floating on the water table was removed using the physical "in situ" traditional treatments like pump & treat, and specifically "slurping".

In two cases further preliminary treatments were applied to heavily polluted unsaturated soil, to limit odors and other possible environmental impacts: the chosen techniques were Soil Vapor Extraction (SVE) and Bioventing (BV).

## DEEP LANDFARMING METHODS

Conceptually, a combination of two treatments was used: "Landfarming" and "Pump & Treat", both applied in the same area where pollution was present. These were designed to maximize soil air permeability and maintain moisture content and so doing optimizing biodegradation conditions, especially aerobic degradation induced by unsaturation. Water drainage was carried out in order to keep groundwater level low enough and constant to maintain a large unsaturated soil thickness.

The soil was initially mixed and arranged in a series of mounds and adjacent trenches, reaching the deepest polluted layer (Step 1).

Infiltration of groundwater was pumped out during the initial period of treatment. Supplementary air injection into the core of the mounds maintained a correct level of oxygenation in the soil; mounds were also periodically shifted (i.e. mounds turned into trenches and viceversa, see Step 2a, 2b) during mixing, guaranteeing treatment of the whole volume to be remediated.

The drawing of Figure 2 shows the above sequence.

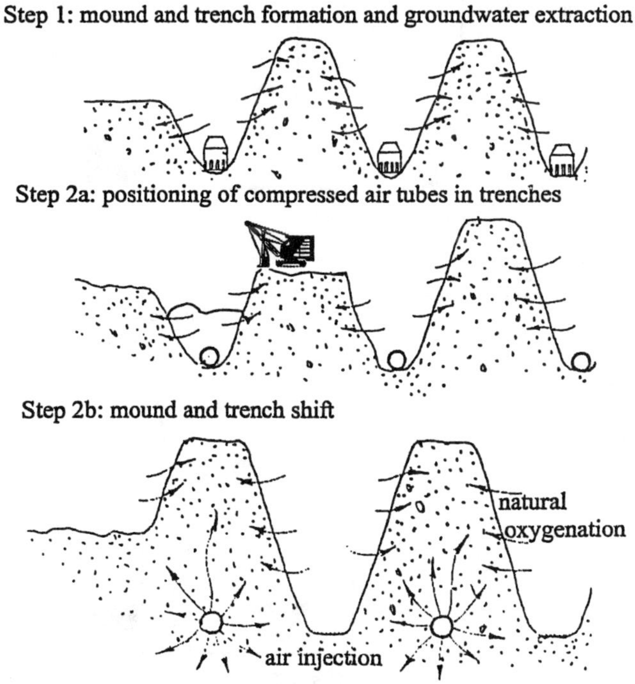


**FIGURE 2.**

## RESULTS

During the initial period of Deep Landfarming, the first two or three movements of mounds, water pumping out was needed; after that, a progressive soil insaturation and microbial growth took place.

All cases showed a considerable reduction in TPH (Total Petroleum Hydrocarbons) (Figure 3a).

From Case Study 1 a reduction in TPH concentration was recorded from several hundred mg/Kg to less than 10 mg/Kg after only 5 months of treatment.

In Case Study 2, after a period of pretreatment by SVE+BV, a reduction in TPH concentration was recorded from about two thousand mg/Kg to near zero in a period of less then 4 months of Deep Landfarming.

In Case Study 3, after SVE+BV pretreatment, a TPH reduction from about one thousand mg/Kg to less than 10 mg/Kg was measured in 8 months of treatment.

In all cases benzene was also present in the soil; at the end of treatment this substance had been considerably reduced, as shown in Figure 3b.

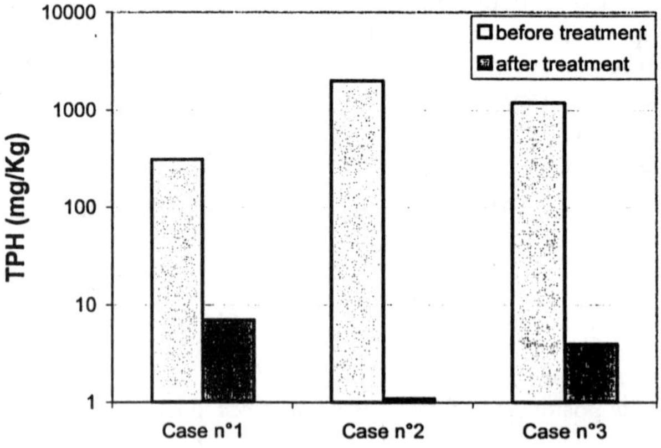


(a)

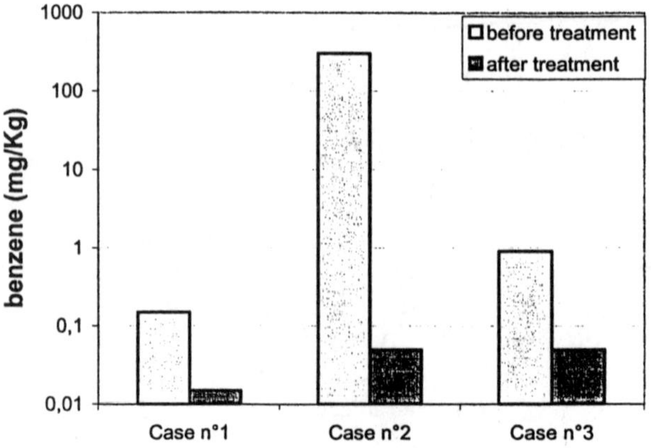


(b)

**FIGURE 3. Deep Landfarming yield in terms of (a) TPH and (b) benzene.**

## ADVANTAGES AND PROSPECTS for the FUTURE

The main advantages of this type of In Situ Deep Landfarming for polluted soil remediation can be summarized as follow:

- applicable "in situ" i.e. without removal, disposal or replacement of impacted soil
- applicable in very fine soils and in aquitards, which both are notoriously resistant to the common "in situ" technique
- applicable in deep pollutions, also affecting groundwater up to a depth of 7 m
- suitable as final stage remediation in industrial sites, after dismantling
- relatively low financial investment needed, time and energy saving; total costs are less than 25 USD per ton.

The main disadvantages are:

- inapplicable to land covered by buildings or process units
- possible odor generation and consequent impact in urbanized areas
- need to extract and treat groundwater in initial period of operation.

## REFERENCES

Please note: no references.

# MEASUREMENT OF TRACE GAS EMISSIONS DURING DIESEL-CONTAMINATED SOIL BIOREMEDIATION

Grant C. Edwards[1], ***Sandra Ausma***[1], Wendy P. Mortimer[2], Terry J. Gillespie[1], Edwina K. Wong[1], Laurie Halfpenny-Mitchell[1]
[1] University of Guelph, Guelph, ON, Canada
[2] Bell Canada, Toronto, ON, Canada

**ABSTRACT:** The use of micrometeorological and chamber methods to monitor trace gas fluxes from bioremediation facilities is described. Flux data collected during two field studies conducted at remotely located diesel fuel contaminated soil bioremediation facilities in northern Ontario are presented along with speciated hydrocarbon data. At the first site, carbon dioxide fluxes were measured using three techniques for a period of 33 days. At the second site, static chamber methods were used to measure total hydrocarbon (THC), $CO_2$ and $N_2O$ fluxes. Experiments showed that THC fluxes increased 16-fold upon cultivation. Nitrous oxide emissions were larger than those typically observed from agricultural soils.

## INTRODUCTION

The loading of important trace gases such as hydrocarbons, carbon dioxide and nitrous oxide to the atmosphere during the bioremediation of petroleum hydrocarbons is not well understood. Knowledge about atmospheric emissions is needed to assess their impact on the environment and human health, to advance the understanding of mechanisms leading to high gaseous emissions, and to develop effective management strategies to optimize biodegradation and minimize the atmospheric release of compounds that have a significant negative impact.

Compounds of interest, from an environmental and human health perspective, which are emitted from bioremediation facilities, include volatile hydrocarbons, $CO_2$ and $N_2O$. These have been linked to both ground-level ozone production and elevated levels of greenhouse gases. Additionally, emitted hydrocarbons may be harmful to humans.

Volatilization may be an important loss mechanism during hydrocarbon degradation in soil. Laboratory scale studies have demonstrated a wide range of hydrocarbon volatilization rates, although few studies have directly measured volatile hydrocarbon losses. Frequently volatile losses are assumed to be negligible, ignored or estimated indirectly through a mass balance.

**Objectives.** The goals of the research efforts described in this paper were to develop methods to evaluate trace gas fluxes from diesel fuel contaminated soils undergoing bioremediation, and to implement these methods to collect data from a variety of landfarm soil surfaces. This was accomplished by developing micrometeorological and chamber methods for monitoring trace gas fluxes at active bioremediation facilities.

## METHODS

Several field methods are available for measuring the air surface exchange of trace gases from soils, including micrometeorological (Fowler and Duyzer, 1989) and chamber techniques (Hutchinson and Livingston, 1993).

**Micrometeorological.** Micrometeorological methods are generally preferred for flux measurements. They are ideal for the collection of long time series of data over large plots of soil. They are continuous, unobtrusive techniques in which the trace gas flux is quantified over an area of emission in real-time. The average emissions from an area can then be determined. Micrometeorological methods have been extensively applied to the measurement of trace gas fluxes from a variety of surfaces including soils. Several trace gas analyzers can be incorporated into a micrometeorological measurement system design to allow measurement of multiple trace gas fluxes.

Two different micrometeorological methods will be described: the flux gradient (FG) and the aerodynamic mass balance methods (AMB). With the FG method, turbulence over a surface is assumed to be horizontally homogeneous and the flux is calculated from the product of the vertical concentration gradient and the eddy diffusivity:

$$Flux = -K\frac{\Delta C}{\Delta z}, \text{ and } K = \frac{u_* k(z_2 - z_1)}{\ln\left(\frac{z_2 - d}{z_1 - d}\right) - \psi_{H_2} + \psi_{H_1}} \qquad 1$$

where *Flux* is the calculated flux (μg m$^{-2}$ s$^{-1}$), $K$ (m$^2$ s$^{-1}$) is the integrated trace gas eddy diffusivity, and $\Delta C/\Delta z$ (μg m$^{-3}$ m$^{-1}$) is the vertical concentration gradient. $K$ is a function of the friction velocity, $u_*$ (m s$^{-1}$), the sample intake heights ($z$), and the integrated stability function for heat transfer ($\psi_H$) at the sampling heights (Businger et al., 1971). $k = 0.40$ is the dimensionless von Karman constant, $z_1$ and $z_2$ are the intake heights (m), and $d$ is the zero plane displacement (m).

The FG method is well suited for situations where a large uniform fetch is available. Since this is often not the situation with bioremediation facilities, this method should be applied with caution at small bioremediation facilities to ensure that appropriate sampling heights are selected to answer the scientific questions posed.

The AMB method differs from the FG method in that fluxes are directly quantified using a mass balance approach. The horizontal flux from an emission area is calculated using the vertically integrated product of wind speed and concentration divided by the fetch length ($l$) (m):

$$Flux = \frac{\sum u\Delta C\Delta z}{l} \qquad 2$$

A concentration difference, $\Delta C$ (μg m$^{-3}$), is measured between several sampling points above the soil surface (height = $z$) and the ambient background. The background can either be measured off-site, or on-site at a sufficiently high height to represent background concentration levels. $u$ is the wind speed (m s$^{-1}$) as measured with cup anemometers at each concentration measurement height, and $l$ is the along-wind distance from the intakes to the edge of the source (m). $\Delta z$ is the thickness of each air layer (m) to which the mass balance is applied. The sum over all the layers gives the total flux.

Fluxes can be measured using a wind speed profile, a concentration profile and knowing the wind direction. This technique is ideal for the measurement of fluxes from small finite sources that are surrounded by land that is not a source: generally the case with bioremediation facilities. While $l$ can be determined for any site geometry using a combination of a wind vane and accurate site dimensions, the ideal geometry is a circle at the center of which profile measurements are performed: eliminating the need for a wind vane.

**Chamber Methods.** The principle of chamber operation is to isolate a surface from the environment and measure concentration changes in the overlying air with time. Chamber methods are preferred when quantifying spatial variability in trace gas fluxes over an area or when examining fluxes from small sources where micrometeorological methods cannot be applied. However, if an average flux from a surface is desired, a statistically valid sampling plan should be used to ensure representative chamber sampling has been performed.

The use of chambers can be limiting because their presence alters the local environment under study. Chambers have been widely used to estimate a variety of trace gas fluxes from soil surfaces and to estimate hydrocarbon emissions from refinery landfarm soils. Hutchinson and Livingston (1993) describe chamber designs, discuss the limitations of chambers and provide references to illustrate their application versatility.

Chamber methods include static and dynamic designs. Static chambers are enclosures that are inserted into the soil to capture emitted gases. Air samples are either collected discretely over time and subsequently analyzed or air is drawn from the chamber through a non-destructive analyzer and returned to the chamber so that the net loss of air from the chamber is minimal. The flux ($\mu g\ m^{-2}\ s^{-1}$) can then be calculated directly from the rate of change in trace gas concentration ($\Delta C/\Delta t$) within the chamber, and knowledge about the chamber volume ($V$) and the enclosed area ($A$):

$$Flux = \frac{V}{A}\frac{\Delta C}{\Delta t} \qquad 3$$

Static chambers offer precision when detecting small fluxes and they need be in place for only a short period of time, limiting site disturbance. They are generally inexpensive and easy to construct and use. In their simplest form they require no electricity; collected samples can be analyzed at a later time.

Since static chambers isolate an area completely, gas, heat and water vapour exchanges are restricted between the inside and outside of the chamber; leading to limitations which must be taken into consideration when using this method.

Dynamic chambers have air flowing continuously through them, thereby maintaining environmental conditions closer to those of an uncovered area. They are the preferred chamber method when performing long-term monitoring of gaseous fluxes. The flux of a trace gas from the enclosed surface is calculated from the product of the trace gas concentration difference between the incoming ($C_{in}$) and exhaust air ($C_{out}$), and the volumetric flow ($Q$) rate to area ($A$) ratio:

$$Flux = \frac{Q}{A}(C_{out} - C_{in}) \qquad 4$$

Dynamic chambers require equilibration time while gas concentrations in the soil and chamber air come to steady values. They may also create artificially high fluxes by inducing airflow through pressure deficits within the chamber. A modified microclimate may also be created within dynamic chambers.

**Method Selection.** Considerations to be made when selecting a method for a site include site geometry, the types of trace gases to be monitored and the objectives of the measurements. If a bioremediation site is in the planning stages and the measurement of trace gas fluxes is anticipated, site geometry can be optimized to allow application of a particular measurement method. The results from several field studies conducted by our research group are presented to illustrate the utility of the different methods in characterizing a variety of trace gas emissions.

**Ambient Speciation Methods.** Solid phase micro-extraction (SPME) was used to analyze the headspace of buckets filled with landfarm soil collected at the MacDowell Lake site described below. SPME is an equilibrium concentration technique that utilizes a coated fused silica fibre to extract compounds from a headspace. Sixty-five μm PDMS-DVB (Supelco, Oakville, ON) and 65 μm PDMS (Supelco) fibres were used to concentrate analytes within headspaces.

Samples were speciated using a gas chromatograph with a mass selective detector (GC/MSD). SPME fibres were inserted directly into the injection port of the GC/MSD where the analytes were thermally desorbed when the fibre was exposed. Compounds were identified through the library associated with the GC/MSD.

## FIELD STUDY RESULTS

Our research team has applied the described methods at several bioremediation sites located in remote areas of northern Ontario. In addition we have used a flux gradient method to measure total hydrocarbon (THC) emissions from active refinery landfarms (Ausma et al., 2001a)

**MacDowell Lake 1996.** During the summer of 1996, two FG micrometeorological field studies were successfully performed at a diesel fuel contaminated soil bioremediation facility near MacDowell Lake ON (latitude 52° and longitude 94°). A landfarm was constructed in 1995 to treat approximately 50 $m^3$ of contaminated soil. A total of 18 days of total hydrocarbon flux measurements were collected. The results of this study are detailed elsewhere (Ausma et al., 2001b).

**Deer Lake 1997.** In 1997, a field study was performed at a diesel fuel contaminated soil bioremediation facility in the First Nations community of Deer Lake ON (latitude 52.5° and longitude 94°). The 530 $m^3$ landfarm was constructed in 1996 to treat soil accidentally contaminated with diesel fuel during generator refuelling and fuel transfer. THC and $CO_2$ flux measurements were performed using a static chamber method along with a FG and an AMB method.

It was unseasonably wet during the field study; as a result THC fluxes were below detection limits. On the other hand, $CO_2$ fluxes were successfully measured using all three techniques. Fig. 1 presents the $CO_2$ flux results. The average daytime $CO_2$ fluxes were 0.207 (± 0.018 s.e.), 0.244 (±0.019 s.e.) and 0.215 (±0.022 s.e.) mg $CO_2$ $m^{-2}$ $s^{-1}$ when using a FG, AMB and static chamber method, respectively.

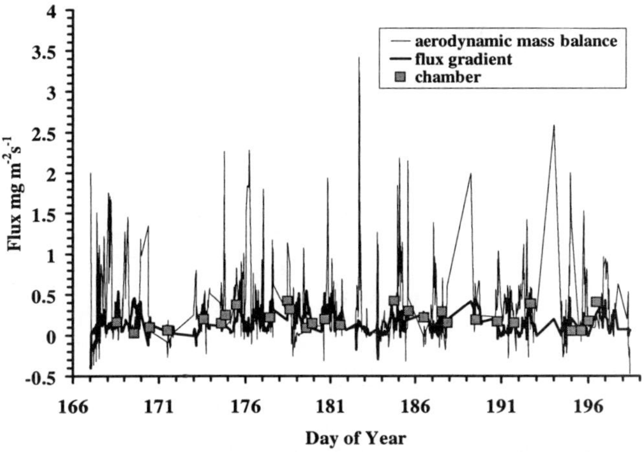


**FIGURE 1. $CO_2$ flux results from the Deer Lake landfarm using three different flux measurement techniques.**

**MacDowell Lake 1999.** In the fall of 1999, a one-day static chamber study was performed at the MacDowell Lake site to quantify THC, $CO_2$ and $N_2O$ fluxes from a landfarm constructed in 1999 to treat approximately 140 $m^3$ of soil contaminated with diesel fuel. The landfarm was located within a forest clearing that had formerly been occupied by a generator facility and telephone support equipment. Fertilizer was applied in August 1999; however, due to wet soil conditions it was not fully incorporated into the soil.

$CO_2$ and $N_2O$ fluxes were determined using a static chamber method in which the chambers were deployed for approximately 0.5 hr and samples were collected in evacuated glass vials sealed with septa. Analyses were performed via gas chromatography upon return to the lab. A modified static chamber method was used to estimate THC fluxes since whole air samples containing hydrocarbons can be difficult to store due to the propensity of some hydrocarbons to adhere to storage container walls. Air was continuously drawn from the chamber at a low flow rate and the concentration determined using a custom-built FID-based THC detector. In this way, the THC flux was estimated in real-time at the site. Less than 10 minutes of concentration data from the chamber were required to estimate the THC flux.

A series of experiments were performed to evaluate the impact of simulated cultivation on trace gas fluxes. A chamber was first deployed on undisturbed soil; approximately 0.5 hr later, the soil was disturbed to a depth of about 30 cm and a chamber was again deployed. Flux results along with temperature measurements from these experiments are presented in Fig. 2.

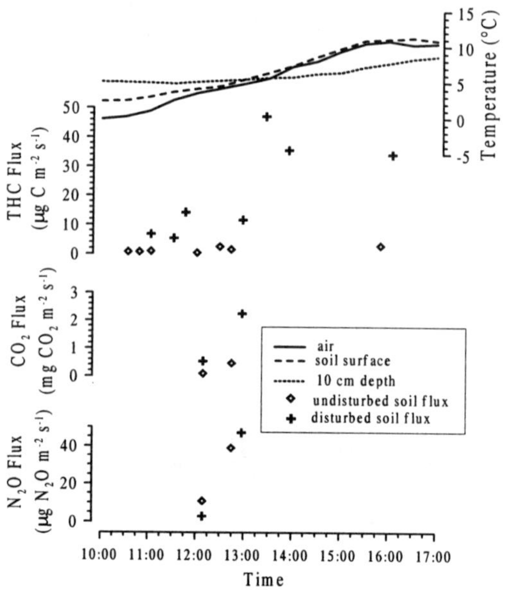


**FIGURE 2. Trace gas fluxes measured using a static chamber method at the MacDowell Lake site.**

On average, disturbing the soil surface resulted in a 16-fold increase in the flux of THCs (Fig. 2). Cultivation loosens up the soil surface exposing buried soil to the air, introducing air to the subsurface and decreasing the resistance to trace gas movement. This leads to higher emissions of gases and volatile species.

For both disturbed and undisturbed soils, the emission rate of THCs increased over the 8 hours of measurements (Fig. 2). The cause of this is likely the increase in soil temperature over this time, however; this is complicated by spatial variations in soil diesel fuel concentrations. THC fluxes increased as sampling moved across the landfarm, correlating with increases in soil diesel fuel concentrations.

A limited number of chambers were deployed to measure $CO_2$ and $N_2O$ fluxes since a longer period of time was required to conduct a full experiment. Disturbing the soil surface resulted in a 5-fold increase in $CO_2$ emissions, while the $N_2O$ data was inconclusive.

The $N_2O$ fluxes were high relative to emissions from agricultural soils, which are typically in the order of tens of ng $m^{-2}$ $s^{-1}$. Fertilizer had been applied late in the season, the fertilizer had not been fully incorporated into the soil, and the soil was wet. These factors contributed to an abundance of mobile nitrogen species (avg. $NH_4$-N 701 mgN $kg^{-1}$ soil: avg. $NO_3$-N 502 mgN $kg^{-1}$ soil: avg. $NO_2$-N 68 mgN $kg^{-1}$ soil) and ensured denitrification conditions.

**Speciation Results.** A variety of hydrocarbons were found in the headspace above soil samples collected in 1999 from the MacDowell Lake landfarm (Table

1). The detected compounds were typical of those found in petroleum mixtures and were comprised primarily of aromatics and aliphatics 7 to 15 carbons in length.

The concentration of THCs downwind from an area source can be approximated using a Gaussian plume model for areal sources at ground level. The maximum emission rate from the 1996 MacDowell Lake Field study of 131 $\mu g\ C\ m^{-2}\ s^{-1}$ (Ausma et al., 2001) was used to estimate a THC concentration of approximately 80 $\mu g\ C\ m^{-3}$ at a height of 2 m slightly downwind of the landfarm.

**Table 1: Speciated hydrocarbons identified in headspace samples above the MacDowell Lake soil.**

| Detected Compounds | |
|---|---|
| p-xylene | 1,2,4-trimethylbenzene |
| ethylbenzene | limonene |
| methylbenzene | octadecane |
| ethylhexane | eicosane |
| pentadecane | docosane |
| heptadecane | tetradocosane |
| toluene | 1-ethyl-2,3-dimethylbenzene |

Human inhalation exposure limits are not established for THC volatiles. When evaluating risks associated with petroleum products in the terrestrial environment, BTEX and PAHs generate the greatest concern due to their associated toxic, carcinogenic and/or mutagenic properties. BTEX is generally only found in trace quantities in diesel fuel due to its flash-point specifications, therefore they were not expected to be found in significant concentrations at the diesel fuel contaminated soil landfarm at MacDowell Lake.

Available American Conference of Governmental Industrial Hygienists (ACGIH) TWA exposure limits for straight, branched and cyclic alkanes, and aromatics range from 52 to 1720 mg $m^{-3}$. The TWA values are at least an order of magnitude larger than the estimated maximum downwind concentration at the MacDowell site. The TWA limits for benzene are more stringent at 3.5 mg $m^{-3}$. Even if THC fluxes were composed of 100% benzene, TWA levels would not be approached. This is a very unlikely scenario at a diesel fuel contaminated soil landfarm. Due to the low levels of THCs in the air above and downwind of the landfarm, it is expected that the effect on workers servicing the landfarm and nearby residents is minimal and no greater than exposure to other local sources of volatile hydrocarbons.

## CONCLUSIONS

Micrometeorological and chamber methods that can be used to measure trace gas fluxes from bioremediation facilities were described and results from several micrometeorological and chamber field studies were presented. These methods allow the direct measurement of trace gas losses from bioremediation facilities. Information about volatile losses can contribute to assessments of environmental and human health impact; aid in evaluating routes of contaminant reduction within the soil; and, help to develop site management protocols to optimize biodegradation and minimize volatile losses of trace gases of concern.

When these methods are implemented in conjunction with comprehensive measurements of environmental parameters such as temperature, humidity, soil moisture content, and nutrient levels, insights into mechanisms controlling volatile releases can be made.

Site geometry, measurement objectives and trace gas characteristics should be taken into consideration when selecting a method for a particular site.

A flux gradient, an aerodynamic mass balance and a static chamber method were used to monitor $CO_2$ fluxes from a diesel fuel contaminated soil landfarm.

Static chamber methods were used to evaluate the impact of cultivation on emissions of total hydrocarbons, $CO_2$, and $N_2O$. They were simple to apply and allowed characterization of site heterogeneity. Cultivating the landfarm soil surface resulted in an approximately 16-fold increase in THC fluxes and a 5-fold increase in $CO_2$ fluxes.

Solid phase micro-extraction (SPME) was used to characterize THC emissions from soil collected at a diesel fuel contaminated soil landfarm. The emissions were comprised primarily of aliphatics and aromatics 7 to 15 carbons in length.

Using a Gaussian areal source model, it was concluded that off-site concentrations would not pose a significant health risk to nearby humans.

**ACKNOWLEDGEMENTS**

Bell Canada and the Natural Science and Engineering Research Council financially supported this research.

**REFERENCES**

Ausma, S., G.C. Edwards, E.K. Wong, T.J. Gillespie, C.R. Fitzgerald-Hubble, L. Halfpenny-Mitchell, W.P. Mortimer. 2001a. "A Micrometeorological Technique to Monitor Total Hydrocarbon Emissions from Landfarms to the Atmosphere." *J. Environ. Qual.* 30:776-785

Ausma, S., G.C. Edwards, C.F. Fitzgerald-Hubble, T.J. Gillespie, L. Halfpenny-Mitchell, W.P. Mortimer. "Volatile Hydrocarbon Emissions from a Diesel Fuel Contaminated Soil Bioremediation Facility." Submitted to *J. Air & Waste Manage. Assoc.* Feb. 2001

Businger, J.A., J.C. Wyngaard, Y. Izumi, E.F. Bradley. 1971. "Flux Profile Relationships in the Atmospheric Surface Layer." *J. Atmos. Sci.* 28:181-189

Fowler, D., J.H. Duyzer. 1989. "Micrometeorological Techniques for the Measurement of Trace Gas Exchange." p. 189-207. *In*: *Exchange of trace gases between terrestrial ecosystems and the atmosphere*. Dahlem Workshop Proceedings. John Wiley & Sons, Berlin

Hutchinson, G.L., G.P. Livingston. 1993. "Use of chamber systems to measure trace gas fluxes." p. 63-78. *In* L.A. Harper et al. (ed.) *Agricultural ecosystem effects on trace gases and global climate change.* ASA Spec. Publ. 55. ASA, CSSA, and SSSA, Madison, WI.

# PREDICTING BIOTREATABILITY END-POINTS FOR CRUDE OIL IN SOIL

***Sara J. McMillen*** (Chevron Research and Technology Co., Richmond, CA)
Robert E. Hoffmann (Chevron Canada Resources, Calgary, Alberta)[1]

**ABSTRACT:** Laboratory scale biotreatability and/or field pilot studies have generally been used to determine whether ex situ bioremediation (composting or land treatment) can achieve total petroleum hydrocarbon (TPH) or oil and grease (O&G) cleanup goals for a given site soil. In this study, data from bioremediation studies (laboratory to full scale) involving weathered crude oil contaminated soil were used to develop and validate predictive models. Chemical data characterizing the starting hydrocarbon composition were used to predict bioremediation end-points. Two equations were found to adequately describe biodegradation end-points for crude oil-impacted soil (one based on O&G content and the second based on TPH quantified by gas chromatography). The O&G predictive equation was validated using bioremediation results from aged hydrocarbon spill areas. The degree of correlation between equation-predicted and observed biodegradation end-points for the equation suggests that they are suitable for making screening-level decisions about the bioremediation potential of a particular site soil, thereby eliminating the need for costly treatability and pilot studies in some cases.

## INTRODUCTION

Bioremediation often represents a cost-effective approach to site cleanup for crude oil impacted soil. Composting and land treatment of oily soils and wastes have been successfully deployed at full scale at crude oil exploration and production sites (Fyock et al., 1991; Zimmerman and Robert, 1991; McMillen et al., 1996; Hoffmann and Chaw, 1999).

Crude oils vary widely in composition, depending on factors such as source bed type and maturation conditions (Hunt, 1979). The composition, which governs the relative biodegradability of a crude, can be characterized by hydrocarbon class distribution (saturates, aromatics, polars) and the abundance of *n*-alkanes relative to branched and cyclic alkanes within the saturated hydrocarbon class (Cook et al., 1974; León et al., 1998).

API gravity is a measure of the specific gravity of a given crude oil. A previously published method for predicting the biodegradation potential of crude oil based on API gravity indicated that 10 to 70% of a topped crude oil may be susceptible to biodegradation (McMillen et al., 1995). However, the utility of this API gravity predictive model is limited since it was based on the amount of biodegradation achieved after only four weeks. Furthermore, this model is not suitable for historical, weathered crude oil spill sites or wastes. Another published method for predicting biodegradation end-points (Huesemann, 1995)

[1] Currently with P.T. Caltex Pacific Indonesia, Sumatra, Indonesia

cannot be easily deployed because of the complexity of the analytical requirements and the degree of interpretation required.

In addition to the relative biodegradability of a crude oil, the extent of bioremediation in the field will also be dependent on the degree of physical and microbiological weathering that has occurred since release (natural attenuation). The degree of attenuation can be quantified via the concentration of a biomarker in the source oil relative to its concentration in the oil in the soil. The concentration of triterpanes (a class of compounds generally resistant to biodegradation) in the residual increases as the extent of biodegradation increases. Therefore, the ratio of the triterpane, 17α(H), 21β(H)-hopane, in the initial and biodegraded extracts has been used as a biomarker to quantify oil depletion after marine or terrestrial spills and bioremediation activities (Moldowan, 1992; McMillen et al. 1993; Douglas et al., 1994; Prince et al, 1994). This paper will describe how two improved API gravity predictive models, coupled with hopane data, can be used to predict biodegradation end-points for weathered crude oil spill sites and wastes.

## MATERIALS AND METHODS

Eighteen laboratory and pilot bioremediation projects were conducted using unweathered crude oil wastes (e.g. oily sludges, oil-spiked soils, and recent spill sites) in order to first develop the models, and then eleven site soils collected from aged hydrocarbon spill areas in Canada were used to validate one of the models. All bioremediation activities were conducted under optimal conditions (nutrient concentration, moisture content, and aeration [tilling or forced aeration]), and were continued until an asymptotic end-point was achieved for TPH and O&G. The bioremediation projects used to derive the predictive equation included a wide spectrum of crude oil types (API gravity range of 10° to 49°), soil textures, and ecological regions (Venezuela, Canada, Indonesia, and the United States).

API gravity is the measurement of specific gravity of a crude oil at 16°C, as defined by the American Petroleum Institute (API), and the reported value is converted to an integer. The O&G concentrations were obtained by tecator-solvent extraction with dichloromethane. The extract was dried at 40°C under a nitrogen stream to a constant weight. Crude oils are topped (removal of compounds $<C_{12}$) using the same procedure. Hydrocarbon class distributions were characterized by asphaltene precipitation followed by alumina-silica gel chromatography. Gas chromatography with flame ionization detection was used to obtain saturate and aromatic "fingerprints" and to quantify both resolved and unresolved complex mixtures of these fractions. The sum of the fractions equals the TPH detected by gas chromatography (TPH-GC). The fractions were also analyzed by GC/MS to determine concentrations of sterane and triterpane biomarkers, e.g. 17α(H), 21β(H)-hopane (Douglas et al., 1992).

## RESULTS AND DISCUSSION

Bioremediation end-point data from 18 laboratory and field projects worldwide were used to develop a predictive equation based on API gravity. As reported previously, API gravity is a surrogate for crude oil composition because of the strong positive correlation between API gravity and the amount of saturates, *n*-alkanes, and the ratio of saturates to aromatic hydrocarbons, as well as the negative correlation to the amount of more recalcitrant fractions such as asphaltenes and resins (Figure 1, from McMillen, et al. 1995).

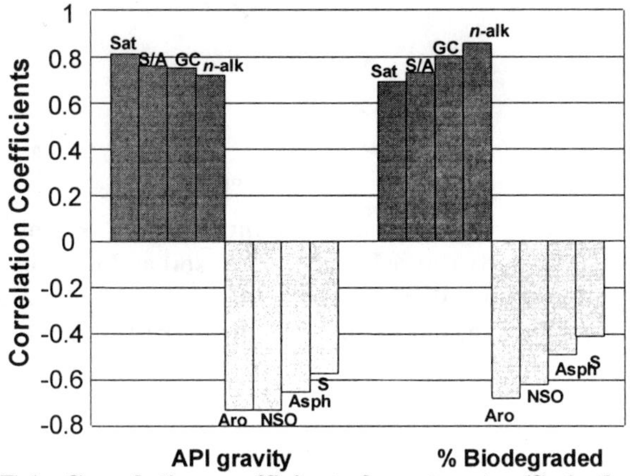


**FIGURE 1. Correlation coefficients for saturates (Sat), the saturate to aromatic ration (S/A), the total resolved and unresolved hydrocarbons detected by GC (GC), the *n*-alkanes (*n*-alk), aromatics (Aro), asphaltenes (Asph), and the % sulfur (S).**

The bioremediation end-points, calculated as maximum percent loss of O&G plotted against the API gravity of the starting crude oil are depicted in Figure 2. Figure 3 illustrates the TPH-GC results from the same studies plotted versus API gravity. Linear regression analysis reveals that API gravity alone can be used to predict biodegradability of freshly released crude oil in soil:

$$\text{max \% O\&G removed} = 2.24\text{API}° - 19.28 \quad (1)$$

$$\text{max \% TPH-GC removed} = 1.77\text{API}° + 9.31 \quad (2)$$

The API 10° Boscan oil from Venezuela, denoted by the star symbol in Figures 2 and 3, was omitted from the correlation because it is compositionally immature relative to the other oils in the study. Most low API gravity crude oils have been biodegraded in reservoir. In contrast, the Boscan oil still possesses significant amounts (14.4%) of labile n-alkanes, in spite of its low API.

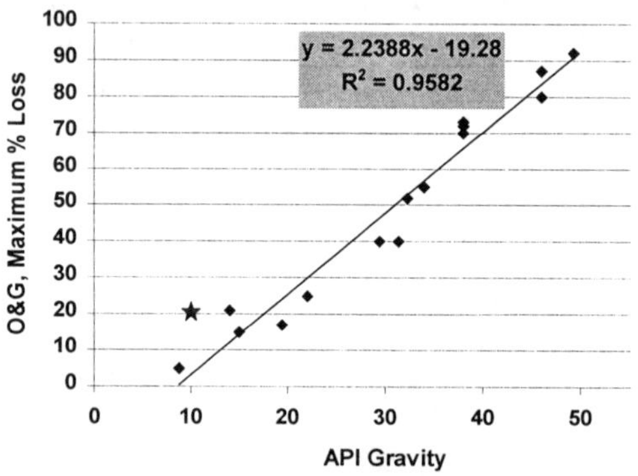


**FIGURE 2. The maximum amount of O&G removed during bioremediation for unweathered crude oils in soil and composts versus the API gravity of the crude oil. The star denotes the immature crude oil from Boscan that was not included in the regression analysis.**

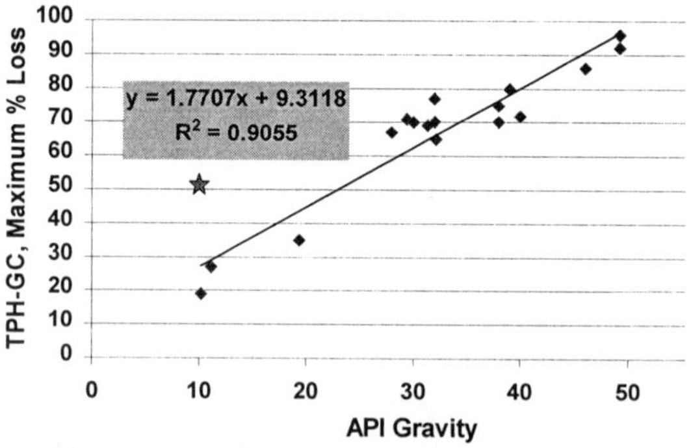


**FIGURE 3. The maximum amount of TPH-GC removed during bioremediation for unweathered crude oils in soil and composts versus the API gravity of the crude oil. The star denotes the immature crude oil from Boscan that was not included in the regression analysis.**

It is not surprising that the predictive equations for O&G and TPH-GC are substantially different, since it is well recognized that quantitative analytical measurements of hydrocarbons can be method-dependent in terms of solvent used, extraction method, and detection method (Blackburn et al., 1993). The variability between the two analytical methods increases with decreasing API gravity (Table 1) because gas chromatography does not detect high molecular weight hydrocarbons ($>C_{44}$) and polar compounds such as asphaltenes which do contribute to gravimetric O&G measurements. Therefore, it is critical to use comparable analytical procedures to those described herein in order to use the predictive equations.

**TABLE 1. Comparison of hydrocarbon concentrations in soil or waste by the analytical method used**

| | O&G , % | TPH-GC, % | Difference |
|---|---|---|---|
| 10° API oil in soil | 4.7 | 1.57 | 67% |
| 12.7° API oil in soil | 11.5 | 5.2 | 55% |
| 14° API oil in soil | 5.3 | 2.32 | 56% |
| 29.4° API oil in soil | 3.7 | 2.48 | 33% |
| 31.4° API oil in soil | 3.8 | 2.42 | 36% |
| 38° API oil in soil | 3.45 | 2.98 | 14% |
| 49.2° API oil in tank bottoms | 23.0 | 22.5 | 2% |

To validate the O&G predictive model, hydrocarbon data from eleven lab, pilot and field bioremediation projects in Canada involving weathered crude oil impacted soils were analyzed. In order to utilize the models, the amount of oil removed by natural attenuation prior to active biodegradation was first quantified. This was accomplished by analyzing the amount of C30-17α(H), 21β(H)-hopane (other biomarkers may also be used) in the hydrocarbon extracts from the site soils and comparing this amount to the amount of hopane in the topped, source crude oil (McMillen et al., 1993; Prince et al., 1994). The amount of natural attenuation that had occurred at the historical crude oil impacted sites is summarized in Table 2, along with a brief description of the sites. The amount of oil removed in the weathered samples, prior to active bioremediation, ranged from 1 to 66.5%.

**TABLE 2. Description of weathered site soils and amount of prior oil removal due to natural attenuation**

| Project | API° | (ug hopane/g soil extract) | (ug hopane/g topped oil)* | Oil Removal via Natural Attenuation |
|---|---|---|---|---|
| Tank battery soil #1 | 46.0 | 64.3, 65.0 | 21.4, 21.9 | 66.5% |
| Tank battery soil #2 | 25.8 | 153.3, 157.7 | 112.7, 118.2 | 25.7% |
| Flare pit #1 | 42 | 78.8 | 54.3 | 31.1% |
| Flare pit #2 | 42 | 82.6 | 54.3 | 34.3% |
| Tank battery soil #3 | 41.0 | 73 | 60 | 17.8% |
| Land treatment area #1 | 40.9-42.5 | 53 | 40 | 24.5% |
| Land treatment area #2 | 40.9-42.5 | 55 | 40 | 27.3% |
| Flare pit surface soil | 40.9-44.9 | 56 - 68 | 39 - 41 | 35.5% |
| Oily soil #1 | 40.9-44.9 | 51 | 39 - 41 | 21.6% |
| Oily soil #2 | 40.9-44.9 | 51 | 39 - 41 | 21.6% |
| Oily soil #3 | 35.7-36.8 | 107 | 106 | 1% |

*Mean values used to calculate % biodegradation

The portion of oil removed by natural attenuation in the weathered site soils was then used to estimate the initial concentration of crude oil present in the soils (concentration of oil in soil/(100-%removed)). Using this estimated initial concentration, the API gravity predictive model was applied to predict the O&G end-points that would be realized after aggressive bioremediation. The results of the model predictions versus the observed final concentrations are shown in Figure 4 for O&G. A strong correlation exists between predicted and observed [O&G] endpoints ($r^2$=0.92), though it would appear the O&G equation may over predict the degree of removal (slope = 1.3) realized. However, since monitoring was terminated in several cases before the bioremediation endpoint was achieved (either because the regulatory closure limit [1000 ppm in Alberta] was achieved, or because it was obvious that it would not be achieved) further biodegradation may have been possible.

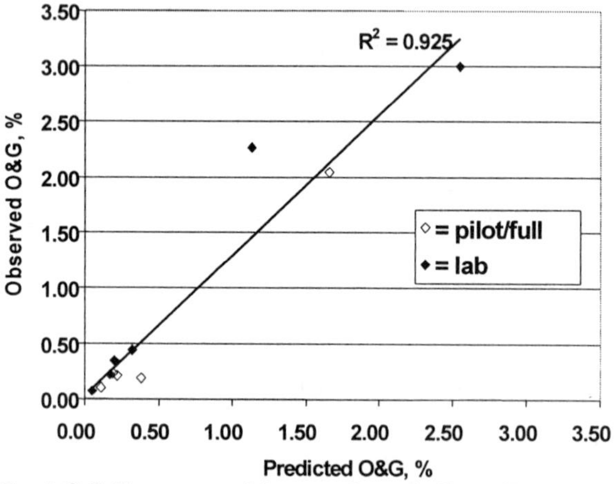


**FIGURE 4. Final O&G measured in weathered oily soils versus those predicted by the API gravity model.**

TPH-GC data were available for 9 of the 11 site soils; however, the measured TPH-GC concentrations in soil were highly variable as illustrated in Figure 5. Therefore, adequate predictions for TPH-GC removal could not be made using these data. TPH-GC results tend to be more variable than O&G data due to the greater complexity of the TPH-GC analytical method, laboratory modifications to the standard TPH method, and different operating parameters for the GC instrument.

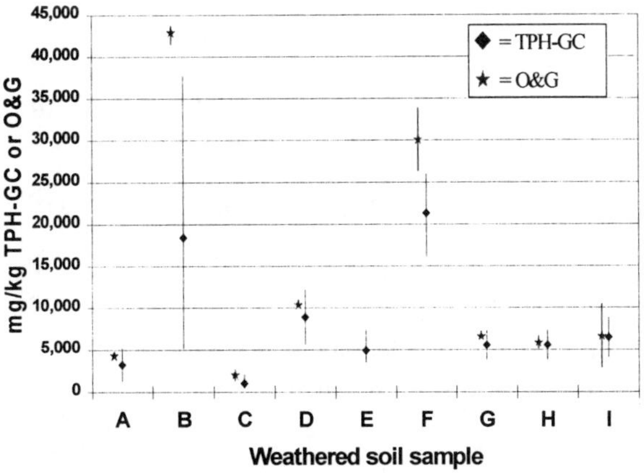


**FIGURE 5. The range of initial TPH-GC and O&G measurements in weathered, oil impacted soils from Canada**

## CONCLUSIONS

The hydrocarbon composition of crude oils, as reflected in their API gravity, coupled with hopane ratios to reflect the degree of natural attenuation in situ, is a good predictive tool for potential O&G biodegradation in crude oil impacted soils. The degree of correlation between equation-predicted and observed biodegradation end-points suggests that the predictions are suitable for making screening-level decisions about the bioremediation potential of a particular site soil, and can be used to prevent costly treatability and pilot studies for those site soils that have low potential for further biodegradation. The limitations of the predictive equations are: (1) the predictive equations are not appropriate for immature crude oils (infrequently encountered); (2) not all crude oils contain hopane; however, other biomarkers such as oleanane can be used; (3) the API gravity/hopane predictive model is limited to those sites where only one crude oil has been spilled and the source of the oil is known; and (4) similar analytical methods must be used. This last conclusion would appear particularly important for TPH-GC analyses, for which a large degree of inter-lab variability was observed.

Work is continuing to evaluate compositional data to develop predictive models for spill sites where multiple sources have impacted the site or where the API gravity of the source is unknown. As with any predictive equation, the practitioner must exercise caution and identify site specific variables that could potentially inhibit the rate or extent of oil removal (e.g. high salinity, presence of inhibitory chemicals). Biotreatability testing is the logical follow-up for any such case to confirm the rate and degree of removal that may be achieved at full scale.

## REFERENCES

Blackburn, J. W. E. J. Harner, W. K. Robbins, R, C. Prince, J. R. Clark, R. M. Atlas, and J. B. Wilkinson. 1993. "Experimental Linkages Issues of Petroleum Site Bioremediation." *Biodegradation*. 4: 207-230.

Cook, F. D., A. Jobson, R. Philippe, and D. W. S. Westlake. 1974. "Biodegradability and Crude Oil Composition." *Canadian Journal of Microbiology* 20: 915-928.

Douglas G. S., R. C. Prince, E. L. Butler, and W. G. Steinhauer. 1994. "The Use of Internal Chemical Indicators in Petroleum and Refined Products to Evaluate the Extent of Biodegradation." In R. E. Hinchee, B. C. Alleman, R. E. Hoeppel, and R. N. Miller (Eds.), *Hydrocarbon Bioremediation,* pp. 219-235. Lewis Publishers, Ann Arbor, MI.

Fyock, O. L. S. B. Nordrum, S. Fogel, and M. Findlay. 1991. "Pilot scale Composting of Petroleum Sludges." In Proceedings of Treatment and Disposal of Petroleum Sludge. University of Tulsa, Center for Environmental Research and Technology, Tulsa, OK.

Hoffmann, R.E.and D. Chaw. 1999. "Composting Crude Oil-Impacted Soil: Performance Comparison with Land Treatment and Soil Productivity Implications." In B.C. Alleman and A. Leeson (Eds), *Bioreactor and Ex Situ Biological Treatment Technologies*. pp. 189-196.

Huesemann, M. H. 1995. "Predictive Model for Estimating the Extent of Petroleum Hydrocarbon Biodegradation in Contaminated Soils", *Environmental Science and Technology, 29*(1):7-18.

Hunt , J. M. 1979. Petroleum Geochemistry and Geology, W. M. Freeman, San Francisco, CA.

Leon N. C. Infante, M. Marquez, A Gorrin, McMillen, S. J. and Smart R.S. 1998. "Biodegradability of Venezuelan Crude Oils," presented at *The 1998 SPE International Conference on Health, Safety, and Environment in Oil & Gas Exploration* in Caracas, Venezuela, Society of Petroleum Engineers.

McMillen, S. J., J. M. Kerr, and N. R. Gray. 1993. "Microcosm Studies of Factors That Influence Bioremediation of Crude Oils in Soil." SPE paper 25981 presented at the *1993 SPE/EPA Exploration & Production Environmental Conference*, San Antonio, TX, Society of Petroleum Engineers.

McMillen, S.J., A. G. Requejo, G. N. Young, P. S. Davis, P. D. Cook, J. M. Kerr, and N. R. Gray. 1995. "Bioremediation Potential of Crude Oil Spilled on Soil,"

In: *Microbial Processes for Bioremediation*, Eds. R. E. Hinchee, G. S. Douglas, and S.K. Ong, (Battelle Press: Columbus, OH) *3*(8): 91-99.

McMillen, S. J., J. M. Kerr, P. S. Davis, J. M. Bruney, M. E. Moir, P. Nicholson, C. V. Qualizza, R. Moreau, and D. Herauf. 1996. "Composting in Cold Climates: Results of Two Field Trials," SPE paper 35878 presented at *The 3rd International Conference on Health, Safety, and Environment in Oil & Gas Exploration* in New Orleans, LA, Society of Petroleum Engineers.

Moldowan, J. M., C. Y. Lee, P. Sundararaman, T. Salvatori, A. Alajbeg, B. Gjukic, G. J. Demaison, N. Slougui, and D. S. Watt. 1992. "Source Correlation and Maturity Assessment of Select Oils and Rocks from the Central Adriatic Basin." in *Biological Markers in Sediments and Petroleum*, Eds. J. M. Moldowan, P. Albrecht, and R. P. Philp (Prentice Hall: Englewood Cliffs NJ), pp. 370-401.

Prince R. C., D. L. Elmendorf, J. R. Lute, C. S. Hsu, C. E. Haith, J. D. Senius, G. J. Dechert, G. S. Douglas, and E. L. Butler. 1994. "17α(H), 21β(H)-Hopane as a Conserved Internal Marker for Estimating the Biodegradation of Crude Oil." *Environmental Science and Technology*, *28*: 142-145.

Zimmerman, P. K. and J. D. Robert. 1991. "Oil-based Drill Cuttings Treated by Landfarming." *Oil & Gas Journal 89*(32):81-84.

# DECOMMISSIONING DIESEL GENERATING STATIONS IN SUB-ARCTIC CANADA: A SIX-YEAR PROGRESS REPORT

***Wendy P. Mortimer*** (Bell Canada, Toronto, Ontario, Canada)

**ABSTRACT:** In 1995, Bell Canada and a government partner embarked on a joint program to decommission, assess and remediate their 16 remote, air-access only, generating stations in Northern Ontario, as well as 15 other Bell-owned facilities in the same area. Most sites were on, or adjacent to, First Nations' lands. Seven sites were clean. Twenty-four others had diesel-contaminated soil, with quantities estimated from 50 to 1500 $m^3$, and minor, if any, groundwater contamination. Soils ranged from sand to clay. At only one site was free product found (two years after the initial assessment). Bioremediation was selected as the soil treatment technique.

Remediation objectives were determined to be 100 ppm total petroleum hydrocarbons. A commitment to act in an environmentally sustainable manner meant using natural materials rather than plastics for liners (e.g. locally collected peat moss), using rototillers for aeration rather than blowers, designing the treatments to use no electricity, no activated carbon, no piping, and no requirements for decommissioning following bioremediation.

The environmental impacts of site remediation are often greater than the original problem, and conventional approaches must be carefully reconsidered in the special circumstances of these remote locations.

## INTRODUCTION

In the mid-1970's, Bell Canada undertook to provide phone service to remote First Nation communities in Northern Ontario, in collaboration with the Ontario Ministry of Transportation and Communications. As these communities were without electricity, Bell operated Ministry-owned diesel-powered electricity-generating stations to power the telephones.

These communities are hundreds of kilometers north of the end of the road. Diesel fuel was, and still is, brought in by seasonal snow road, or by air. As with other installations of this kind, incidental, or minor spills were common. Occasionally, more serious leaks, overflows or accidents occurred, resulting in soil contamination. A common practice was to toss used engine oil on the ground, rather than fly it out.

During the 1980's, large-scale diesel generating stations were built in many communities, and Bell Canada retired its power-generating equipment, often abandoned in place and subsequently vandalized. In 1995, Bell Canada and the Ministry embarked on a joint program to decommission, and remediate their 16 stations in Northern Ontario, as well as 15 other Bell-owned facilities. They are located within the area bounded by the Manitoba border, Hudson Bay, James Bay and the 51st parallel, as shown in Figure 1.

**FIGURE 1. Locations of Former Generating Stations in Northern Ontario.**

## APPROACH

**Project Initiation:** The Bell Canada property manager and the author, an environmental consultant, traveled by chartered aircraft to 31 sites of concern over a six-week period in June and July, 1995. At each site, the inspection day began with a meeting with the Chief and Council of each of the local First Nation. (Four sites were remote repeater stations on provincial crown land.) The reasons for our visit, our interest in hearing of any fuel spills or other environmental problems, and our willingness to communicate the results of our study were explained, with the help of Oji-Cree translators, if necessary. Responses to our presentations ranged from indifference to enthusiasm, with the First Nations' interests centering on the possibilities of employment and our rental of their heavy equipment, as well as the potential health and environmental risks posed by spilled fuel.

**Phase I Assessment and Classification.** At the site of the former diesel generating station, any remaining junked equipment was inventoried, evidence of soil contamination was gathered, a limited number of samples were taken, and other environmental issues were noted. In many cases, fuel had been brought to the tanks by overland hose from floatplanes, so the route to the dock was also checked for evidence of spills. Each installation typically consisted of five 4500 L above-ground steel tanks, and a 3 m x 4 m steel building which housed the generators. Most stations also had small accommodation buildings equipped with oil-fired furnaces and 900 L fuel tanks. A tank up-grade program was coordinated with the decommissioning program where possible, and two demolition contractors were used to cut up and haul out the remaining tanks and equipment on the winter roads or by chartered aircraft.

Seven of the locations did not have any environmental problems, as these locations had received phone service after community power was available. Six other sites were either still operational, or had other reasons for delaying clean-up activities.

**Phase II Assessments.** In September, 1995, two consulting companies were contracted to conduct detailed site assessments of soil and groundwater contamination at 18 "high priority" sites. We used the most stringent soil quality guideline of 100 mg/kg Total Extractable Petroleum Hydrocarbons (TPH), according to the Ontario Ministry of the Environment Guidelines Table A – Residential, with potable groundwater. At only one site was free product found (but not until two years after the initial assessment). This site is described in detail in a poster in this Session (Reinecke, Mortimer and Livingstone). Others had diesel-contaminated soil, with quantities estimated from 50 to 1500 $m^3$, and minor, if any, groundwater contamination. Soils ranged from sand to clay. All sites were deemed to require remediation and plans were developed over the winter.

**Phase III Remedial Action Plan.** Bioremediation was selected as the soil treatment technique, and a total of eight (so far) consulting companies, with varying levels of experience, and using various approaches, were involved throughout this work. All proposals outlined the standard "biopile" design, in accordance with Environment Canada's 1993 Guidelines on the Ex-Situ Bioremediation of Petroleum Hydrocarbon Contaminated Soils on Federal Crown Land including liners, covers, blowers, water treatment, etc. Due to the remoteness of the sites, the huge expense of the proposals, an example of a failed implementation of just such a design in a neighboring northern community, and the perception that, from a social and environmental perspective, the cure was worse than the disease, A modified "landfarm" design was developed to replace the standard biopile approach. Advice was sought from Environment Canada as to the application of a more appropriate technology for these northern sites, but they were not willing to approve or endorse any approach other than that given in their Guidelines. Understandably, the First Nations did not want to be stuck with any soil treatment facility that did not have the approval of Environment Canada.

Working on First Nations' lands means that certain political protocols must be respected, including seeking project approvals in the form of Band Council Resolutions. Since the federal government's Department of Indian Affairs holds Reserve lands "in trust," their concurrence with the projects' design and execution were seen as important as well, and they look to Environment Canada for guidance. And, as several of Bell's sites were on provincial crown land adjacent to the Reserve, the provincial Ministry of the Environment was the third government level with requirements to satisfy.

In spite of the government's reservations about the landfarm technique, a commitment was made by Bell Canada to act in an environmentally sustainable manner, which included using natural materials rather than plastics for liners (e.g. locally collected peat moss), using manual labor for aeration rather than blowers, designing the landfarms to use no electricity, no activated carbon, no piping, and, most significantly, no requirements for decommissioning following bioremediation.

Having made a decision to use a landfarm approach, it was necessary to monitor potential impacts through intensive air monitoring, groundwater monitoring, soil monitoring, and soil sampling under the landfarms, and the dissemination of monitoring information annually, or as requested.

**Phase IV Implementation.** Before starting to dig, we met again with the Chief and Council of each First Nation, and explained the problems and our proposed solutions. As Bell Canada was the first agency in many communities to start to clean up old fuel spills, the Councils had little or no experience with this type of activity. Few, if any, restrictions were placed on the operations, and little attention was given to issues of landfarm location, etc. Employment of local people was their major concern, and we hired locally for all aspects of the project. We found that the support of the Chief and Council was imperative for the success of these projects.

Excavation equipment, if available, is usually owned by the First Nation, or its economic development corporation. In some cases, repairs to the equipment were necessary prior to its use. Backfill material, needed to fill the excavations following removal of contaminated soil, is frequently unavailable, or is extremely expensive, given the thin, clay soils and limited road networks of many communities. It was often necessary to re-grade the site considerably following excavation, to minimize the requirement for clean fill material.

Our biological treatment facilities were constructed on the same properties as the accommodation buildings, which were, in most cases, the same properties where the generating facilities had been located. This avoided the need to lease additional property, and had added advantages, including reduced costs in trucking, availability of water supply, proximity to storage facilities for rototiller, fuel and fertilizer, ease of monitoring, and enhanced weed and brush control under the towers, and, unexpectedly, site landscaping.

Peat moss collection in the local forests is a traditional fall activity of the women of the First Nations. In the days before disposable diapers, the moss was dried by draping it in trees, and, stored over the winter, it was used to line babies' diapers. To collect moss for landfarm "liners," local people were hired where possible. For example, twelve people working for five hours, collected enough moss for one of the smaller landfarms.

The peat moss was laid over bare soil to a thickness of about 15 cm. Depending on availability, a 15 cm layer of sand was spread over the moss to protect it from damage by the rototiller tines. Contaminated soil was then spread over the sand (or directly on the moss) as thinly as the available treatment area would allow. In most cases, we were able to limit the thickness of contaminated soil to about 0.5 m. Commercial fertilizer was added to the soil, rocks picked out by hand, and an 8 hp rototiller was used to mix and aerate the soil, taking care not to disturb the liner. A contract with the First Nation to have the landfarm rototilled regularly was made wherever possible.

Excavations were generally done to a clean background, to the extent possible, given the available machinery. The deepest excavation was about 4 m. Floor and wall samples were taken prior to backfilling, and the first round (time zero) sampling of the landfarm occurred after the last load of soil was leveled.

The season for this work is short. The ground thaws in May (or early June) and snow can be expected by the beginning of September. It makes the most sense to dig in the spring, so that the warm summer weather can result in the maximum possible amount of biodegradation in the first season. With temperatures frequently in the –40°C range, there is no point in trying to keep the soil warm over the winter.

## RESULTS AND DISCUSSION

**Disappearance of Diesel-Range Organics.** At three sites, the residual TPH level in the soil was close to 100 mg/kg within the first 13 months, from starting concentrations of about 4000 mg/kg as shown in Table 1. These three sites had very clayey soils, and the landfarms were not tilled as often as expected (monthly vs weekly). We were surprised by the positive outcomes. Soil samples taken under the moss liners had no detectable TPH (<10 mg/kg).

**TABLE 1. TPH concentrations at three landfarm sites, from the time of construction to closure.**

| | Site 1 | | Site 2 | | Site 3 | |
|---|---|---|---|---|---|---|
| **Date** | Mean | Std. Dev. | Mean | Std. Dev. | Mean | Std. Dev. |
| July 96 | **5731** | 7508 | **2298** | 2033 | **4063** | 3427 |
| Aug. 96 | **812** | 341 | - | - | - | - |
| Sept. 96 | - | - | **200** | 102 | **273** | 109 |
| Oct. 96 | - | - | - | - | **811** | 393 |
| June 97 | **207** | 175 | **1007** | 846 | **820** | 475 |
| July 97 | **779** | 1157 | **1592** | 1171 | **474** | 239 |
| Aug. 97 | **<17** | 11 | **<81** | <124 | **82** | 74 |

A larger number of sites had a very active first year, but they "stalled" at an average concentration between 1000 and 3000 mg/kg TPH for the next two to three years. When stratified sampling was done to distinguish between the top layer of soil (under 20 cm) and the rest of the landfarm (20 – 50 cm), the upper layer was usually found to be much less contaminated (<100 mg/kg) than the deeper soil layers. In those cases where space was available, the top layers were scraped off and spread around the site.

There may be a number of reasons why the top layer of soil reached significantly lower concentrations than the deeper layers, including: better aeration from tilling, warmer temperatures, better drainage/more rainfall, and more fertilizer, which is applied from the top.

Efforts to characterize the remaining material and to re-start the process are discussed in the poster in this Session by Ford, Child, Vinnai and Mortimer.

Due to the lack of tilling implements in these non-agricultural communities, we were limited to rototillers and backhoes for soil aeration. When possible, the local backhoe is brought to the site to "flip' the soil right down to the peat liner, mixing the less-contaminated surface material in with the soil from greater depths. At two sites where this was done last spring, the soil concentrations declined over the summer sufficiently that we can now close these landfarms.

Fourteen-day worm toxicity tests were conducted on these "stalled" soils. All results were negative (in each test soil, all 25 worms survived). Corn, squash, peas and various grasses grew to maturity in the landfarm soil in several test locations.

The air monitoring is described in a poster in this Session, by Edwards, Ausma, Mortimer, Gillespie, Fitzgerald-Hubble, Wong and Halfpenny-Mitchell.

**Environmental/Social Impacts.** Fuel consumption during each phase of the assessment, remediation and monitoring is high, due to the need to fly consultants and equipment to these remote locations. Fuel consumption during excavation and trucking of contaminated soil, and subsequent excavation and trucking of clean fill (and potential excavation and trucking of remediated soil) can be significant, with hundreds of truckloads of soil moving around the small gravel roads of the community. Rototilling uses little fuel, while flipping soil with a backhoe uses somewhat more, and is much more effective at reaching lower soil depths. Fuel costs in the north are double those in the south, due to the expenses of trucking over ice roads, and bulk storage facilities in the communities. Moving and storing this fuel carries a risk of creating new fuel spills.

Air impacts from hydrocarbons are generated by the earth-moving equipment, from the excavation of contaminated soil and from the operation of the landfarm or biopile. These are far in excess of that which would result from following a "do nothing" or *in situ* approach to remediation.

Due to the low solubility of diesel fuel, surface water and groundwater are not at risk from hydrocarbon contamination, although fertilizer run-off from soil treatment facilities can occur. Floatplanes used to transport people to the sites left more fuel on the water than did all the remediation activities combined.

Solid waste issues include disposal of fertilizer bags, oil changes, monitoring wells and sampling residues. It is particularly during site closure that the advantages of landfarming over biopiles become evident. All that is required is to scatter some seeds and ship the rototiller home. Decommissioning a biopile would involve moving all the soil once again (after finding a willing landlord), demolishing the liners, covers, pipes and filters, disposing of spent carbon, and restoring the biopile area. Disposal of waste materials on reserves is not recommended as the "landfills," or dumps, are usually just designated tipping areas, and many are reaching capacity as garbage only flies in, and never flies out.

## CONCLUSIONS

The environmental impacts of site remediation can be significant, especially when the impacts of all the associated activities are counted. Bell Canada acted as a good corporate citizen in taking action to clean up contaminated soil at its remote facilities. The company persevered with a low-impact (relatively speaking), "low-tech" approach, in spite of reservations on the part of the government and pressure from expert advisers to increase level of expenditure of both financial and environmental capital. Landfarming and the use of a rototiller for aeration has proven to be an economical and effective method of diesel fuel contaminated soil remediation, but, perhaps "no-tech" would be the best approach of all.

This is an issue for our society to ponder, along with the scientific community and the site remediation industry, which seems to be overlooking the impacts resulting from its own activities. Conventional approaches must be carefully considered, particularly so in the special circumstances of remote locations.

**ACKNOWLEDGEMENTS**

The author is grateful to Dr. Louise Millette, Bell Canada Environmental Services, for the opportunity to take on this very interesting project, and for her foresight and commitment to conduct the work with cultural and environmental sensitivity. Mr. Richard Nancarrow, currently with the Ontario First Nations Technical Services Corporation, provided invaluable insight and expertise in the early years of the program. Many helpful discussions were had with Mr. Don Stewart, Nexacor Realty Management, Mr. James Ford, Investigative Sciences, Inc., and Dr. Brent Sleep, Civil Engineering, University of Toronto.

**REFERENCES**

Ontario Ministry of the Environment, 1997. "Guideline for Use at Contaminated Sites in Ontario."

Environment Canada, Conservation and Protection, 1993. "Guidelines on the Ex-Situ Bioremediation of Petroleum Hydrocarbon Contaminated Soils on Federal Crown Land."

Reinecke, S., W. P. Mortimer and S. Livingstone, 2001. "Remediation In A Remote Northern Site: A Case Study." In *Proceedings of the Sixth International In Situ and On-Site Bioremediation Symposium, San Diego, California, June 5-7.*

Edwards, G. C., S. Ausma, W. P. Mortimer, T. J. Gillespie, C. R. Fitzgerald-Hubble, E. K. Wong, and L. Halfpenny-Mitchell, 2001. "Measurement Of Trace Gas Emissions During Diesel-Contaminated Soil Bioremediation." In *Proceedings of the Sixth International In Situ and On-Site Bioremediation Symposium, San Diego, California, June 5-7.*

Ford, J., P. Child, L. Vinnai and W. Mortimer, 2001. "Characterization And Treatment of Recalcitrant Diesel Contaminated Biocell Soil From Northern Ontario, Canada." In *Proceedings of the Sixth International In Situ and On-Site Bioremediation Symposium, San Diego, California, June 5-7.*

# REMEDIATION IN A REMOTE NORTHERN SITE: A CASE STUDY

***Stefan Reinecke***, Franz Environmental Inc., Ottawa, Ontario, Canada
Wendy Mortimer, Bell Canada, Toronto, Ontario, Canada
Stephen Livingstone, Franz Environmental Inc., Ottawa, Ontario, Canada

**Abstract:** MacDowell Lake is the site of a small First Nation community in northwestern Ontario, north of the 52nd parallel. The operation of a diesel-powered telephone station next to the village resulted in diesel contamination in both soil and groundwater, including a 700 $m^2$ liquid-phase petroleum hydrocarbon plume. Contaminated surface soils are being treated by landfarming. Groundwater remediation is currently in the free-product recovery stage, and will eventually proceed to a monitored natural attenuation (MoNA) approach. Remediation activities at the site are constrained by several factors including: lack of road accessibility, aircraft payload limits, absence of utilities, local fauna, climate, and soil conditions. Because of these constraints and because of the otherwise pristine nature of the site, remediation strategies must be technologically simple and more environmentally benign than conventional approaches.

## INTRODUCTION

Diesel generators were used to power telephone equipment, including a microwave tower, for the MacDowell Lake band community. The telephone station began operation in 1976 and was decommissioned in the summer of 1995. The generator building and five 4,550 L above-ground storage were situated in a bermed area, but on otherwise unlined soil. The initial assessment focused on delineating surface contamination in the bermed area. Liquid-phase petroleum hydrocarbon (LPH) was discovered in 1997 in a monitoring well that was installed approximately 7 m outside of the bermed area in 1995. The full extent of the LPH plume was determined in 2000. Currently, the site has 91 active monitoring wells, with approximately 30% of the wells reporting the presence of LPH.

A description of the Site and a review of the current remediation strategies are presented.

## SITE DESCRIPTION

MacDowell Lake is situated in the northern boreal forest on the Canadian shield approximately 150 km northeast of Red Lake, Ontario (see Figure 1). As shown in Figure 2, average temperatures range from approximately –20 to 20 °C. MacDowell Lake is generally frozen from the beginning of November until the beginning of May. This period coincides approximately with the time during which conditions are suitable for bioremediation of surface soils (i.e. no snow cover and temperatures > 4 °C).

**Subsurface Characteristics.** The local overburden stratigraphy consists of a thin layer of peat moss or top soil, followed by a fine to medium sand (approximately 0.5 m thick), a clayey silt (approximately 2 m thick), and a silt and fine sand layer grading to coarser sand at depth. While soil investigations have not extended beyond 6 m depth, it is presumed that the silt and fine sand layer is likely underlain by till and bedrock at approximately 10 m below surface. The water table is between 2.5 and 4 m below surface, within the silt and fine sand unit.

The telephone station was located on a local topographic high point. The topography slopes downward towards a swamp on one side and towards MacDowell on the other. Groundwater flows from the site towards the lake.

**FIGURE 1: Site location**

**Diesel Impacts.** Surface soil diesel impacts are confined to the bermed area. Excavation and borehole investigations suggest that diesel fuel migrated downwards primarily below the bermed area.

Groundwater impacts (i.e. TEH detected in groundwater samples) extend to at least 60 m outside of the bermed area. LPH has been observed in approximately 30% of the 91 active monitoring wells at the site and at distances as far as 45 m outside of the bermed area. The size of the LPH-impacted area is approximately 700 $m^2$. The total volume of LPH in the subsurface is estimated to range from 1800 to 5800 L. This volume was estimated using an analytical solution (Charbeneau and Chiang, 1995) based on soil and diesel properties, and observed LPH thicknesses in monitoring wells. Assuming that recoverable LPH represents 20 to 50% of the total volume (USEPA, 1996), between 360 and 2900 L may be recoverable.

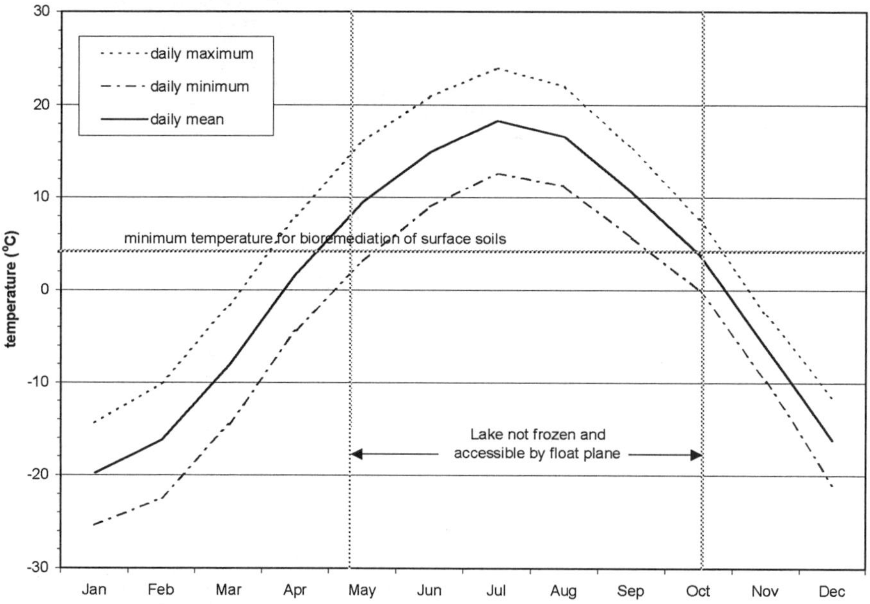


**FIGURE 2: Temperature normals, Red Lake, Ontario**

## REMEDIATION

**Surface Soil.** Contaminated surface soil from the bermed area is being treated by landfarming. Soil was excavated by hand shovel and using a small excavator and placed in cleared areas adjacent to the excavation. Fertilizer (Plant Prod 18-9-27, Plant Products Co. Ltd, Brampton, Ontario, Canada) was added to the soil. Two land farming areas have been constructed: the first in 1996 (and expanded in 1997), measuring 8 m x 13 m; and the second in 1999, measuring 15 m x 20 m. The landfarms are tilled periodically using a gas powered rototiller. The 1999 landfarm, which has a deeper impacted soil layer than the 1996 landfarm, has also been turned over approximately four times using the excavator.

The clay content of the soil in the land farm makes aeration and fertilizer addition difficult, as the soil is either too dry or too wet and forms "clay-balls" when rototilled.

Soil diesel concentrations for the 1996 landfarm are shown in Figure 3. Concentrations dropped rapidly during the first field season, but the rate of mass loss appears to have decreased significantly since that time. Achieving the provincial criteria for surface soils for both potable and non-potable groundwater categories does not appear possible within the next year.

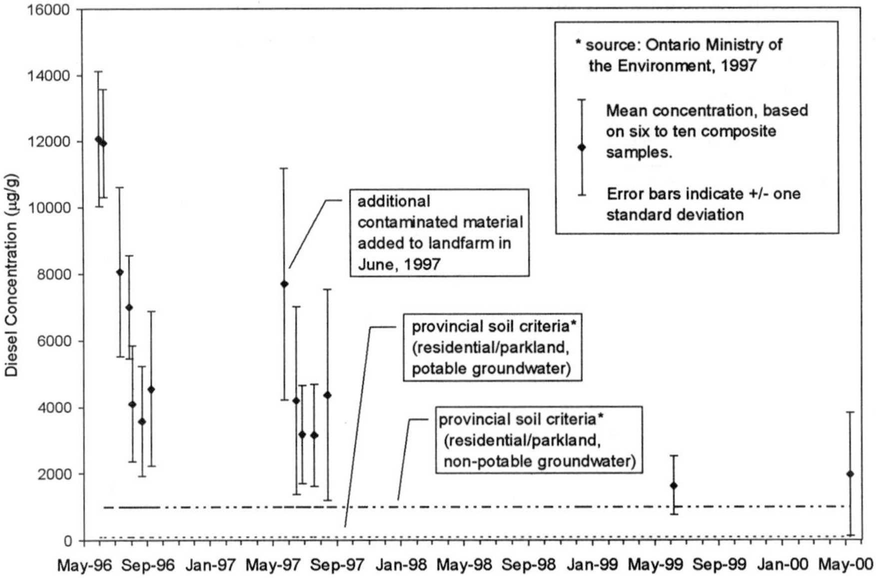


**FIGURE 3: Mean landfarm concentrations, 1996 landfarm**

New risk-based federal standards for petroleum hydrocarbon contaminated soil are now available (Canadian Council of Ministers of the Environment, 2000). All future soil analysis will be conducted in accordance with these standards, which specify limits for four TPH fractions depending on soil type and land-use. The four TPH fractions refer to the following equivalent straight-chain hydrocarbon (nC) boiling point ranges: nC6-nC10, nC11-nC16, nC17-nC34, nC35+. When compared to these standards, soil concentrations at the site may be closer to acceptable levels than when compared with the provincial guidelines shown in Figure 3.

Fertilizer was applied with large volumes of water (~30,000 L) surficially and through injection wells in the bermed area in 1996 over a four month period. Oxygen-releasing compound (ORC) was also placed in injection wells. These in-situ remediation efforts were unsuccessful compared to the performance of the landfarms. This work was done prior to any knowledge of LPH on the water table. It is speculated that nutrient flushing may have mobilized LPH away from the bermed area towards the swamp.

**Groundwater.** Groundwater remediation currently consists of LPH recovery, using three methods: i) passive skimming; ii) pumping LPH from solar-powered automated skimmers (active skimmers); and iii) automated total fluids pumping complete with oil/water separation, and powered by compressed air and a portable diesel generator. The passive and automated skimmers collect LPH through a hydrophobic filter. The filter is 0.3 m long to accommodate small water table fluctuations.

Approximately 425 L of LPH were recovered during the first season of operation (2000). Of this total, 15% was recovered with passive skimmers, 30% with active skimmers, and 55% by total fluids pumping. A large portion of volume recovered by total fluids pumping consisted of LPH that had accumulated in the wells during the winter and was recovered at the beginning of the season. During the rest of the season, the total fluids system experienced numerous failures and could not be successfully automated. The active skimmer system was the most effective system and is now the primary method of LPH recovery.

Following LPH recovery, the remaining groundwater impacts will be addressed using a monitored natural attenuation (MoNA) approach. A preliminary assessment of the natural attenuation potential at the site indicates that biodegradation is occurring in the aquifer. Depleted dissolved oxygen, nitrate, sulfate, and iron III concentrations are associated with groundwater samples having elevated TEH concentrations.

## CONSTRAINTS

The remoteness and climate of the MacDowell lake site places several constraints on remediation activities. Since the site is not serviced by road or airstrip, it is only accessible by float plane. The largest available float plane, a deHavilland Otter, has a payload limit of 900 kg (2000 lbs). Transportation costs are significant, especially for heavy items such as fuel, fertilizer, drilling supplies (e.g. sand and bentonite), and machinery. The payload limit, and the absence of utilities (water and electricity) at the site, severely limits the choice of drilling, excavation, and other remediation equipment.

A detailed discussion of the constraints associated with the decommissioning of other diesel generating stations in northern Ontario is given in the paper in this Session by Mortimer (2001).

**Drilling and Monitoring Well Installation.** Monitoring wells (51 mm ID (2 in)) are installed using a Little Beaver wheel-mounted portable drill rig, equipped with 6 inch hollow stem augers. The drill weighs approximately 315 kg (700 lbs) and can be mobilized through a forest with minimal trail-building effort. Borehole depths are limited to less than 6.5 m (21 ft) due to soil-type and the power limitation of the drill. Because the drill does not allow for in-auger well installations, good-quality well installations were not always possible. While 25 mm ID (1 in) well pipe is easier to install, it is less useful for recovery purposes. To reduce payloads, well sand was obtained from a local beach and from sand deposits found in the forest.

**Excavation.** Initially, contaminated surface soil was excavated with shovels and wheelbarrows. A very small excavator was subsequently brought to the site. Soil was moved using a trailer attached to a four-wheel all-terrain vehicle. While the excavator has made the task less labour-intensive, excavation is very slow compared to what could be achieved using full-scale equipment. Approximately 300 to 400 $m^3$ of soil have been excavated in two field seasons.

**Infrequency of Site Visits.** Site visits are infrequent due to the remoteness of the site. Consequently, the land farm is rototilled monthly, and recovery system failures may go unnoticed for several weeks. Poor weather may rule out tilling and excavation during entire site visits.

**Fauna.** Between site visits, exposed LPH recovery tubing and cables require protection from rodents, caribou and moose. In 2000, three LPH recovery lines were severed by an animal, resulting in the failure of the total fluids pumping system. The problem was addressed by burying tubing and cables, placing them within protective conduit, and covering remaining exposed tubing and cables with tin foil or metallic duct tape.

## DISCUSSION

**Surface Soils.** While diesel concentrations in surface soil still exceed regulatory limits, the landfarm remediation approach has been relatively successful. Contaminant mass has been removed in a technologically simple and non-resource intensive manner. By using landfarms, as opposed to biopiles with aeration pipes and blowers, mechanical breakdowns are avoided and decommissioning is not necessary.

The decrease in the mass loss rate, or "stalling", observed in the landfarms is discussed in detail in the poster in this Session by Ford et al. (2001).

**Groundwater.** The soil conditions at MacDowell lake (clayey and silty soils) reduce the mobility of LPH in the subsurface which inhibits LPH recovery. The automated solar powered skimmers are ideal for this site. Solar power makes the system energy efficient. The pumps inside the skimmers are robust and can be run dry periodically without damage. Fluids handling is minimized, as opposed to total-fluids pumping system which requires oil/water separation.

Unfortunately, site visits are required to adjust skimmer depths in response to water table fluctuations. While the passive skimmers float on the water table and are to some extent self-adjusting, the active skimmers remain at their installed depth.

High water table levels, resulting from heavy precipitation during the summer, significantly reduce LPH recovery rates by trapping LPH within the formation and impeding gravity-assisted LPH flow into recovery wells. Therefore, low water table levels during the winter months may offer the optimum conditions for LPH recovery. However, winter visits are difficult, if not impossible, due to weather conditions and safety concerns associated with landing on the lake with a ski plane.

**Sustainability.** The remediation at MacDowell lake is driven by legal and political considerations associated with a land transfer between two levels of government. The First Nation village has a small number (<10) of seasonal residents who are averse to using groundwater as a source of drinking water.

Diesel exposure levels for human and biological receptors appear to be low, although no risk assessments have been conducted at this time. Remediation activities, however, appear to have a more significant impact on the local environment than the original contamination. The effects on the regional environment resulting from the use of aviation fuel and the manufacture of remediation equipment (e.g. PVC pipe) should also be considered. For these reasons and due to the site constraints discussed above, the use of technologically simple, non-energy intensive remediation methods is especially important.

## REFERENCES

Canadian Council of Ministers of the Environment (CCME). 2000. Canada Wide Standards for Petroleum Hydrocarbons (PHC) in Soil.

Charbeneau, R.J. and C.Y. Chiang. 1995. "Estimation of Free-Hydrocarbon Recovery from Dual-Pump Systems Groundwater", Groundwater. 33(4): 627-634.

Ford, J., P. Child, L. Vinnai and W. P. Mortimer. 2001. "Characterization and Treatment of Recalcitrant Diesel Contaminated Biocell Soil from Northern Ontario, Canada." In *Proceedings of the Sixth International In Situ and On-Site Bioremediation Symposium*, San Diego, CA.

Mortimer, W. P. 2001. "Decommissioning Diesel Generating Stations in Sub-Arctic Canada: A Six-Year Progress Report." In *Proceedings of the Sixth International In Situ and On-Site Bioremediation Symposium*, San Diego, CA.

Ontario Ministry of the Environment. 1997. "Guideline for Use at Contaminated Sites in Ontario."

United States Environmental Protection Agency Office of Underground Storage Tanks OSWER National Risk Management Research Laboratory, ORD. 1996. *How to Effectively Recover Free Product at Leaking Underground Storage Tank Sites, A Guide for State Regulators.*

# *AZOMONAS* IS A NFB CAPABLE TO USE KEROSENE AS A CARBON SOURCE

***J. Pérez-Vargas***; H. M. Poggi-Varaldo; G. Calva-Calva; A. Albores; R. Rodríguez-Vázquez; F. Esparza-García (CINVESTAV-IPN, México D.F., México);
R. Ferrera-Cerrato (Colegio Posgraduados, Montecillos, México)

**ABSTRACT:** A nitrogen-fixing bacterium was isolated from kerosene-contaminated soil. A preliminary taxonomic evaluation suggested that this bacterium belongs to the *Azomonas* genus. This bacterium showed capability to use kerosene as a sole carbon source. When the bacterium was grown in Rennie modified medium added with 5000 mg/L of kerosene, in submerged culture, between 40 to 50% of the hydrocarbon mixture was removed in 13 days. Results from enzymatic analysis for nitrogenase activity strongly suggested that, for this bacterium, a biochemical relationship between the nitrogenase activity and kerosene removal capability may exist. The results from this work, show that this nitrogen fixing bacterium not only withstood high concentrations of kerosene but also could achieve important removal of this pollutant from contaminated soils. These results are promising for bioremediation studies of heavily polluted soils, without the addition of fertilizer.

**Key words:** *Azomonas*, free-living, hydrocarbons, kerosene, nitrogenase, nitrogen fixation.

## INTRODUCTION

Soil contamination produced by undesirable accumulation of hydrocarbons from spills and discharges of the petroleum and petrochemical industries is a common problem in oil producing countries (Walton and Anderson, 1992). Bioremediation can be a convenient and cost-effective alternative as it allows the utilization of autochthonous microflora, eventually acclimated to the pollutants, for site remediation. (Strayer *et al.,* 1983; Shen y Bartha 1996). Microbial consortia play a key role in the biodegradation and eventual removal of hydrocarbons and other pollutants. (Poggi-Varaldo *et al.*, 2000; Rojas-Avelizapa *et al.,* 1999). The use of specialized consortia offers several advantages over pure cultures (Dehorter *et al.,* 1992, Siso *et al.,* 1988), namely, synergism among strains leading to a more effective removal of pollutants and complex substrates. However, for the same reason, the role played by particular strains or microbial groups cannot be easily evaluated (Dehorter *et al.,* 1992). In recent phytoremediation studies using bean plants for the treatment of a soil polluted kerosene, the complex plant-microflora-abiotic factors could remove up to 100% of the hydrocarbons (Hernández-Acosta, 1997). Also, it has been reported that free-living nitrogen-fixing bacteria (NFB) are the most abundant bacteria in that microflora soil. However the role and significance of this bacterial group on hydrocarbon removal and hydrocarbon uptake processes has not been studied yet. In an attempt to understand this role, and to investigate if a relationship between the nitrogen fixation and the kerosene removal capability exists, in a previous report (Pérez-Vargas *et al.*, 2000), a microbial consortium of free-living NFB was isolated from a kerosene contaminated soil. The consortium was able to use kerosene as the sole carbon source and removed up to 75% of an initial concentration of 5000 mg/L in submerged culture assays. In this work we are reporting *i)* further characterization of the microbial consortium and its components, and *ii)* insight on a possible association between the nitrogen fixation and hydrocarbon removal of the NFB.

## METHODS

The microorganisms were from a nitrogen–fixing microbial consortium isolated from a soil contaminated with kerosene (Pérez-Vargas *et al.*, 2000).

Microbial growth, nitrogenase assays, and kerosene removal tests were performed following the procedures described elsewhere (Posgate, 1971; Pérez-Vargas *et al.*, 2000; Pérez-Vargas, 2001; García-Esquivel, 2001). The kerosene concentration was determined by GC as the sum of peak areas in the chromatograms, according to Mondello (1989). The kerosene removal (R in %) was calculated using the equation [1], and subtracting the abiotic control removal (Ramírez-Canseco, 2000).

$$R(\%) = \left[ \frac{(\Sigma A_i - \Sigma A_f) - (\Sigma A_{ic} - \Sigma A_{fc})}{\Sigma A_i} \right] * 100 \qquad [1]$$

Where; i = initial conditions; f = final conditions; c = abiotic control; A = peak area. It is worth emphasizing that equation (1) gives a reasonable approximation to the biological removal of hydrocarbons.

## RESULTS AND DISCUSSION

**Microbiological characteristics.** The microbial consortium CM12K consisted of four types of bacteria (Table 1). Three bacteria cultures were Gram-negative, and the pink culture was, an ovoid-form Gram-variable bacterium, depending on the age of the culture. All strains could grow in complete Rennie medium, which is specific for free-living NFB (Rennie, 1981). When the cultures were grown in the modified Rennie medium (where the mannitol and sucrose were substituted by kerosene as sole carbon source, Pérez-Vargas *et al.*, 2000), the orange culture did not show any growth. The pink culture resulted a free-living NFB, with great motility in the first hours of culturing and a Gram negative stain. However, after the first 36 hours, the Gram stain becomes variable. It was catalase-positive. All these features resemble the *Azomonas* genus characteristics. Thus, the pink bacterium could be tentatively identified as belonging to the *Azomonas* genus.

**Table 1: Main characteristics of the cultures**

| | Cultures | | | | |
|---|---|---|---|---|---|
| Characteristic | CM12K[a] | P[b] | Y[c] | O[d] | W[e] |
| Colony Color | Yellow-white | Pink | Yellow | Orange | White |
| Colony Size (mm) | nd | 2 | 1 | 1 | 3 |
| Shape and borders | nd | Smooth | smooth | Flat | Drop-like |
| Slimy material | +++ | + | + | + | +++ |
| Gram Stain | V | V | - | - | - |
| Motility | V | + | - | Nd | nd |
| Microscopic Morphology | V | Round and ovoid cells | Cocobacilli | small bacilli | cocobacilli |
| Growth M+S | +++ | +++ | +++ | + | +++ |
| Growth Ker | +++ | +++ | ++ | - | +++ |

Notes: [a] microbial consortium; [b] pink culture (*Azomonas*); [c] yellow culture; [d] orange culture; [e] white culture; nd not determined; V variable; M+S Mannitol+Sucrose; Ker kerosene; - negative; + low; ++ moderate; +++ abundant.

**Kerosene removal.** The microbial consortium displayed larger extent of kerosene removal and faster kinetics than the pink pure culture (Figure 1). The kerosene reduction after 13 days of culturing was 75% and 45% for CM12K and pink culture respectively, whereas the kerosene disappearance kinetics for the first 60 hours (fast period) was 33.3 and 16.7 mg Kerosene removed/L-h respectively. This difference could be due to the generic advantage reported for mixed cultures over pure cultures, pointing out a possible synergistic relationships in the consortia (Dehorter *et al.* 1992, Siso *et al.* 1988). Interestingly, there was not any significant difference between the kerosene disappearance kinetics for the second period (slow): 6.9 and 6.2 mg Kerosene removed/L-h for the mixed and pink cultures respectively. Although the pink culture removed less kerosene than consortium, the amount of hydrocarbon removed was very significant (Figure 2). The hydrocarbons profile for the kerosene in the pink culture (Figure 2), showed that this bacterium was able to remove the hydrocarbons in the whole range of the kerosene components, but few peaks of lower molecular weight than the original compounds were produced (Figure 2B). These peaks were also observed in the consortium cultures grown in kerosene (Pérez-Vargas *et al.*, 2000).

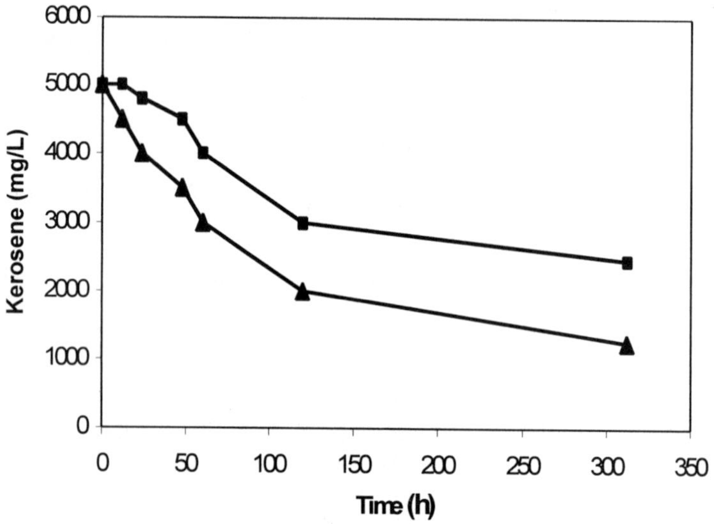


**FIGURE 1: Time course of kerosene removal by submerged cultures of the microbial consortium CM12K (▲), and the Pink culture (*Azomona*) (■).**

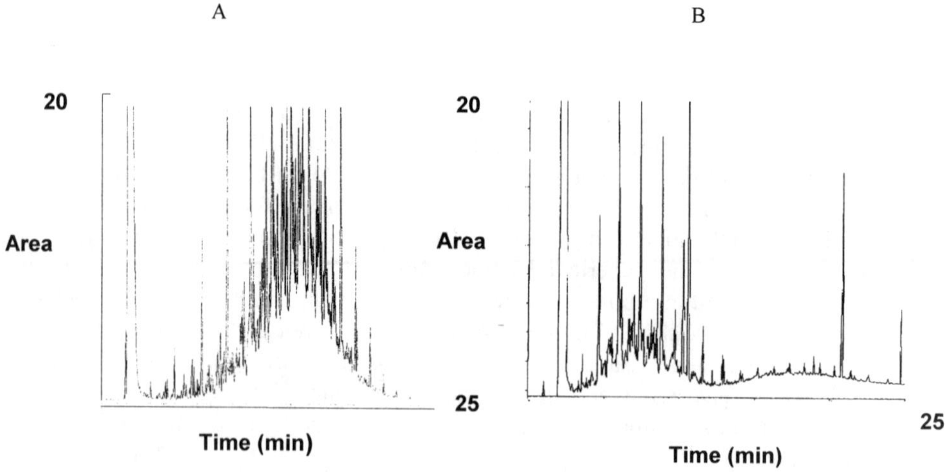


**FIGURE 2: Typical chromatograms of kerosene hydrocarbons for the pink culture grown in presence of 5000 mg/L kerosene at time zero (A); and after 13 days of incubation (B).**

**Nitrogen fixation and hydrocarbon removal.** The nitrogen fixation capability in relation with the hydrocarbon removal was studied by evaluating the nitrogenase activity and the kerosene removal by each culture used in this work. The results for the pink culture after 312 h of incubation (Figure 3), shown an average high value of nitrogenase activity in the fist period of incubation (the first 60 h) with a value of 80 nMol/mg-h. In this period them the nitrogenase activity increased rapidly for first 30 hours reaching a peak value of 160 nMol/mg-h; afterwards, the activity decreased suddenly for the subsequent 30 hours. In the first period of 60 h, the average rate kerosene removal was also maximum, reaching a value of 16.7 mg kerosene removed/L-h. In the second period (60 to 312 h), both the nitrogenase activity and kerosene removal rate were significantly lower: below 10 nMol $N_2$/mg-h and 6.2 mg kerosene removed/L-h respectively.

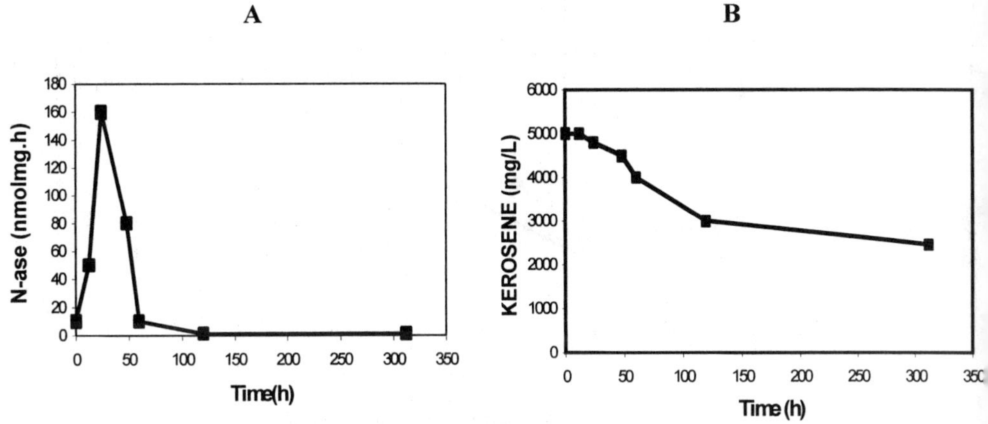


**FIGURE 3 : Time course of nitrogenase activity (A), and Kerosene removal (B) for the pink culture.**

Further experiments, showed a similar association between the overall kerosene removal and the peak nitrogenase activity for all the cultures used in this work (Figure 4). The pure cultures, isolated from the consortium, showed that those with the lower nitrogenase activity growing in presence of kerosene also had the lowest kerosene removal capability (orange and yellow cultures). This general characteristic of these cultures, strongly suggests the existence of a biochemical association between the nitrogen fixation and hydrocarbon removal capability.

It is known that nitrogen fixation is an endergonic process, where 12 moles of ATP are needed for the fixation of 1 mole of $N_2$ (Brock and Madigan, 1991). Therefore, a plausible explanation for the above results could be that the high rate of hydrocarbon uptake would provide the energy required for the nitrogen fixation. Thus, it is logical that after the rate of nitrogen fixation declined, the hydrocarbon uptake rates were decreasing, or viceversa. These results point out that a relationship between the nitrogen-fixing activity and the kerosene removal should exist as suggested by the Equation 2 and the results shown in Figure 4.

$$R(\%)=(0.784 \pm 0.152) * (N\text{-ase}) + (8.24 \pm 7.34);\ r^2= 0.9301 \qquad [2]$$

Where R (%) = kerosene removal, %; N-ase = nitrogenase activity, nmol/mg.h; the numbers following the symbol ± in the parentheses stand for the standard errors of each coefficient.

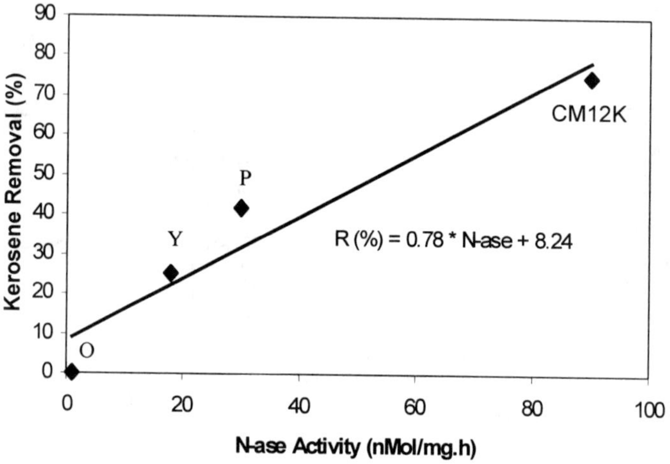


**FIGURE 4: Correlation between the nitrogenase activity and kerosene removal for several nitrogen-fixing cultures: CM12K (nitrogen-fixing consortium); P (pink culture); O (orange culture); Y (yellow culture).**

## CONCLUSIONS

- Four bacteria were isolated from a nitrogen-fixing consortium, which was capable of using kerosene as the a carbon source. The strains were also capable of fixing dinitrogen and utilizing kerosene. However, their nitrogenase activity and kerosene removal capability were inferior to those from the consortium (258 nmol/mg-h and 75% removal for the consortium, and 10 to 168 nmol/mg-h and 0 to 45% removals for the pure cultures).

- One of the isolated bacteria was tentatively identified as belonging to the *Azomonas* genus. This strain was able to remove kerosene up to 45% and use it as a sole carbon source.

- For both the consortium and the *Azomonas* cultures, it was found that kerosene disappearance was maximum during the first period of incubation (60 h), when nitrogenase activity was also maximum.

- Pooling the data from the microbial consortium and from the isolated strains, there was a positive correlation between kerosene removal and the nitrogenase activity [R (%) = 0.784 * N-ase + 7.34; $r^2 = 0.9301$]. However, the underlying mechanistic relationship remains to be unveiled.

This work has shown that free-living nitrogen fixing bacteria not only withstand high concentrations of kerosene, but also could remove important amounts of these pollutants from contaminated soil. In addition, these results are promising for bioremediation of heavily polluted soils.

**ACKNOWLEDGEMENTS**
Financial support of the IMP (FIES project, and CONACYT (graduate scholarship) is gratefully acknowledged. The authors wish to thank the excellent technical assistance of Professors E. Ríos-Leal and J. Martínez of CINVESTAV-IPN.

**REFERENCES**

Anderson T.A., Guthrie E.A., and Walton B.T. (1993). Bioremediation in the rhizosphere. *Environ. Sci. Technol.* **27** (13): 2630-2636.

Brock, T.D., and Madigan, M.T. (1991). *Biology of Microorganisms.* 6th. edn., Prentice-Hall Inc., Englewood Cliffs, N.J., USA.

Dehorter, B., Kontchou, C.Y., and Blondeau, R. (1992). C NMR spectroscopic analysis of soil humic acids recovered after incubation with some white rot fungi and actinomicetes. *Soil Biol. Biochem.* **24** (7): 667-673.

García-Esquivel, G. (2001). Características fisiológicas y bioquímicas de la biodegradación de queroseno y fijación de nitrógeno por un cosorcio aislado de la rizosfera. M.Sc. Tesis Dept. Biotecnología y Bioingeniería, CINVESTAV-IPN, México D.F., México. In Spanish. Extended Abstract in English.

Hernández-Acosta, E. (1997). Influencia de una mezcla compleja de hidrocarburos sobre las poblaciones rizosféricas y el crecimiento de plantas de frijol var. Michoacán 12-A3. Tesisde maestría. Colegio de Postgraduados, Montecillo, México.

Mondello, F. J. (1989). Cloning ad expression in *Escherichia coli* of *Pseudomonas* strain LB400 genes encoding polychlorinated biphenyl degradation. *J. Bacteriol.* **171**(3):1725-1732.

Pérez-Vargas J. (2001). Degradación microbiana de queroseno. Tesis Doctoral. Dept. Biotecnología y Bioingeniería, CINVESTAV-IPN, México, D.F., México.

Pérez-Vargas, J., Poggi-Varaldo H. M., Calva-C. G., Ríos-L. E., Rodríguez-V. R., Ferrera-C. R., Esparza-G. F. (2000). Nitrogen fixing bacteria capable of utilising kerosenen hydrocarbons as a sole carbon source. *Water Sci.Technol.*, **42** (5-6): 407-410.

Poggi-Varaldo, H. M., Ríos-Leal, E., Hernández-Vera, R., Caffarel-Méndez, S., Rinderknecht-Seijas, N. (2000). Activated Soil Bioreactors: an effective alternative for the remediation of heavy soils. *Proceedings 3rd Congress of the Mexican Society for Soil and Aquifer Remediation (SORESCA).* Puebla, Pue., November 2000. In Spanish. Abstract in English. CD-ROM.

Posgate, J.R.(1971). The acetylene reduction test for nitrogen fixation. In: Methods in Microbiology" J.R. Norris & D.W. Ribbons. Eds. Academic Oress. Vol. 6B, 343-357.

Ramírez-Canseco, S. (2000). *Post-treatment of anaerobically-treated Kraft effluents in an aerobic fluidized bed reactor with immobilized ligninolitic fungi.* M.Sci. Thesis. Dept. Biotechnology and Bioengineering, CINVESTAV-IPN. México DF, México. In Spanish. Extended Abstract in English.

Rennie, R.J. (1981). A single medium for the isolation of acetylene-reducing (dinitrogen-fixing) bacteria from soils. *Can. J. Microbiol.* **27** (1): 8-14.

Rojas-Avelizapa, N., Rodrìguez-Vazquez R., Martìnez-Cruz J., Esparza-Garcìa F., Montes de Oca-Garcìa A., Rìos LealE., Fernández Villagomez G.F., (1999). Isolation and characterization of bacteria degfrading polychlorinated biphenyls from transformer oil. *Folia Microbiol.* **44**(3):317-321.

Shen, Y.; and R. Bartha (1996). Metabolic efficiency and turnover of soil microbial communities in biodegradation tests. *Appl. Environ. Microbiol.* **62** (7): 2411-2415.

Siso, M.I.G., Jurado, J.M., Franco, J.M., Miron, J., and González, M.P. (1988). Microgungus-yeast mixed cultures in the degradation of maylaceous wastes I: Interactions affecting amylolytic activity. *Biotech. Letters* **10**(6): 431-436.

Strayer R.F., Edwards N.T., Walton B.T. (1983). Potential terrestrial fate and effects on soil biota of a coal liquification product spill. *Environ. Toxicol Chemistry.* **2** (2): 217-224.

Walton , B.T., and Anderson T.A. (1992). Plant-microbe treatment systems for toxic waste. *Current Opinion Biotechnol.* **3**:267-270.

Yee, D. C., Maynard J. A., and Wood T. K. (1998). Rhizoremediation of tricloroethylene by a recombinant, root-colonizing *Pseudomonas fluorescens* strain expressing toluene ortho-monooxygenase constitutively. *Appl. Environ. Microbiol.* **64** (1): 112-118.

# REMEDIATION OF A FORMER TANKS AREA THROUGH THE BIOPILE TECHNIQUE

P. Carrera*, ***P. Mocciaro****, M.Gavinelli*, J. Hoffmann** and T. Niemeyer**
*Ambiente Spa ENI Group, S. Donato Milanese, Italy
**Hochtief Umwelt Gmbh, Essen, Germany

**ABSTRACT:** In the context of a project to requalify dismissed industrial sites in the Venice area, an extensive site assessment was carried out in 1998 on an 11-hectare area formerly occupied by an oil tank farm.

The characterization of the area included a detailed assessment of hydrogeological, chemical-physical, and microbiological characteristics of soil and groundwater. Together with field activities, a laboratory study was conducted to verify the possibility of utilizing bioremediation techniques to restore hydrocarbon-contaminated soils. Due to the peculiar geology of the site, and the encouraging results of the biodegradation simulation tests, the selected technique was the use of biopiles for on-site treatment of contaminated soils. Design and management of a pilot-scale biopile completed the data required for the design of the full-scale remediation of the area.

In the future, a total of 54,000 tons of soil with Total Petroleum Hydrocarbons (TPH) concentrations as high as 12,000 mg/kg (average TPH concentration 3,000 mg/kg) will be biotreated in batches of four biopiles for a total of nine, having the capacity of 6,000 tons each and dimensions of 110 m x 20 m.

The enhancement of microbial degradation is obtained by oxygenating the soil, previously pretreated and mixed with perlite and nutrients.

The process is monitored over time by evaluating different chemical and physical parameters such as temperature, oxygen and water content, nutrient consumption, contaminant degradation, and microbial flora.

Up to now, the soil present in the first batch of four biopiles has been completely treated. Of particular interest is the degradation trend of the PAH fraction together with the disappearance of different heavy and light hydrocarbon fractions, according to the Italian regulatory standards.

## INTRODUCTION

The re-use of former industrial areas has become a major goal in the policy of most industrialized countries. Huge sites often are located in valuable city areas, and have to be decontaminated and redeveloped for other purposes (parking lots, leizure and cultural centers, large commercial areas).

The heavy industrialized area of Venice lagoon in Italy offers a wide spectrum of such examples. In the region, the problem is even more important due to the need of new infrastructures for touristic development.

This paper presents the remediation project of a former tanks area abandoned since 1988 that has been occupied by more than 60 tanks containing up to 25,000 $m^3$ of petroleum derivatives. In accordance with Italian legislation, the area has been fully investigated both for soil and groundwater. Conventional drilling operations together with soil gas survey and test pits have been used to characterize the site.

The site investigation results showed diffused hydrocarbon contamination ranging from 1,000 to 12,000 mg/kg of dry soil. Gas-chromatographic (GC) data showed the presence of diesel oil and fuel oil as major contaminants of concern. A volumetric extrapolation of the pollution indicated the presence of more than 55,000 tons of contaminated soils.

The hydrogeology of the site according to its geographical location, showed a perched aquifer at 2.5 m from the surface almost in equilibrium with the closed lagoon. The result was poor water quality because of a high concentration of salts. The hydrocarbon presence (concentrations lower than 1 mg/L) was therefore of minor concern.

Together with the chemical characterization of the site, a detailed microbiological study was performed, including microbial counts, end-point tests and bench scale biodegradation trials. The encouraging lab biodegradation data allowed the construction of a pilot biopile of 15 $m^3$ of soil with an average contamination of 6,000 mg/kg. The correlations between hydrocarbon degradation and microbial populations in full scale biopile are not discussed in the present paper.

The soil, amended and fertilized, was aerated to enhance microbial degradation. An automatic monitoring device was installed to follow the most important parameters affecting the process. In particular, direct measurement of pH, redox, humidity and temperature was possible, together with the collection of the off-gas continuously analyzed for oxygen, carbon dioxide, and volatile organic carbons (VOC). The pilot test was managed for 6 months and resulted in a 90% degradation of Total Petroleum Hydrocarbons (THP).

All the information collected allowed the set up of a basic process design based on the biopile technique to remediate the entire tank area. In particular, two sets of four biopiles were to be constructed, each with a capacity of 6,000 tons of soil. A time schedule of 36 months included the preliminary dismantling of the former oil tanks, the demolition of concrete containment tanks, an additional monitoring campaign of soil under the tanks, and the biopile remediation of contaminated soil.

The project was approved by the local authority who fixed the remediation target to 500 mg/kg of Total Petroleum Hydrocarbons (THP), 50 mg/kg of BTEX, and 60 mg/kg of total PAHs (10 mg/kg of more toxics).

## PROCESS DESCRIPTION

The full scale project included the following main activities: area preparation, soil pretreatment, biopile treatment and monitoring.

**Area preparation**. The site covered an extension of 11 ha, of which 2.2 ha were contaminated. An area of 16,800 $m^2$ was identified as "treatment area" for the biopile locations, pretreatment activities and soil contaminated storage area (Figure 1). According to local authority requirements, the whole "treatment area" was sealed with two different layers of asphalt, indicated as the base and binder layer. In addition to these layers, the four areas designated for the biopile were treated with an waterproofing polymer blend, and bounded by asphalt curbs. Differently, the pretreatment area and the buffer storage area were sealed with a concrete layer and surrounded by concrete boundary-walls. The above-mentioned areas have been graded with a fitting slope, collecting drain and underground pipes for rainfall and leachate drainage, the latter inside the biopile area. The biopile leachates were subsequently collected in leachate shafts, placed at the end of the biopiles.

**Soil pretreatment.** Based on the level of contamination, the nutrient content in the original soil was not enough to support the microbial activity. Hence, a fertilizer with 20% nitrogen and 20% phosphorus has been added to the soil at a concentration of almost 0.1% w/w. The presence of fine grain size in the soil, silt and clay fraction, required the addition of a bulking agent to ensure a proper soil particle porosity. The selected bulking agent was perlite, a mineral expanded in grain shape. The main advantages of such compound were: a) no secondary microbial degradation occurred (as in the case of organic material, such as wood chips or peat); b) small soil volume increase, facilitating the refilling of excavated areas. The perlite amount added varied from 2% to 7%, depending on the soil grain size from sandy to silty.

The soil pretreatment was carried out using the following equipment: 1) a star screen unit, which consisted of several rows of tooth marked stars (spaced 3 cm), disposed in parallel and hydraulically engineered; 2) two perlite silos; 3) a dosing unit, composed by a vibrating pipe connected to the perlite silo and the fertilizer tank; 4) a fertilizer tank.

The pretreatment activity, including soil excavation, processed the soil matrix to make it homogeneous, more porous and less aggregate.

**Soil Treatment.** The biopiles were set up on a 20 cm base gravel layer, where a drainage piping for the biopile leachate collection and several branches of slotted pipes for the biopile aeration were fitted. Each biopile was constructed and managed in four sections. Dedicated pipes connect each section to a manifold, connected to the aeration system, including a water knockout vessel, blower, valves, control panel and biofilter (for outlet gas purification). The base layer is covered with a geotextile film to protect the underlying piping. The four biopile dimensions were 110 m length x 20 m width x 2 m height. The pretreated soil was placed in a trapezoid shape, with a 2% slope in width and a 1% slope in length on the top, for discharge of the rainfall water. Then, the pretreated soil was covered with a geotextile layer, on the top of which dripper line branches were arranged. This system was connected to a manifold and then to an irrigation system constituted by water tank, irrigation pump, valves, flowmeter, and control panel. The biopile constructions was completed by covering them with waterproof breathing sheets, of a semi permeable membrane.

**Monitoring.** Each section of the biopiles was monitored using the following equipment: 1) an oxygen meter device, 2) a humidity meter device, 3) a

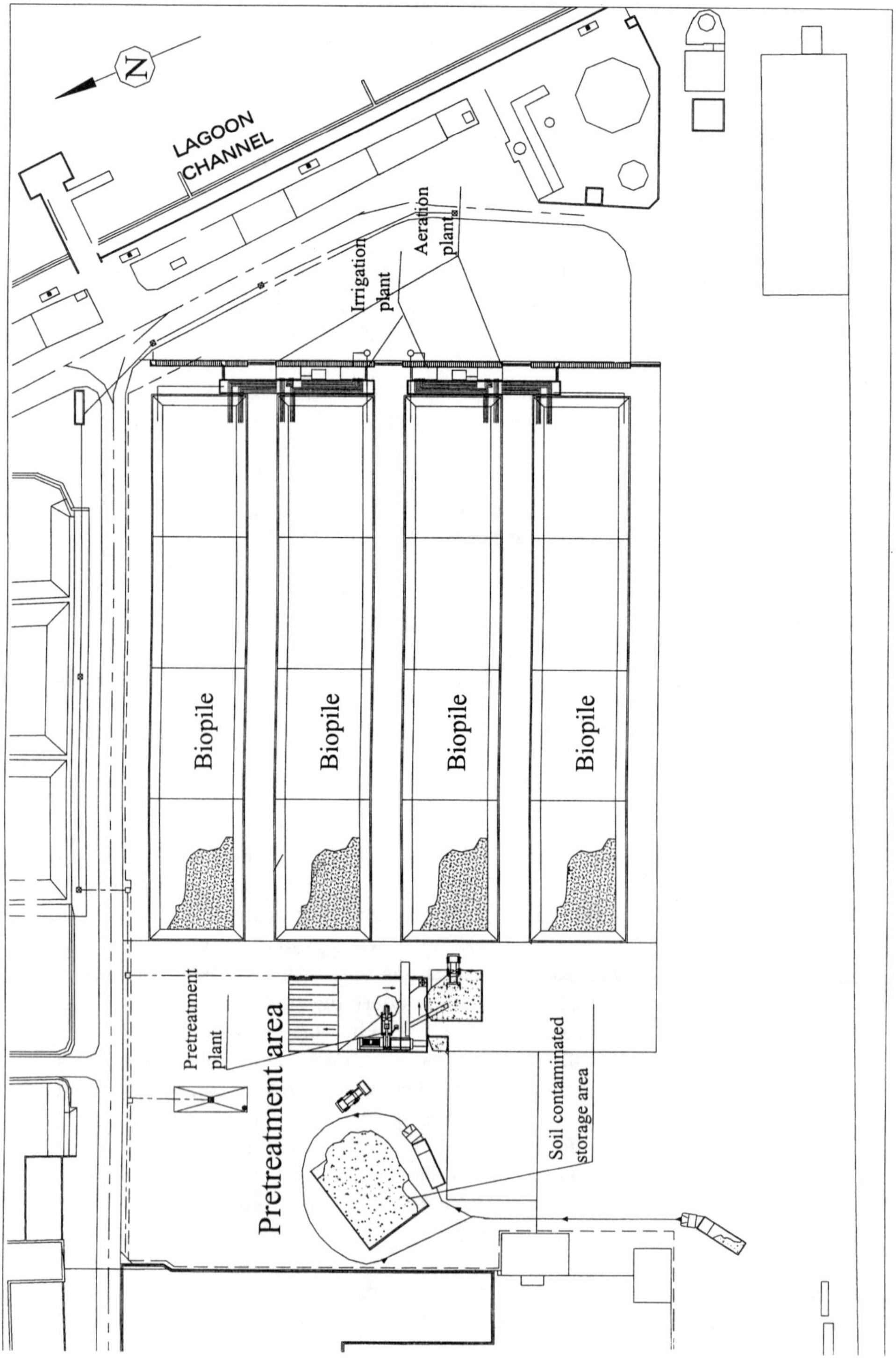


Figure 1. Site aeral view

thermocouple for field temperature measurements, 4) three steel hollow tubes for soil gas measurement, positioned at three different depths at 0.5 m, 1.0 m and 1.5 m.

The oxygen and humidity meter devices operated remotely and were connected to the control panel of the aeration and irrigation system. These systems regulated the valves of air and water flow according to the oxygen and humidity soil content.

In each biopile, 19 sampling points were arranged corresponding approximately to 200 $m^3$ of stockpiled soil. At regular intervals, from each sampling point, three soil samples were collected at 0.5 m, 1.0 m and 1.5 m. The three samples were mixed to create a composite representative sample.

Monitoring included monthly soil sampling and respirometry testing, and daily external and internal temperature measurements. The chemical analyses executed on every soil sample were: TPH, BTEX, PAHs, humidity and nutrients.

## RESULTS AND DISCUSSION

The first biopile batch (24,000 tons of soil) was treated for 6 months. After treatment, the first batch was dismantled and the soil re-used in the same area. The preparation of the second batch is scheduled for April 2001. The following results pertain to the first biopile batch.

**Hydrocarbon Degradation.** Among the contaminants of concern, due to the age of the contamination found in the site, BTEX never presented a problem for the soil quality. In contrast, the concentration and degradation of PAHs and TPH followed in detail.

Even though, according to the project quality standards, the PAH degradations was monitored over time, a low initial PAH concentration was revealed, in some biopile close to the target limits. The monitored PAH are the 16th terms of the EPA 610 set.

The PAH amount in the biopiles at the start-up was estimated by analyses under 100 mg/kg. The most representative PAH were Fluoranthene, Pyrene, Benzo(b)fluoranthene and Indeno(1,2,3-cd)pyrene. After 4 months, the measured level of every single PAH term was under the detectable limit (0.05 mg/kg).

In two biopiles of the first batch, TPH were analyzed by infrared spectroscopy (IR) and, in the same samples, by GC for specific hydrocarbon content: distribution of hydrocarbon fractions, from C6 to C12, from C12 to C24, from C24 to C32. Table 1 shows the hydrocarbon distribution of these two biopiles, shared in three different hydrocarbon classes, at the start-up, T0, and at the end of the treatment after 6 months, T6. The following data represent the relative percentage variation of each hydrocarbon classes.

**Table 1. Hydrocarbon composition vs. time.**

| Biopile | C6 to C12 $T_0$ | C6 to C12 $T_6$ | C12 to C24 $T_0$ | C12 to C24 $T_6$ | C24 to C32 $T_0$ | C24 to C32 $T_6$ |
|---|---|---|---|---|---|---|
| 3 | 22.0% | 13.0% | 75.8% | 55.6% | 2.2% | 31.5% |
| 4 | 38.5% | 4.3% | 54.4% | 50.5% | 7.0% | 45.2% |

The most considerable hydrocarbon class is the intermediate one from C12 to C24, all over the treatment period, even though an increase of the heavier species jointed to the decrease of the other two, in percentage terms is noticed.

Figure 2 shows TPH decrease over time in a sample biopile. The hydrocarbon removal reached up to 87% percent from an initial average concentration of 1800 mg/kg over 6 months.

In the following figures, time points, settled in scale, correspond to sampling or measuring events.

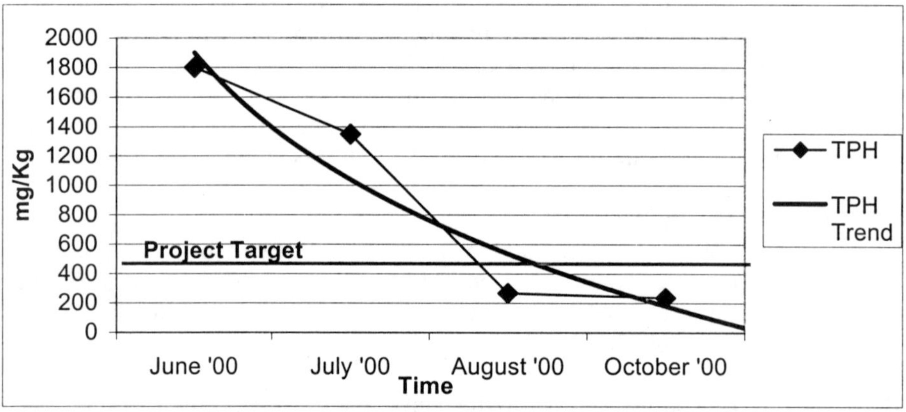


**Figure 2. TPH abatement vs. time (Biopile 3).**

**Respirometry and degradation rates.** The respirometry tests clearly indicate a correlation between hydrocarbon degradation and the microbial activity measured through the consumption of oxygen and carbon dioxide production (Figure 3). It is interesting to notice that the respiration rate decreased over time according to the loss of organic carbon to be degraded.

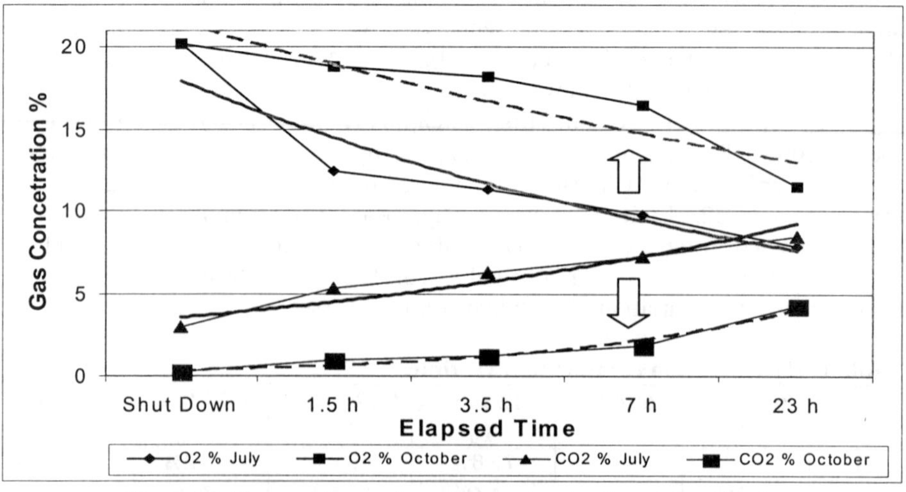


**Figure 3. Comparison of respirometry testing (Biopile 3).**

In order to evaluate the hydrocarbon degradation rate vs. time, Figure 4 shows the rate calculated for Biopile 3. The degradation velocity decreased with the amount of degraded hydrocarbons.

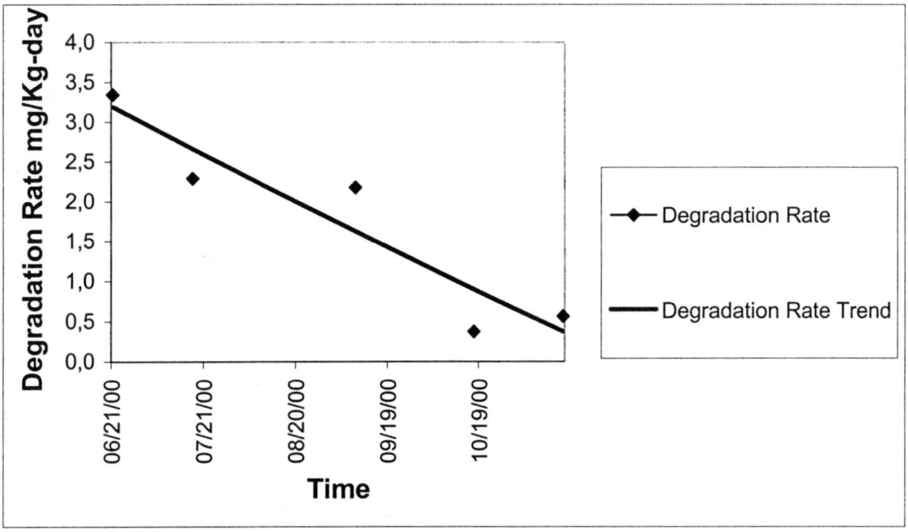


**Figure 4. Degradation Rate Trend (calculated on Biopile 3)**

**Temperature measurements.** The ambient temperature ranged from 10 to 27°C from May to October. Temperatures in the body of the biopiles ranged from 15 to 37.4°C, indicated temperature increase of 5 to 10 °C (with a spike of 12°C after 1 month from the start-up), due the intense microbial activity.

## REFERENCES

Von Fahnestock F.M., Wickramanayake G.B., Kratze R.J., Major W.R. 1998. *Biopile design, operation and maintenance handbook for treating hydrocarbon contaminatd soils.* Battelle Press.

Vinciguerra G., Gavinelli M., Miori M. and Carrera P. 1999. *Assessment and remediation strategy of former MGP sites in Italy.* In situ and on site remediation symposium. San Diego (CA) April, 19-22.

Tamburini D., Porta A., Ambert J. and Di Luise G. 1997. *Evaluation of optimum hydrocarbon degradation conditions: a biotreatability study.* In situ and on-site bioremediation Vol. 1 Battelle Press.

# ENHANCING HYDROCARBON BIODEGRADATION RATE ON LANDFARMING USING EXOGENOUS MICROORGANISMS: A FIELD STUDY

***Claudio Belloso*** (Centro de Investigación y Estudios Superiores, Facultad Católica de Química e Ingeniería, Rosario, Argentina)

**ABSTRACT:** A field study was conducted at a landfarming site belonging to the Petroleum Refinery San Lorenzo located in San Lorenzo City, Santa Fe Province, Argentina, to evaluate the effectiveness of added exogenous microorganisms. Two existing cells were used for the study (Cell 1 and Cell 2). The soil was characterized as sandy loam (texture: 56% sand, 18% silt, 26% clay), the mean soil porosity was 0.38, and the mean bulk density was 1,270 kg/m$^3$. Initial total hydrocarbon concentration in the soil averaged approximately 6% w/w, including saturates 51%, aromatics 12%, resins 12%, and asphaltenes 25%. In both cells, heavy metals were present: 23,000 mg/kg Fe; 90 mg/kg V; 348 mg/kg Ni; 44 mg/kg Pb; 55 mg/kg Cu; and 1.3 mg/kg Hg. Exogenous microorganisms (hydrocarbon degraders) were added (at the beginning of the study) to compare degradation rates in Cell 2 with those in Cell 1 in which indigenous microorganisms were allowed to develop. The exogenous microorganisms were isolated from the same landfarming area and were cultivated in the laboratory before inoculation. Landfarming operations began in September 1997 and concluded in August 1998. The operations in both cells included tilling and irrigation. Moisture content, pH, total hydrocarbons, total aerobic heterotrophic bacteria, and hydrocarbon-degrading microorganisms in each cell were determined monthly during the study. No fertilizers were applied in this study. Cell 2 (which contained exogenous microorganisms) achieved a higher percentage of total hydrocarbon removal than Cell 1. At the end of the study, the removal efficiency was 84% for Cell 2, but only 49% for Cell 1. The results of the field study showed that bioaugmentation was effective, in spite of the presence of high concentrations of heavy metals and the fact that fertilizers were not used.

## INTRODUCTION

Landfarming, also known as land treatment or land application, is an above-ground remediation technology for soils that reduces concentrations of petroleum constituents through biodegradation. This technology usually involves spreading excavated contaminated soils in a thin layer on the ground surface and stimulating aerobic microbial activity within the soils through aeration and/or the addition of minerals, nutrients, and moisture. The enhanced microbial activity results in degradation of adsorbed petroleum product constituents through microbial respiration. Landfarming has been proven effective in reducing concentrations of nearly all the constituents of petroleum products typically found at underground storage tank (UST) sites (EPA 510-B-94-003).

The economic situation in Argentina results in many industries with serious economic problems associated with their waste disposal. Nowadays, the treatment of wastes by incineration costs 100 dollars/ton. For this reason, petrochemical and petroleum refineries often consider landfarming to treat their wastes. Fifty years ago, these industries began treating their solid wastes by simply spreading them over the soil. Soils with hydrocarbon concentrations between 20 to 80% w/w were commonly spread over the soil. Over the years, these soils have lost considerable capacity to degrade hydrocarbons, and the total bacteria concentration has decreased over time to 1,000 CFU/g dry soil. We are trying to recover these soils by bioaugmentation and periodically monitoring the physical and chemical parameters of the soil.

**Objective:** The objective of this field study was to compare degradation rates in two landfarming cells when exogenous microorganisms were added.

## MATERIALS AND METHODS

This field study was conducted in two existing landfarming cells (Cell 1 and Cell 2) belonging to Petroleum Refinery San Lorenzo located in San Lorenzo City, Santa Fe Province, Argentina. Each cell was 80 m long by 40 m wide.

**Initial Conditions.** Both cells contained soil which was characterized as sandy loam (texture: 56% sand, 18% silt, 26% clay), the mean soil porosity was 0.38, and the mean bulk density was 1,270 kg/m$^3$, as measured on-site at the start of the study. Initial total hydrocarbon concentrations in the soil averaged approximately 6% w/w (expressed as ethylic ether extract). The hydrocarbon distribution was: saturates 51%, aromatics 12%, resins 12%, and asphaltenes 25%. The initial total organic carbon (TOC) of the landfarming soil was 2.1% w/w. The initial ammonia-nitrogen concentration in both cells was 120 mg/kg and the inorganic phosphorus concentration was 45 mg/kg. In both cells, heavy metals were present at high concentrations: 23,000 mg/kg Fe; 90 mg/kg V; 348 mg/kg Ni; 44 mg/kg Pb; 55 mg/kg Cu; and 1.3 mg/kg Hg.

Exogenous microorganisms (hydrocarbon degraders) were added (at the beginning of the study) to compare degradation rates in Cell 2 with those in Cell 1 in which indigenous microorganisms were allowed to develop.

Details of the field system setup, system design conditions (such as nutrient and soil moisture requirements), operation and monitoring details (such as tilling and irrigation frequency), and water management were predetermined in a work plan. A sampling plan was predefined to collect samples from the landfarming cells, the soil parameters that were to be measured, and the analytical methods to were to be used to evaluate the samples.

**Summary of Activities.** Landfarming operations and field activities began in September 1997 and concluded in August 1998. Cell 1 was periodically irrigated and tilled only, with no biostimulating agents added. Cell 2 was supplemented with exogenous microorganisms at the beginning of the study. Based on a preliminary study (Belloso, 1999), it was decided to add 1 L of liquid microbial culture ($1.8 \times 10^{13}$ CFU/mL) per square meter. Microorganisms were added to Cell 2 using a spreader to ensure uniform distribution. Exogenous microorganisms (hydrocarbon degraders) were isolated from the same landfarming area and were cultivated in a liquid mineral medium (Table 1) with landfarming soil ethereal extract as sole carbon source.

**TABLE 1. Liquid mineral medium composition.**

| | |
|---|---|
| NaCl | 5.0 g |
| $MgSO_4.7H_2O$ | 0.2 g |
| $KNO_3$ | 3.0 g |
| $K_2HPO_4$ anhydrous | 1.0 g |
| $(NH_4)_2SO_4$ | 1.0 g |
| $(NH_4)H_2PO_4$ | 1.0 g |
| Distilled water | 1000 mL |

Each cell was tilled twice a week to aerate and mix the soil. Moisture added to the cells came from rainfall and an irrigation system. The moisture was 15 to 20% during the study. Moisture content and pH were measured once a week in the field on the day before tilling and irrigation were scheduled. The average temperature was between 10-18°C in the winter and 20-30°C in the summer.

**Sampling Methods.** A two-dimensional grid was established for each test cell to create discrete sampling locations for performance monitoring. Intervals between grid nodes were equal in length and width (10 m). Internal grid nodes for each cell were individually numbered, and eight nodal points were selected for sampling. These eight samples collected per sampling event were blended to yield a single sample to be analyzed.

**Analysis.** At the beginning of the study and every 30 days the following parameters in each soil were determined: total hydrocarbons, total aerobic heterotrophic bacteria (TAHB), and hydrocarbon-degrading microorganisms (HDM).

The soil pH was measured in suspensions of soil (40% w/v) in $CaCl_2$ 0.01M solution. Moisture in soils samples was determined by drying to 105°C until constant weight. The hydrocarbon content was determined by extraction in Soxhlet with Ethylic Ether for 6 h. TAHB in soils were enumerated by the pour plate method (Clesceri et al., 1992) using Merck plate count agar. All plates were incubated at 30°C for 24 hours. HDM in soils were enumerated by the most probable number (MPN) method (Clesceri et al., 1992) using Bushnell-Hass mineral broth (Bushnell and Hass, 1941) with n-hexadecane as the sole carbon source. A five-tube MPN technique was used. Three sets of five screw-cap tubes, each containing 5 mL of Bushnell-Hass mineral broth were inoculated with 1, 0.1, and 0.01 mL of a sample dilution. After inoculation, 50 μL of UV-sterilized n-hexadecane (Merck) and 100 μL of autoclaved resazurin solution

(Merck, 50 mg/L) were added to each tube. After incubation at 30°C for 7 days, tubes giving positive reactions were counted. The color of positive tubes changed from blue to pink.

**RESULTS AND DISCUSSION**

During the 360-day study, the moisture varied between 15% and 20%. The pH varied between 6.3 and 7.5.

The field study showed that Cell 2 (which contained exogenous microorganisms) achieved a higher percentage of total hydrocarbon removal than Cell 1 (Figure 1). At the end of the study, the removal efficiency was 84% for Cell 2, but only 49% for Cell 1.

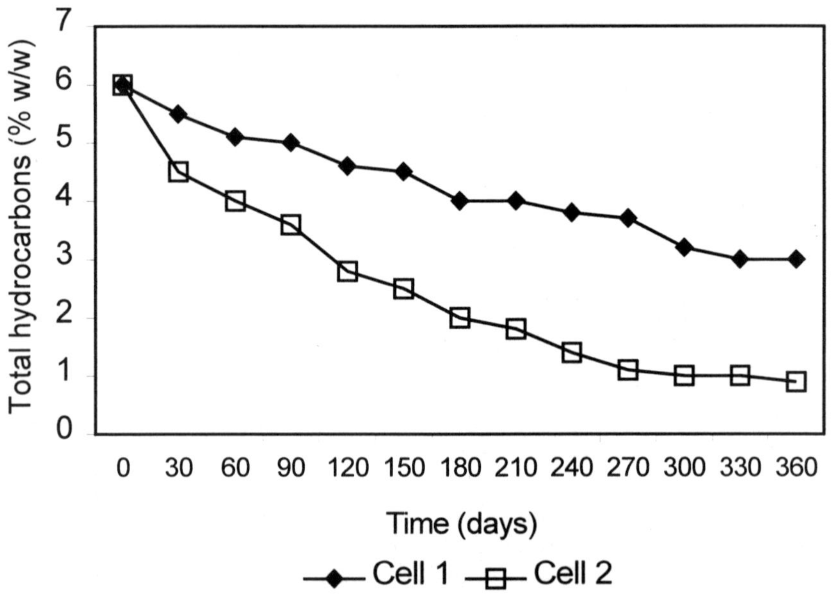


**FIGURE 1. Hydrocarbons concentration (mean) in Cell 1 and Cell 2.**

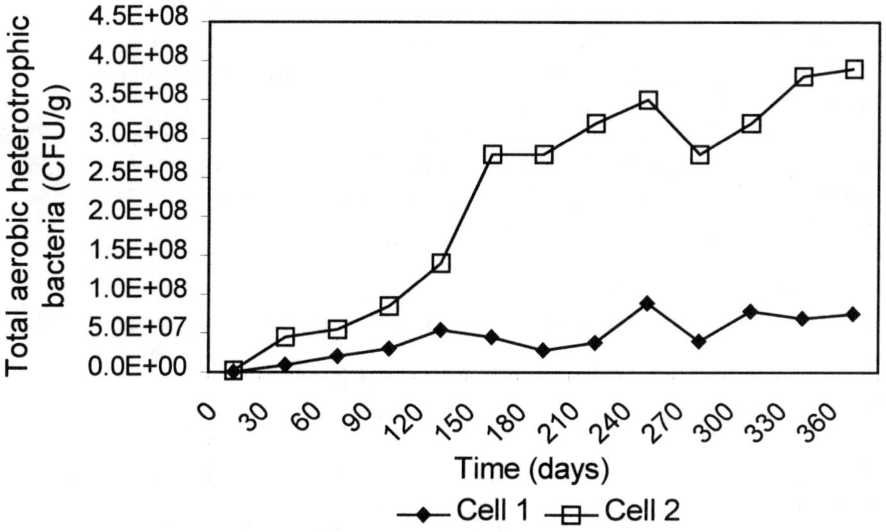


**FIGURE 2. Total aerobic heterotrophic bacteria (TAHB) in Cell 1 and Cell 2.**

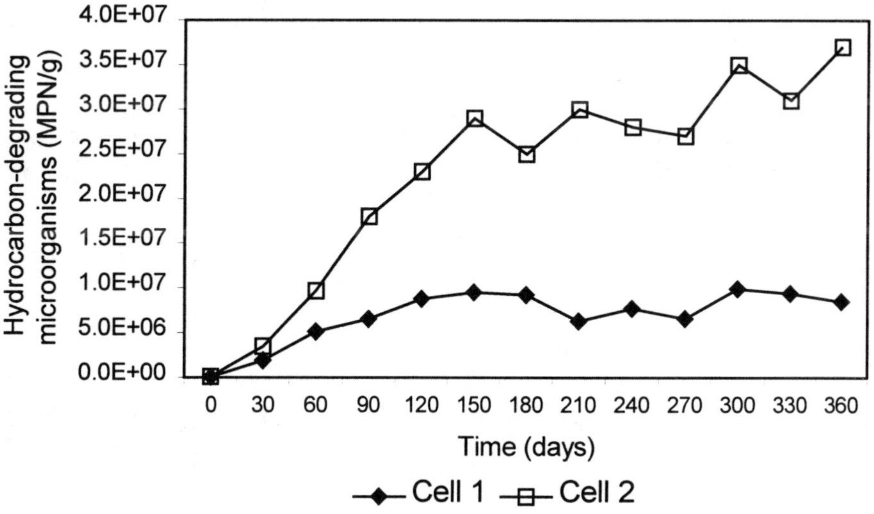


**FIGURE 3. Hydrocarbon-degrading microorganisms (HDM) in Cell 1 and Cell 2.**

## CONCLUSIONS

The higher extent of hydrocarbon removal observed in Cell 2 is consistent with the higher increase in the concentration of TAHB and HDM relative to Cell 1 (Figures 2 and 3). The presence of high concentrations of heavy metals in both cells has not harmed the bioremediation process.

For the refinery, this field study was a success, and it encouraged the refinery to use the same plan for three addition cells. After one year of treatment, similar positive results were obtained in these bioaugmented cells. Since then, the refinery has treated more solid waste in less time

## ACKNOWLEDGMENTS

I wish to thank Francisco A. Casiello Ph.D. for their assistance with this project.

## REFERENCES

Belloso, C. O. 1999. "Hydrocarbons Biodegradation in Landfarming Soils Contained in Containers" In B. C. Alleman and A. Leeson (Eds.), *Bioreactor and Ex Situ Biological Treatment Technologies*. pp. 75-80. Battelle Press: Columbus, Ohio, USA.

Bushnell L. D., and H. F. Hass. 1941. "The Utilization of Certain Hydrocarbons by Microorganisms". *J. Bacteriol. 41*: 653-673.

Clesceri, L. S., A. E. Greenberg, and R. Trussel Rhodes. 1992. *"Standards Methods for the Examination of Water and Wastewater"*. 18$^{th}$. ed., American Public Health Association, N.Y.

USEPA. 1996. *How to Evaluate Alternative Cleanup Technologies for Underground Storage Tank Sites: A Guide for Corrective Action Plan Reviewers.* (EPA 510-B-94-003).

# SOIL WASHING AND THERMAL SOIL TREATMENT AT THE TRECATE OIL SPILL SITE

Jack Ambert, Roland Mage, ***Klaus Müller,*** and Augusto Porta
(Geneva Research Centre, Battelle Memorial Institute, CH-1227 Carouge)

**ABSTRACT:** About 20 000 tons of oil had spilled into rice fields during a 1994 blowout of an oil well in Trecate, Northern Italy. While major part of the Trecate oil spill site was remediated by biotechniques, the two hectares surrounding the former well have been excavated, the soil was treated by washing with aqueous detergent solutions, and the soil containing excessive pollutant was thermally treated on site. The treated soils were back-filled. The process, results, experience, advantages and disadvantages as well as the analytical work accompanying the effort are described.

**Introduction.** South of Milan's airport Malpensa, oil wells yield crude from strata three miles beneath the Po plains. One of them, Trecate 24, erupted during exploration and produced a rare incident of large-scale land-locked oil pollution. Well owner Agip, administrations, local laboratories, and research staff from various Battelle labortories have collaborated in a large program of monitoring, risk assessment, and various bioremediation campaigns starting with the emergency measures taken in early March 1994. (Tamburini et al., 1997; Porta et al., 1997; Porta et al., 1999; Porta and Duart, 1999; Mage and Porta, 2001)

For about six years after the spill, the two hectares of land surrounding the dismantled well served as an operating base for remediation work in surrounding rice fields. Contaminated debris was buried, polluted soil stored in shacks, machinery for active remediation (like bioslurping, bioventing) stationed, and a field laboratory operated there. Between 1995 and 1997, one of Europe's largest biopiles was set up on the site. When these activities had ended and the biopile soil returned to the adjacent fields (Andreotti et al., 2001), the platform site itself was ready for remediation.

**TR 24 remediation scheme.** Agip issued a tender with specifications formulated by Battelle on the basis of preliminary Geoprobe®, trial-bore, open-trench, and geophysical investigations. In view of the high levels and large average depths of pollution and of a large and inhomogeneous fraction of landfill and coarse gravel at this site, bioremediation techniques could not have provided a short-range solution. Down to 3 m, the platform site was expected to contain about 18,500 $m^3$ (estimated to represent about 30,000 tons) of soil in need of a treatment. An Italian-Dutch consortium proposing the combination of on-site soil washing and thermal treatment performed the work between November 1999 and July 2000. Battelle performed analytical work on and off-site and steered the day-to-day operating decisions. The scheme shown in Fig. 1 illustrates the approach taken.

Boundary conditions of restricted space, short transport routes, and timely backfill had to be respected. The excavations proceeded by opening trenches 4 m wide, which were subdivided horizontally and longitudinally according to the estimated soil pollution levels. A limited number of trenches were open at any given time, and limited amounts of partly treated soil fractions stored.

The excavated soil, if assessed to be dirty (>500 ppm of total petroleum hydrocarbons), underwent classification and washing following the procedures described in the scheme of Fig. 2, which should be examined while referring to the typical particle size distribution reported in Table 1.

SOIL EXCAVATED > ***analysis***
*clean:* **backfill**
*dirty:* ↙
*waste:* **controlled dump**

SOIL DIRTY > ***classification, washing, analysis***
*clean:* **backfill**
*still dirty:* ↙
*waste:* **controlled dump**

SOIL STILL DIRTY > ***thermal treatment, analysis***
*clean:* **backfill**
*still dirty:* **repeat**
*waste:* **controlled dump**

**FIGURE 1. Scheme of soil remediation around the Trecate 24 well site**

After washing, large part of the soil was clean enough to be back-filled. The part still dirty was subjected to a thermal treatment in an oil-fired rotary furnace. This operation is summarized in the scheme shown in Fig. 3.

**TABLE 1: Trecate 24 soil particle size distribution** (humidity: 6.5 %)

| **Fraction** (mm) | ≥ 4.75 | 2.36/ 4.75 | 1.18/ 2.36 | 0.60/ 1.18 | 0.30/ 0.60 | 0.15/ 0.30 | 0.075/ 0.150 | ≤0.075 |
|---|---|---|---|---|---|---|---|---|
| **Percent** (dry weight) | 54.2 | 11.3 | 8.6 | 11.6 | 6.8 | 3.0 | 2.3 | 2.2 |

**Operating experience.** The excavations revealed a rather important inhomogeneity of soil pollution, which meant that lengths of as little as 4 m, and horizontal slices between 25 and 100 cm, were dug up, analyzed, and entered into the treatment schemes mentioned above. This reflects a number of points characterizing the situation: nonuniform penetration of the crude into the ground, the presence of subsurface structures related to the former oil rig, and debris introduced after the accident. Vertical sections such as that shown in Figure 4 were prepared from the analytical results and field geometry for each of the more than 80 trenches excavated. The excavation activities were timed so as to provide sufficient materials at all times for subsequent activities.

The operations of soil classification and washing, like those of the thermal treatment, were run 24 hours a day, seven days a week except for repairs and maintenance. The washing occurred with aqueous detergent solutions. Their concentrations were adjusted to the soil pollution levels encountered. The "clean" fractions produced by washing were monitored for their residual hydrocarbon content. The sand, for instance, regularly analyzed at less than 200 ppm of TPH.

SOIL EXCAVATED, ANALYZED, FOUND DIRTY:<br>
⇨ ***hopper, grating***<br>
> **fraction >120 mm:** rocks, big objects to dump<br>
> **fraction <120 mm:** ↙<br>
FRACTION <120 mm<br>
⇨ ***washing drum (trommel), mud remover***<br>
> **fraction 40/120 mm:** clean gravel to backfill<br>
> **fraction 10/40 mm:** clean fine gravel to backfill<br>
> **fraction <10 mm:** ↙<br>
FRACTION <10 mm<br>
⇨ ***sifter (screen)***<br>
> **fraction 2/10 mm:** ↙<br>
> **fraction <2 mm:** ↙<br>
FRACTION 2/10 mm<br>
⇨ ***jig (shaking washer)***<br>
> **light fraction:** to thermal treatment<br>
> **medium fraction:** clean sand to backfill<br>
> **heavy fraction:** to thermal treatment<br>
FRACTION <2 mm<br>
⇨ ***hydrocyclone***<br>
> **fraction >63 μm:** clean sand to backfill<br>
> **fraction < 63 μm:** ↙<br>
FRACTION <63 μm<br>
⇨ ***thickener (polyelectrolyte)***<br>
> **thick mud:** to thermal treatment<br>
WASHING LIQUORS<br>
⇨ ***oil trap***<br>
> **recycle water:** back to washing operations<br>
> **oil sludge:** to waste disposal

**FIGURE 2. Scheme of soil classification and washing**

Minor problems in operation were encountered only during prolonged periods of freezing temperatures, but even then the process proved to be rather robust.

Furnace operation was more delicate. Thus, if the ingoing soil was too wet, throughputs could not be maintained at the projected level of 10 tons per hour. Also, under these conditions the soil produced was black. Probably steam prevented access of sufficient air oxygen to the oil-contaminated soil particles and thus the oil was converted to soot rather than burned. This problem was eased when, in the warmer months, ingoing soil could be predried in the open.

Physical operating problems arose from excessive levels of fines in the ingoing soil. This problem was alleviated by adding coarser fractions from soil clean after washing, which unfortunately translated in higher thermal treatment volumes.

DIRTY SOIL
with 500 - 15,000 ppm of hydrocarbon contamination

- ***predrying***, granulometric adjustment if needed
- ***rotary furnace primary unit***, thermal desorption/combustion
  ⇨heat-treated soil: cooling, mixing with some water ⇨ backfill
  ⇨gases: to post-combustion unit
- ***post-combustion unit***, thorough combustion with preheated air
  ⇨dust-laden gases: to scrubber
- ***scrubber***
  ⇨dust-laden gases: quenching, washing, absorption in venturi, packed tower
  ⇨clean gases: analysis, venting
  ⇨heated cooling air: to primary unit and post-combustion unit
  ⇨washing liquors: sludge separation

**FIGURE 3. Scheme of thermal soil treatment**

Because of the high percentage of fines in the Trecate soil, the product of thermal treatment also had a high dust content and had to be conditioned with the addition of water to bind this dust.

**Analytical work.** The analytical work was an essential part of the operations:

- after excavation of a layer of earth from a trench, material heaped up alongside the trench was analyzed to determine the need for treatment.
- after the production of clean material by washing, regular checks were made.
- after the production of heat-treated soil in the furnace, checks were made every four hours as well as on samples from the 12-hour heaps.

The field tests were performed with Petroflag$^R$ kits from SDI Europe (Alton, Hampshire, UK), which involve the extraction of 10 grams of soil (or as little as 1 g when highly contaminated) with methanol, filtration, addition of proprietary reactant to produce turbidity, and quantitative instrumental reading of the turbidity against standards. Calibration of this field-test procedure was performed by comparison with the Trecate hydrocarbon levels as determined by GC; this was an important point, inasmuch as different oils behave differently in these tests.

Duplicate samples were routinely reanalyzed by GC (about 10 % of all samples). Agreement usually was good, with GC tending to give lower residual TPH.

Treated materials on an average analyzed 100 to 200 ppm for the "clean" sand (a value probably exaggerated due to residual detergent), and 0 to 150 ppm (GC) or 20 to 400 ppm (kit) for the furnace-treated soils. Residual TPH exceeding 500 ppm were found during a single week in furnace-treated soils; these soils were recycled into the furnace. Coarser material always analyzed in the lower tens of ppm after washing. To the contrary, washed fines destined for the furnace analyzed between 1500 and 4000 ppm (with 25 to 40 % humidity).

It should be mentioned here that emissions from the furnace (CO, $SO_2$, $NO_x$, $O_2$) and emission flows (CO, $SO_2$, $NO_2$) were closely monitored by the furnace operators and verified by the local authorities.

**Results of remediation.** In view of future industrial use of the well site, the clean-up limit had been set at 500 ppm of TPH. The material used for backfill was very safely below this limit. The total amount of material excavated was about 55,000 tons, nearly double the original estimate.

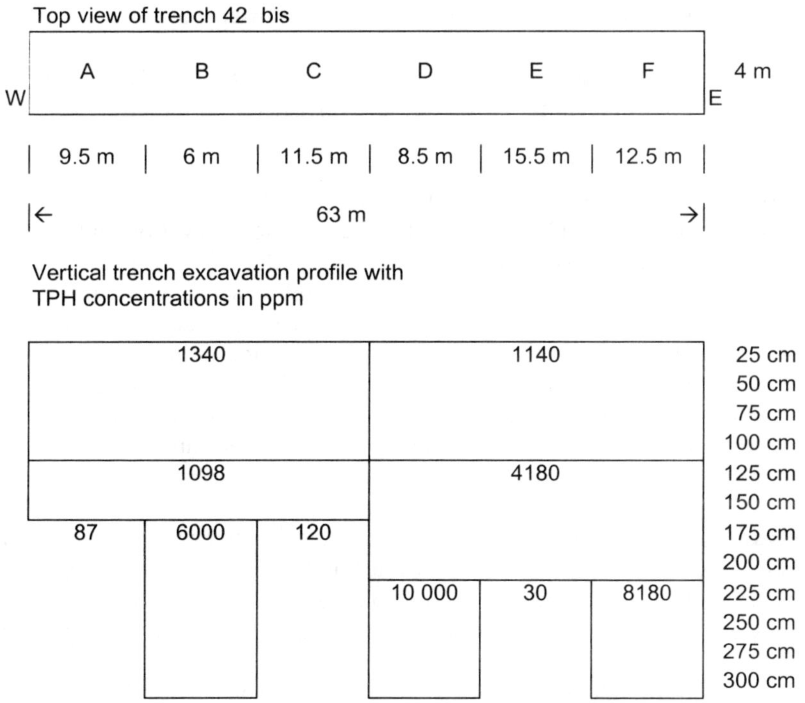


**FIGURE 4: Excavation and contamination of Trench 42 bis.** Letters in the top view (west to east) designate the different trench subdivisions; numbers in the vertical profile are TPH concentrations (ppm) in the corresponding trench volume.

The limiting depth of excavation had been fixed at 3 m. In part of the area, pollution was absent or restricted to a limited range of depths (see subdivisions A,

C, E in Fig. 4). Trenches were discontinued when clean strata were encountered. To the contrary, in an appreciable part of the excavated trenches, pollution levels were still in the "dirty" range at that depth, of 300 or 350 cm. Thus, a coherent zone of about 30 x 60 m in the center of the site and an incoherent zone of about 40 x 100 m in the NW corner retain contamination levels exceeding 500 pm at depths of 300 or 350 cm. The presence of this contamination will have to be remembered for short-term uses. It should be monitored for long-term degradation, which is likely to occur by the action of soil bacteria (natural attenuation).

Excavating to a depth of more than 3 m was problematic for a variety of reasons:

- ordinary excavation equipment and nonfortified trench walls were no longer feasible
- the cost-benefit ratio became very high
- the projected, ordinary industrial use did not require more profound clean-up under the Trecate conditions, where volatile hydrocarbons (which could constitute a health risk at the site) had long since disappeared.

More profound clean-up in greater depths could be stimulated by air sparging, bioventing, or by injection of an oxygen release agent at any later time if necessary.

**Advantages.** The combined soil washing and partial heat treatment turned out to be an economic solution (180,000 ITL/ton) relative to excavation and disposal in special waste dumps (estimated to cost at least 300,000 to 500,000 ITL/ton, depending on distance) and relative to on-site soil washing with solvents. Since less than 40 % of the soil had to be heat-treated, biological activity remains in the back-fill. The heat-treated soil is also not "dead-burned", since hydrocarbons do not require the same high temperatures as some toxics such as dioxins.

With only a minute fraction of excavated material sent to dump sites, the back-filled site has experienced practically no loss of elevation, which is important for a site in neighborhood with rice fields flooded in spring.

**Disadvantages.** Part of the work was affected by freezing temperatures and torrential rains occurring in parts of winter and spring. This was a problem in particular for the furnace, when ingoing soil was highly caked or even frozen, or overly wet. Operation of the furnace with wet soil produced unsightly black product.

The fines content of the excavated Trecate soil (see Table 1) and even more so that of the washed soils was rather high, and required special measures as mentioned in the section on operating experience.

The very heterogeneous distribution of contamination (in three dimensions, see Fig. 4) was responsible for a large subdivision of the trenches and far greater analytical effort than originally envisaged. It also made high demands on the skills of the bulldozer operators.

Apart from causing certain delays and additional costs, these problems are minor from a remediation point of view, and do not detract from the basic success of the approach taken.

## REFERENCES

Tamburini, D., A. Porta, J. Ambert, and G. Di Luise. 1997. "Evaluation of optimum hydrocarbon degradation conditions: A biotreatability study." *Fourth International In Situ and On-Site Bioremediation Symposium*, B. C. Alleman and A. Leeson (Eds.), Vol. 1, Battelle Press, Columbus OH. pp. 467-472.

Porta, A., A. Trovato, K. McCarthy, A. Uhler, and G. Andreotti. 1997. "Degradation of saturated and polycyclic aromatic hydrocarbons and formation of their metabolites in bioremediated crude oil-containing soils." *Fourth International In Situ and On-Site Bioremediation Symposium*, B. C. Alleman and A. Leeson (Eds.), Vol. 1, Battelle Press, Columbus OH. pp. 505-510.

Porta, A., R. Mage, D. Tamburini, J. Ambert, and G. Andreotti. 1999. "Natural attenuation study of the aquifer underlying the Trecate 24 oil field." *Fifth International In Situ and On-Site Bioremediation Symposium*, B. C. Alleman and A. Leeson (Eds.), Vol. 1, Battelle Press, Columbus OH. pp. 159-164.

Porta, A., and P. Duart. 1999. "Ecological and human health risk assessment of the Trecate 24 oil well blowout." *Fifth International In Situ and On-Site Bioremediation Symposium*, B. C. Alleman and A. Leeson (Eds.), Vol. 1, Battelle Press, Columbus OH. pp. 359-364.

Porta, A., A. Trovato, J. M. Neff, and G. Andreotti. 1999. "Metabolic products from petroleum hydrocarbons in oil polluted soils." *Fifth International In Situ and On-Site Bioremediation Symposium*, B. C. Alleman and A. Leeson (Eds.), Vol. 5 Battelle Press, Columbus OH. pp. 69-74.

Porta, A., N. Filliat and N. Plata. 1999. "Phytotoxicity and phytoremediation studies in soils polluted by weathered oil." *Fifth International In Situ and On-Site Bioremediation Symposium*, B. C. Alleman and A. Leeson (Eds.), Vol. 6, Battelle Press, Columbus OH. pp. 51-56.

Mage, R., and A. Porta. 2001. "Long-term biodegradation of underground and aquifer contamination at Trecate." *Sixth International In Situ and On-Site Bioremediation Symposium*, Battelle Press, Columbus OH.

Andreotti, G., N. Plata, A. Porta, and K. Müller. 2001. "Phytoremediation of hydrocarbon-polluted agricultural soils." *Sixth International In Situ and On-Site Bioremediation Symposium*, Battelle Press, Columbus OH.

# BIOREMEDIATION OF OIL-CONTAMINATED SOIL IN KUWAIT (Part 2)

***Hiroyuki CHINO***, Hirokazu TSUJI,
Yoji ISHIKAWA, Takashi MATSUBARA
(Obayashi Corporation, Tokyo, Japan)
Nader AL-AWADHI, M. Talaat BALBA, Reyad AL-DAHER
(Kuwait Institute for Scientific Research, Safat, Kuwait)

**ABSTRACT:** A bioremediation field experiment was carried out in the Kuwait desert to remediate oil contaminated soil. Cost reduction by improving the efficiency of the large-scale demonstration was required,resulting in the remediation of composting piles using a pile-turner to remediate 5500 $m^3$ of oil contaminated soil. The treatment duration was 15 months and total petroleum hydrocarbon (TPH) had been degraded 80 %. Mutagenicity test results showed that soil became non-toxic after 3 months of treatment. The criteria for the treatment of the contaminated soil in Kuwait was based on the end points of several proposed experimental results. The protocol for the actual field treatment depened on the contamination level and area proposed for treatment. This study was carried out under the Joint Research Program between the Petroleum Energy Center under the Ministry of Economy, Trade and Industry, Japan, and the Kuwait Institute for Scientific Research, Kuwait.

## INTRODUCTION

Bioremediation was selected for the treatment of contaminated soil in the Kuwait desert, caused by the Gulf War. Three different technologies, namely, landfarming, windrow composting, and static bioventing, were able to reduce oil concentration significantly, as reported by Chino et al (1999) [1]. The effectiveness of bioremediation had been verified, but the economic feasibility of the technology had to be assessed. We therefore carried out field experiments in Kuwait to improve the efficiency of bioremediation. A pile turning machine, which could mix soil effectively was introduced and compared to other technologies, including landfarming, windrow composting, and static bioventing.

## MATERIALS AND METHODS

**Site Preparation and Construction.** A field demonstration experiment was carried out at the site near Oil Lake No. 102, located about 70 km south of the city of

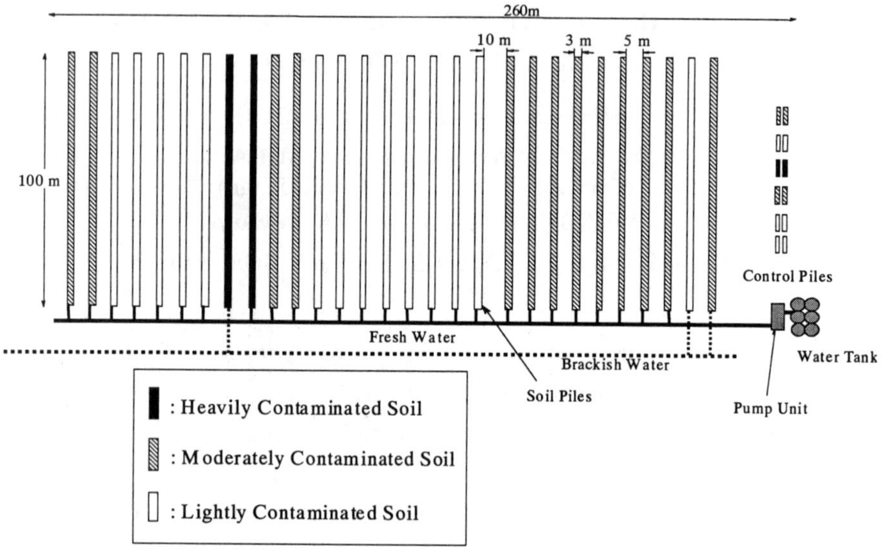


**FIGURE 1. Site Layout**

**TABLE 1. Soil Characteristics**

| Item | unit | Lightly Contaminated Soil | Moderately Contaminated Soil | Heavily Contaminated Soil |
|---|---|---|---|---|
| pH | | 8.0 | 7.9 | 8.1 |
| EC | mS/m | 555 | 4450 | 4703 |
| Moisture | % | 0.5 | 1.62 | 1.84 |
| Combustible Matter | % | 4.5 | 9.1 | 10.4 |
| TEM (Total Extractable Matter) | % | 2.33 | 4.32 | 5.55 |
| TPH | % | 1.95 | 3.29 | 4.43 |
| $Cl^-$ | mg/kg | 2230 | 23330 | 24580 |
| $SO_4^{2-}$ | mg/kg | 2190 | 2840 | 2590 |
| Total Bacterial Count | Log cfu/g | 6 | 6 | 6 |

Kuwait. Moderately and heavily contaminated soil was excavated from Lake No. 102, and lightly contaminated soil was excavated from Lake No. 148. The excavated soils were moved to the treatment yard, then homogenized. Figure 1 illustrates the conceptual design and layout of the treatment yard. TPH concentration in heavily and moderately contaminated soil was about 3.5 - 4.5%, while that in lightly contaminated soil was about 2%. Table 1 provides some information on the characteristics of the contaminated soil. Soil type was sand and field capacity was around 15 %. The large volume and subsequent evaporation of seawater used to extinguish the oil well fires caused the very high salinity observed in the soil.

Nitrogen and phosphate fertilizers were added to the soils. Additionally, 2.5% of wood chips and 0.8% of sheep manure were mixed into the soils. A pile-turner was introduced for the mixing of the soil piles. This machine has rotary knives that can turn over the soil

**PICTURE 1. Experimental Yard**

**TABLE 2. Analyses and Methods**

| | Item | Methods |
|---|---|---|
| Moisture | Moisture | 105 °C, 1 night |
| | Amount of Irrigation | Flowmeter |
| | Ambient Temperature | Automated measurement by data logger |
| | Soil Temperature | Thermometer |
| Oil | TPH | Freon extraction, measured by FTIR (EPA 418.1) |
| | TEM | Dichloromethane extraction, gravimetrically |
| | Fractional Analysis | Column chromatography |
| | Aliphatic Compounds | Freon extraction, GC (FID) |
| | Aromatic Compounds | Dichloromethane extraction, HPLC (FLD) |
| Chemical | pH | Soil :Water =1:2, by pH meter |
| | Combustible Matter | 550 °C 3 h |
| | EC | Soil :Water =1:2, by EC meter |
| | Chloride | AgCl, titration method |
| | Sulphate | BaSO4, nephlometric method |
| | Phosphate | Molybdenum yellow colorimetric method |
| | Ammonium Nitrogen | Nessler's reagent, colorimetric method |
| | Nitrate Nitrogen | Cadmium reduction method |
| | Kjeldahl Nitrogen | Kjeldahl decomposition, titration |
| Microbial | Bacteria | Agar plate (pH 7.3 nutrient agar media) |
| | Fungi | Agar plate (pH 5.6 PDA media) |

with at than 200 rpm. This machine was originally developed for production of compost from sewage sludge, wood chips, and so on, and was applied for oil contaminated sandy soil. 29 piles (cross section 1.8 $m^2$ and length 100 m) were prepared and more than 5,000 $m^3$ of soil was treated using the pile-turner as shown in Picture 1. Water was supplied through the pile-turner and the moisture content of the soils was maintained at between 8 and 10%.

Monitoring. The parameters monitored in the tests are listed in Table 2. Samples were collected from a total of 20 locations on both sides of each pile, and tested after being composited.

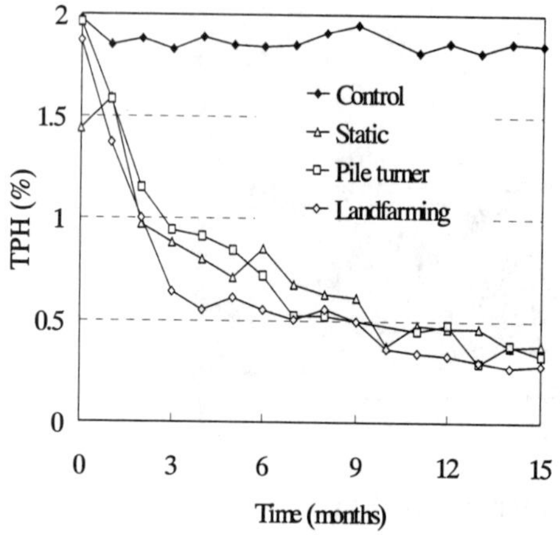


**FIGURE 2. TPH Concentrations Overtime for Three Treatments and Control**

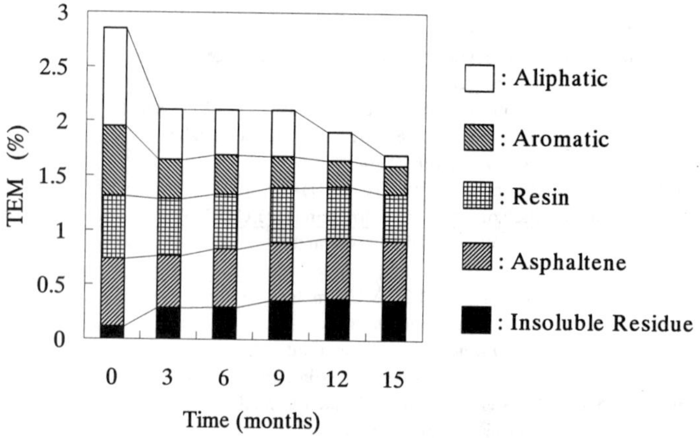


**FIGURE 3. Result of Fraction Analysis in Lightly Contaminated Soil**

**Mutagenicity Tests.** Mutagenicity test (Ames test) was carried out after every 3 months of treatment, according to the Japanese Ministry of Health, Labor and Welfare's notification method. Five different Salmonella strains were used.

## RESULTS AND DISCUSSION

**Degradation Behavior of Oil Components.** The changes over time in TPH values are shown in Figure 2. TPH values decreased rapidly and after 15 months, over 80% of the total TPH had been degraded. Pile-turner technology effectiveness was similar to andfarming.

The fractional amounts of TEM components are shown in Figure 3. Prior to the experiment, the oil consisted of roughly equal thirds of aliphatic compounds, aromatic compounds, and other substances. As is shown in Figure 3, the aliphatic and aromatic compounds were broken down, but the resin and asphaltene components were barely degraded. In the case of lightly contaminated soil, more than 50% of both the aliphatic and the aromatic fraction had been degraded after 9 months, and these rates increased to over 90 and 60%, respectively, after 15 months.

**Soil-Water Balance.** The cumulative amount of water used per unit volume of soil in each technology is shown in Figure 4. Landfarming required more than three times as much water per unit volume of soil, pile-turner required relatively little water, that is 1 -1.5

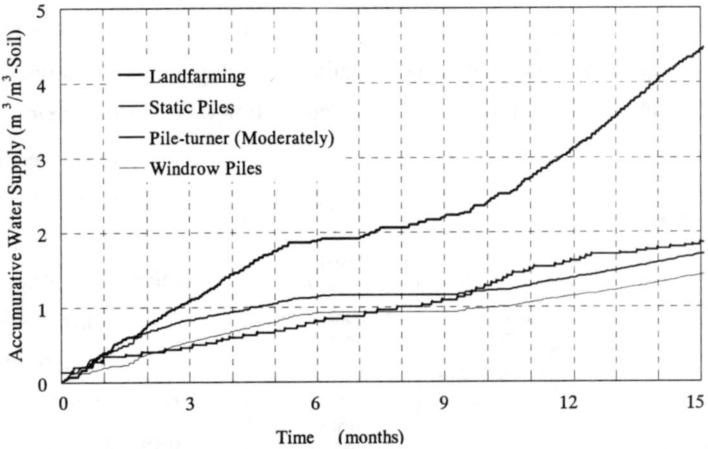


**FIGURE 4. Cumulative Water Supply**

**TABLE 3. Result of Mutagenesis Test**

| Moderately Contaminated Soil S9 mix - Pile-turner | 0 month | 3 months | 6 months | 9 months | 16 months |
|---|---|---|---|---|---|
| TA100 strain | - | - | - | - | - |
| TA1535 strain | - | - | - | - | - |
| WP2 uvrA strai | - | - | - | - | - |
| TA98 strain | + | - | - | - | - |
| TA1537 strain | + | - | - | - | - |

| Moderately Contaminated Soil S9 mix - Landfarming | 0 month | 3 months | 6 months | 9 months | 16 months |
|---|---|---|---|---|---|
| TA100 strain | - | - | - | - | - |
| TA1535 strain | - | - | - | - | - |
| WP2 uvrA strai | - | - | - | - | - |
| TA98 strain | + | - | - | - | - |
| TA1537 strain | + | - | - | - | - |

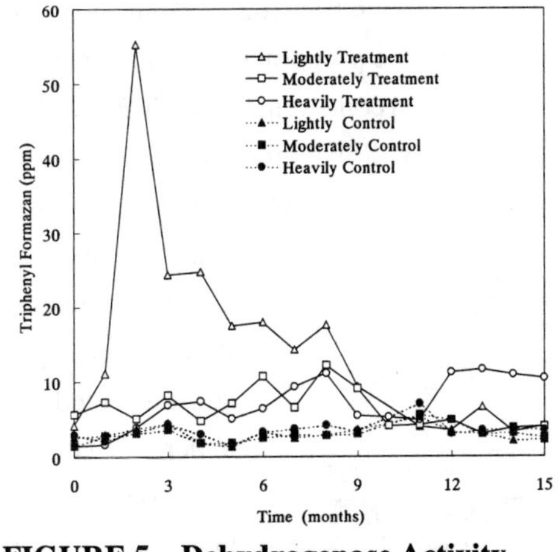


**FIGURE 5. Dehydrogenase Activity of the Soil**

$m^3/m^3$-soil/year.

**Mutagenicity Tests.** Table 3 shows the mutagenicity test results for moderately contaminated soil. TA98 and TA1537 strains initially present mutagenicity. After three months treatment, degradation had removed the existing mutagenicity. The results of the heavily and lightly contaminated soils were similar.

**Microbial Monitoring.** Many kinds of microbial monitoring methods were tested, and dehydrogenase activity is shown in Figure 5. Dehydrogenase activity was correlated well with TPH degradation.

**Effect of Salt in the Soil and Water.** Salinity of the heavily and moderately contaminated soil was extremely high. Degradation of

TPH shown in Figure 6 was stagnant, and about 2 % of TPH remained after 15 months treatment. Compared with the result of landfarming technology, done with low salinity and equal oil concentrations (Chino et al. (1999)) [1)]degradation was inferior. The inpact of salt was tested in the laboratory using washed soils to decrease salinity. and degradation was inferior.

Figure 7 shows the result of a respiration test, used as an indicator of microbial activity. The result show that carbon dioxide production increased after desalination.

On the other hand, there was no difference between the fresh water and the brackish water tests (Figure 6).

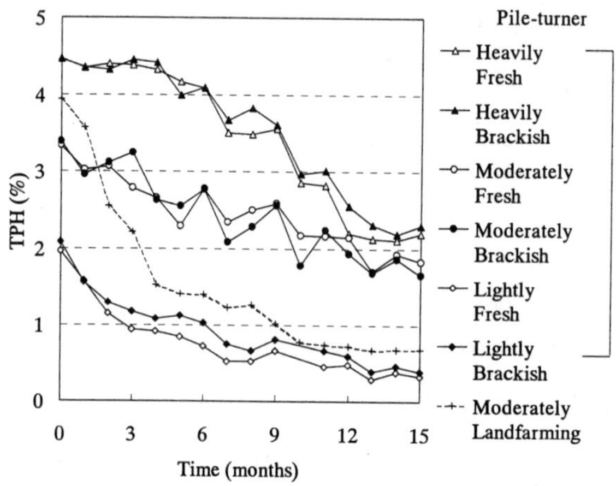


**FIGURE 6. Effect of the Brackish Water**

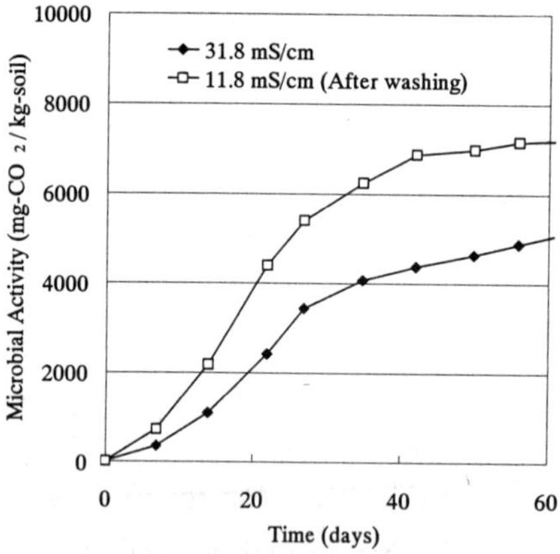


**FIGURE 7. Effect of Soil Washing**

## PROPOSED TREATMENT CRETERIA

Based on our analytical result, Table 4 lists the treatment endpoints based on contamination level. The proposal treatment goal of the oil contaminated soil in Kuwait is 1 %. Soil contaminated below level does not have an oilly odor or an oil film, and could be used for vegetation. If there is very lightly contaminated soil under 1 % TPH, 2 - 3 months of bioremediation may

**TABLE 4. End Point of the Each Monitoring Results**

| | Oil Degradation | Vegetation | Mutagenicity |
|---|---|---|---|
| Moderately to Heavily Contaminated Soil | TPH = 0.7-0.9%<br>*PAH = 0.5 mg/kg-soil | TPH <1% | More than Three months Treatment |
| Lightly Contaminated Soil | TPH = 0.3-0.4%<br>*PAH = 0.3 mg/kg-soil | TPH =0.4% | More than Three months Treatment |

*Total amount of 16 kinds PAHs designated by EPA

**TABLE 5. Comparison of Bioremediation Technologies**

| Technology | | Landfarming | Windrow Composting | Static Bio-venting | Pile-turner |
|---|---|---|---|---|---|
| Illustration | |  Moderately & Lightly |  Moderately |  Moderately & Lightly |  Heavily, Moderately & Lightly |
| Feed Method | Water | Rotary Irrigation System | Leachy Pipes and Top Water | Leachy Pipes and Top Water | Supplied through the Pile-turner |
| | Air | Tractor and Tiller | Mixing by Front loader | Supplied from the Bottom of the Pile by Air Compressor | Supplied by the Pile-turner during Mixing |
| Result after 12 Months Treatment | | 80% of TPH had been degraded, Moderately 0.7 %, Lightly 0.3%, 40-50 % Aromatics had been degraded. | 70% of TPH had been degraded, Moderately 0.95 %, 40% of Aromatics had been degraded. | 70% of TPH had been degraded, Lightly 0.46 %, 50% of Aromatics had been degraded. Possible only for Lightly | 80% of TPH had been degraded, Lightly 0.3%, 60 % Aromatics had been degraded. |
| Water Consump-tion | Summer | 17-20 $L/m^3/d$ | 4-7 $L/m^3/d$ | 4-8 $L/m^3/d$ | 4-8 $L/m^3/d$ |
| | Autumn, Winter, Spring | $5L/m^3/d$ | $1L/m^3/d$ | $1L/m^3/d$ | $1L/m^3/d$ |
| | Total Amount | 6 $m^3/m^3$-soil | 1.4 $m^3/m^3$-soil | 1.6 $m^3/m^3$-soil | 1.2-1.8 $m^3/m^3$-soil |
| Required Equipmen | | Tractor and Tiller, Irrigation System | Front loader, Leachy Pipes | Compressor, Leachy Pipes | Pile-turner |

be enough for the removal of residual oilly odor and oil film.

Table 5 shows the comparison of each bioremediation technology. Landfarming technology was effective but the water consumption was very high. The pile-turner technology was also effective and the water consumption was relatively low and did not require as much treatment area. Therefore; the pile-turner technology is recommended. If contamination depth is less than 50 cm, in situ landfarming technology is also recommended. If soil contamination levels are very low (TPH<1%), static bioventing pile technology is proposed.

## CONCLUSION

Composting piles using a pile-turner was introduced in a bioremediation field experiment conducted for the treatment of 5500 $m^3$ of oil contaminated soil in Kuwait. The pile-turner was effective and recommended for future bioremediation work. The criterion for the treatment of the contaminated soil in Kuwait based on the end points of several experimental results, was proposed as TPH < 1%. The protocol for an actual field treatment depends on the initial contamination level and of interest.

## ACKNOWLEDGMENTS

We are grateful to Petroleum Energy Center, Japan and Kuwait Institute for Scientific Research, Kuwait for financial support, and to Kuwait Oil Company for their cooperation and valuable contribution to the project. We would like to express our sincere appreciation to Professor Matsumoto and Professor Oyaizu of the Department of Applied Biological Chemistry, the University of Tokyo, Professor Omori of the Biotechnology Research Center, the University of Tokyo, Mr. Makishima of Analytical Research Center, Japan Energy Corporation for their generous guidance and analysis of oil compound in this research.

## REFERENCES

1) H. CHINO, H. Tsuji, T, Matsubara, N, Al-Awadhi, M. Talaat Balba, R. Al-Daher, Bioremediation of Oil-Contaminated Soil in Kuwait, The Fifth International In Situ and On-Site Bioremediation Symposium, (5) Bioreactor and Ex Situ Biological Treatment Technologies, pp13-18 (1999).

2) H. CHINO, H. Tsuji, Y. Ishikawa, M. Yotsumoto, T, Matsubara, N, Al-Awadhi, M. Talaat Balba, R. Al-Daher, Bioremediation of Oil Contaminated Soil in Kuwait (2000).

# IN SITU BIOSTIMULATION: A KEY FACTOR IN BIOREMEDIATION

Carmen Infante (PDVSA Intevep, Los Teques, Venezuela)
Marianela Arias (PDVSA Intevep, Los Teques, Venezuela)
***Gisela Sánchez (PDVSA Intevep, Los Teques, Venezuela)***
Nydia León (PDVSA Intevep, Los Teques, Venezuela )
Luis Sánchez (PDVSA Intevep, Los Teques, Venezuela)

**ABSTRACT:** Bioremediation technology has been successfully used in several places of the world to recover oil-impacted soils and to treat organic wastes. However, the selection of the process to be applied as well as the evaluation criteria, and handling of abiotic and biotic factors require knowledge of the technology to guarantee success. Bioremediation technology in tropical countries has been highly effective due to high and constant temperatures throughout the year. In other words, the environmental conditions promote the natural dynamic of the microbiological communities in soils, favoring hydrocarbon biodegradation. To stimulate the microbial population, without introducing exogenous microorganisms (bioaugmentation), abiotic factors are conditioned *in situ* by adjusting C/N, C/P ratios, and adding organic materials (bulking agents), which enhance the physical properties of the wastes being treated. The present paper shows a summary of laboratory and field experiments in which *in situ* biostimulation effectiveness vs. bioaugmentation is demonstrated. Field experiences are described showing how bioremediation technology has helped to minimize oil industry wastes.

## INTRODUCTION

PDVSA Intevep (Center for Research of the Venezuelan Oil Company) has dedicated important efforts toward protection of the environment. Several research projects have been developed in order to maximize the efficiency of bioremediation for the treatment of solid oily waste, in particular, those generated from oil pits, contaminated soil, and drilling activities (cuts). For this purpose, the addition of bacterial inoculum (bioaugmentation) for treatment of contaminated drill cuttings is compared with the stimulation of the indigenous microbial population by the enhancement of abiotic factors (biostimulation). Different bulking agents were also tested in order to increase the efficiency of the process.

**Objective.** The objectives of the work described in this paper were:

- To compare laboratory and field scale biodegradation of oily waste using bioaugmentation vs. biostimulation;
- To evaluate different bulking agents in order to improve the structure of the oily waste and to enhance the biodegradation process.

**Site description**. Field experiments were carried out in two locations of the Venezuelan petroleum industry. The bioaugmentation vs. biostimulation test was performed at the Paraguaná Refinery Center using oily sludge from tank bottoms of a cleaning process. The bulking agents were evaluated in a drilling location situated in Barinas state at the southwest of the country.

## MATERIALS AND METHODS

**Inoculum selection.** Sewage sludge from a water treatment plant, oil contaminated soil and native soil were used as source of inoculum. Bacterial enrichments were obtained using mineral media amended with crude oil as a carbon source. The best hydrocarbon-degrading consortium was selected from a shake flask test by comparing the decrease of oil concentration over time. Sterile controls were prepared to quantify abiotic losses

**Field experiment.** Plots of 4 $m^2$ (triplicates) were constructed and contaminated with oily sludge at the final concentration of 10%. For bioaugmentation purposes the inoculum was applied in such a dose that the bacteria population was about $10^8$ colony-forming units (CFU) per gram of contaminated soil. For biostimulation, plots were amended with fertilizers to stimulate the native microbial population. In order to monitor oil biodegradation and compare both conditions, plots were sampled periodically over six months.

**Organic enhancers test.** Laboratory composts in soil pans were prepared mixing oil-based drilling cuts with litter from tropical trees or grass. Fertilizers were added to give a C/N ratio of 60 and C/P ratio of 800 based on the amount of carbon present in the crude oil. The composts were incubated at approximately 25° C and moisture was maintained at 20%. A set of controls was constructed following the same design without the bulking agents. Aeration was performed by mixing the compost pans with fork garden tool once per week. Samples were taken at time zero and after 30 days to measure total oil content and saturated + aromatic fractions.

The effectiveness of bulking agents on biodegradation was also evaluated at field scale compost to treat 500 $m^3$ of drilling cuts. The evaluation of biodegradation process was followed by measuring oil and grease concentration and $CO_2$ production during the experimental period.

## RESULTS AND DISCUSSION

**Bioaugmentation vs. Bioestimulation.** The most successful bacteria consortium tested at laboratory scale was that obtained from sewage sludge plants (Figure 1). In the field experiment however, there were not significant differences on oily sludge biodegradation between bioaugmentated and biostimulated plots (Table 1). These results agree with other reports that demonstrate the stimulation of native soil microbial population by improving abiotics factors such as nutrients, aeration, and moisture. These factors allow the development of a native oil degrading microbial populations without the needs of adding biological cultures that are not only unsuccessful but also increase the cost of biotreatment (Leavitt et al, 1994).

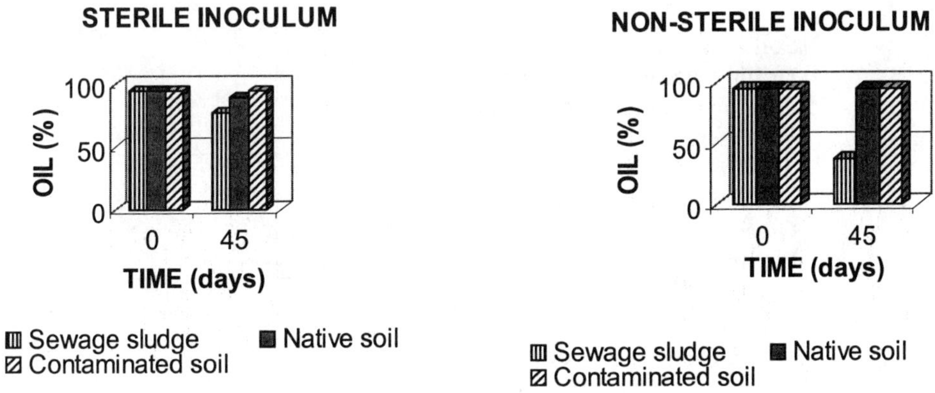


**Figure 1. Oil content in oily-sludge with different inoculum.**

**TABLE 1. Extent of biodegradation of saturates + aromatic fractions**

| Treatment | Sampling time | | | | |
|---|---|---|---|---|---|
| | Day 0 | Day 60 | Day 90 | Day 150 | Day 180 |
| | Biodegradation (%) | | | | |
| Oily sludge + A + F | 100 | 60 | 72 | 81 | 80 |
| Oily sludge + A + F + I | 100 | 68 | 71 | 75 | 78 |

A= aeration; F= fertilizers; I= inoculum

**Organic Enhancers.** The organic bulking agents increased the biodegradation of the oily drilling wastes compared to control (Table 2). However, no significant differences were observed between the two types of vegetable biomass (biomass from tropical trees vs. biomass from tropical grass). These bulking agent are thought to have improved the humidity, structure, nutrients, and porosity conditions of the waste (although these factors were not tested), which in turn resulted in improvement of the biodegradation process.

**TABLE 2. Biodegradation of the oil present on the drilling waste.**

| Treatments | Biodegradation rate expressed as reduction of oil and saturate + aromatic fractions in 30 days | |
|---|---|---|
| | Oil and Grease (%) | Saturate + Aromatic (%) |
| Soil 70%<br>Drilling waste (10% of oil)<br>Organic material type A:20%<br>Fertilizers | 80 | 90 |
| Soil 70%<br>Drilling waste (10% of oil)<br>Organic material type B: 20%<br>Fertilizers | 80 | 84 |
| Control<br>Soil 70%<br>Drilling waste (10% of oil)<br>Without bulking agents<br>Fertilizers | 20 | 28 |

A: vegetable biomass from tropical tress
B: vegetable biomass from tropical grass

The bulking agent results were confirmed at field scale where the hydrocarbon content in the drilling cuts compost decreased from 8 to 0,5% in 130 days (Figure 2). The $CO_2$ production from compost is shown in Figure 3. The highest level was observed in the compost compared to the surrounded native soil. This result indicates that the $CO_2$ production was due to biodegradation activity. The biodegradation technology can be applied to eliminate toxic fractions (saturate and aromatic) from the total amount of oily waste produced (500 $m^3$) during drilling operations.

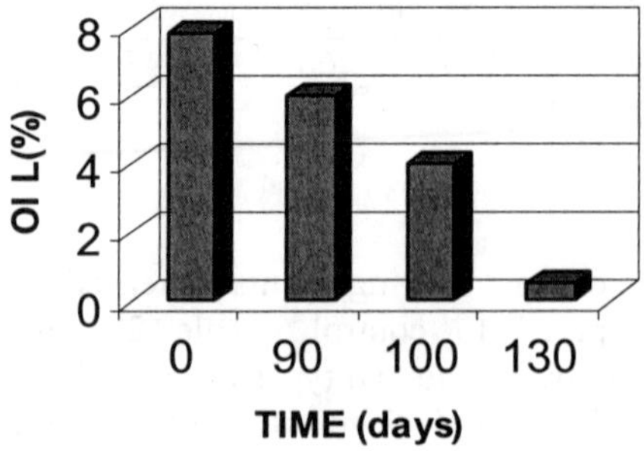


**FIGURE 2. Oil content in drilling cuts compost during the biotreatment**

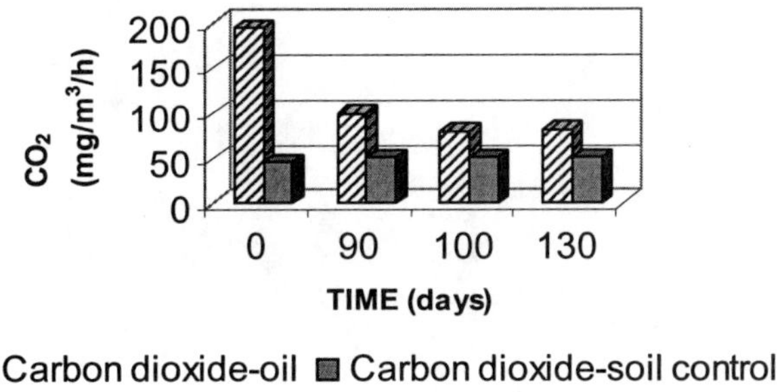


**FIGURE 3. Carbon dioxide production in composts compared with native soil.**

## CONCLUSIONS

Oily-drilling waste and soils containing oily-sludge, generated during exploration, production and refining activities of the Venezuelan petroleum industry can be bioremediated. Nutrient and bulking agents addition resulted in the successful stimulation of indigenous microorganisms to degrade the oily waste. Results from laboratory and field experiences confirm the important role of that biostimulation plays in the biodegradation process, avoiding the use of exogenous microbial populations for the enhancement of the biological process.

## REFERENCES

1. Leavitt, M.E.; K.L. Brow. 1994. Biostimulation versus Bioaugmentatio- Three cases studies. pp 72-79. In: Hydrocarbon Bioremediation. Hinchee, R. et al (editors). Publishers Lewis.

2. Infante, C.; M. Romero; A. Arocha; D. Gilbert and F. Brito. 1999. In situ Bioremediation of pits from Puerto La Cruz Refinery. pp 215 – 221 In: *In Situ Bioremediation of Petroleum Hydrocarbon and other Organic Compounds.*

# MYCOREMEDIATION - COMMERCIAL STATUS AND RECENT DEVELOPMENTS

***Richard T. Lamar, Ph.D.*** and Richard B. White, P.E.
(EarthFax Development Corporation, North Logan, Utah)

**ABSTRACT:** Mycoremediation, or fungal-based soil remediation, has been under development since the mid 1970s. Despite initial high expectations, which were based on the demonstrated ability of white-rot fungi (WRF) to degrade a plethora of hazardous organic chemicals, the technology has been chosen only infrequently as a field remediation solution. From our perspective as technology developers, there are a number of factors that have lead to less then satisfactory field performance by some technology applicators that has, in turn, lead to infrequent selection of mycoremediation as a remediation treatment. The lack of consistent success in the field has resulted from the following: (1) A history of research emphasis on pollutant degradation by WRF without equal emphasis on development of fungal inoculum production techniques or field engineering designs which has led to difficulties in scale-up; (2) The necessity of seeking commercial application prior to completing technology development to fund and provide platforms for further development; (3) Initial overselling of the technology as a result of extrapolation of laboratory results, which were obtained from work in aqueous culture, to potential field performance which has led to attempts at application of the technology to chemicals (e.g. PCBs) whose treatment in a soil environment was only poorly understood; and (4) Attempts at technology applications by personnel who were not familiar with all aspects of the technology. Despite these barriers to commercialization, there have been several successful small-scale commercial projects completed. In addition, we have developed techniques that have greatly enhanced the ability of WRF to decrease the concentrations of extremely hydrophobic contaminants (e.g. high molecular weight PAHs, PCBs and DDT). Future directions in development and applications will be discussed in light of experience in technology field performance and new developments in enhancing fungal pollutant degradation.

## INTRODUCTION

In the early 1970s, T. Kent Kirk and co-workers at the United States Department of Agriculture Forest Service, Forest Products Laboratory (FPL), synthesized $^{14}$C-labeled-lignin and with it developed an unequivocal assay for studying lignin biodegradation (mineralization of the radioactive lignin to $^{14}CO_2$) by white-rot fungi (WRF) (Kirk et al., 1975). The assay was used together with the nutritional conditions that trigger lignolysis in the WRF *Phanerochaete chrysosporium*, namely nutrient nitrogen starvation. They found lignin to be rapidly mineralized by the fungi and in subsequent years described and optimized the various factors affecting its degradation. They also found that lignin was mineralized by WRF even after kraft cooking and bleaching with elemental chlorine (Lundquist et al., 1977). In the same year, Fukuzumi et al. (1977) demonstrated decolorization of a chlorine bleach plant effluent by a WRF.

Based on their results with radioactive chlorinated lignin, Kirk and co-workers in collaboration with scientists at North Carolina State University (NCSU) also examined (without knowing about the Japanese work) the ability of WRF to decolorize bleach plant effluent. They focused on the first extraction stage effluent following chlorination of the pulp, because this effluent contained most of the bleach plant color. This work (Eaton et al., 1980) verified the Japanese work and demonstrated that the factors influencing lignin mineralization had a similar influence on decolorization, implicating the fungal ligninolytic system. The collaborative work between FPL and NCSU continued over a number of years, and resulted in the development of the MyCoR (Mycelial Color Removal) process for decolorizing the effluents; a patent was obtained in 1987 (Chang et al., 1987). Importantly, in these investigations it was discovered that low molecular weight chlorinated aromatic components (i.e. chlorinated phenols, anisoles and guaicols) of the effluents were dechlorinated and mineralized by the fungi (Huynh et al., 1987). The information on the ability of WRF to mineralize low molecular weight chlorinated aromatics ultimately led to the concept of developing and using a soil remediation technology based on the pollutant-degrading abilities of WRF.

Using the assay (i.e., assessment of mineralization of $^{14}C$-labeled pollutants by liquid cultures of WRF) developed by Kirk and co-workers to study lignin degradation, investigations conducted by many researchers on fungal pollutant degradation have resulted in an extensive list of hazardous organics that are transformed, and in most cases mineralized by the fungi (see Barr and Aust, 1994; Paszczynski and Crawford, 1995 for review). The plethora of chemicals able to be degraded by WRF, among them many unusually recalcitrant pollutants (e.g., polychlorinated biphenyls (PCB) and high molecular weight polyaromatic hydrocarbons [PAH]), coupled with the belief that the key components of the lignin-degrading system, which are of necessity extracellular (i.e., because lignin is a large, hydrophobic polymer), non-specific and highly oxidative, would result in superior pollutant-degrading performance in soil, presented a strong, albeit circumstantial case, that the fungi were potentially excellent soil remediation agents.

Despite this potential, much research, and several successful commercial projects, fungal-based soil remediation has not yet become a commercially viable, consistently-used soil remediation alternative. In this paper we discuss factors that have delayed the development of the technology to mature, commercial status.

## BARRIERS TO COMMERCIALIZATION

The overriding factor that has limited the commercial use of fungal-based remediation is the lack of consistent success in field projects. There are a number of factors that have contributed to the inconsistent performance. First, most of the work that has been conducted on fungal-based remediation has focused on the fundamental aspects of fungal pollutant degradation with very little attention to the applied aspects of the technology like inoculum production and field engineering. Lack of information in these areas has resulted in significant problems in scale-up that have led to unsatisfactory field performance. Second, in common with many biotechnologies, the commercial development of fungal-based remediation from initial concept has been dependent on financing from grants and/or private funding. In both cases, the time required for full commercial development of the technology has exceeded both the interests of funding organizations and the patience of investors to realize a return. As a consequence, commercial projects have been sought, prior to full technology development, to provide opportunities for further development sometimes with less than satisfactory results. Third, another common factor shared by biotechnologies that rely on funding for development is initial overselling of the potential market and/or the technical capabilities of the technology, either to obtain funding or because the actual market potential and/or technical capabilities are not known at the beginning stages. When the technology is not able to meet predicted expectations, it is usually labeled as ineffective. In the minds of many within the U.S. environmental community who are familiar with fungal-based remediation, the technology has certainly achieved that status. Finally, there have been attempts at technology application by entities unskilled with the technology. These attempts have sometimes led to failures, which have further exacerbated the view that the technology does not work.

**Lack of attention to applied aspects of fungal-based remediation.** The preponderance of research efforts in the area of fungal-based remediation has focused on elucidating the physiology and biochemistry of WRF pollutant degradation. While the information generated from this work is important and has provided ideas for enhancing the technology performance, it has not provided the kinds of information that are required to apply the technology economically, under large-scale field conditions. There has been very little work published on two extremely important technology components: large-scale inoculum production and field engineering approaches.

<u>Inoculum</u> - A variety of lignocellulosic materials have been used as inoculum substrates to support the growth of WRF. Examples include agricultural and forest industry residues like wood chips (Lamar and Dietrich, 1990), corn cobs (Fernando et al., 1990), straw (Loske et al., 1990), a nutrient-fortified grain sawdust mixture (Lamar et al., 1993; Lamar et al. 1994), and sugar beet pulp (Bradley et al., 1996). Using these lignocellulosic substrates the inoculum can be prepared off-site and transported to the site once it is partially or fully colonized or the substrates can be delivered to the site and used to produce the inoculum on site. Inoculum production using these types of substrates with either off- or on-site production is expensive because it is labor intensive and because of the costs of the primary inoculum (i.e. the liquid or solid fungal inoculum used to inoculate the bulk substrates), the substrates (the cost of even relatively inexpensive substrates like straw can be significant because of the amounts required), and transport of the substrates or inoculum to the site.

If the inoculum is produced off-site, care must be taken to minimize degradation (e.g., from thermogenesis) of the inoculum potential (i.e., ability of the inoculum to support fungal colonization of the contaminated soil) during transport to the site (Lamar et al., 1994). Producing inoculum on-site can be extremely difficult. Contaminated sites are usually located in active industrial areas with an abundance of air-born microbial contaminants. This makes production of fungal inoculum difficult, even with pasteurized substrates. In addition, no quantitative method exists to determine when a fungal inoculum is ready (i.e., when the inoculum potential is greatest). Therefore, the inocula prepared using bulk substrates are usually determined to be "ready" when the substrate(s) appear, visually, to be fully colonized. While these substrates support fungal colonization of contaminated soil, the inocula produced from them usually have extremely low inoculum potentials which necessitates the use of application rates ranging from 20% to 60% (i.e., [dry weight of inoculum/dry weight of soil] x 100) depending on the soil type, contaminant(s) and contaminant(s) concentrations. For example, maximum degradation rates, in soil inoculated with *Phanerochaete chrysosporium* cultivated on ground corn cobs, of pentachlorophenol (PCP), 2,4,5-trichlorophenoxy acetic acid (2,4,5-T) and polychlorinated biphenyls (PCBs) occurred at a ground corn cob:soil ratio of 4:1 (by weight) (Bumpus et al., 1988; Ryan and Bumpus, 1989). While we haven't used application rates this high in the field, an application rate of 20% using a mixture of cottonseed hulls and sawdust has been routinely used. At this rate 200 kg of inoculum are required for each metric ton of soil. On a volume basis, the ratio is close to 1:1 (inoculum:soil).

Spent mushroom compost (SMC) has also been evaluated for use as inoculum (Okeke et al., 1993; Eggen, 1999). If the commercial fungus can effectively degrade the target contaminants, SMC can be used if delivery to the site is cheaper than other methods of producing inoculum. An overall 86% removal of PAHs was obtained in a soil with an initial total PAH concentration of 1909 mg $kg^{-1}$ that was treated with SMC of *Pleurotus ostreatus* (Eggen, 1999). While spent mushroom compost can usually be obtained for free, there are significant costs for materials handling and transport to the site and the commercial fungus must have the ability to degrade the target contaminants effectively.

A very important factor to consider when selecting a lignocellulosic substrate is the effect of the inoculum substrate on fungal pollutant degrading ability. The carbon/nitrogen (C/N) ratio of inoculum substrates can have a significant effect on the pollutant-degrading performance of WRF that varies depending on the fungal species. For example, increasing the C/N ratio from 300/1 to 10/1 significantly increased the extent of PCP degradation in soil by *P. chrysosporium* but not *Trametes versicolor* (Table 1).

One approach that has been investigated but not put into practice is the use of pelleted fungal inocula. Two variations have been developed. In the first, inocula are in the form of pelleted solid substrates coated with a sodium alginate suspension of fungal spores or mycelial fragments and incubated until overgrown with the mycelium of the selected WRF inocula (i.e., overgrown pellets) (Lamar et al., 1999; Lestan and Lamar, 1996; Lestan et al., 1996; Lestan et al., 1998). In the second variation, the pellets consist of mycelial fragments and/or spores of selected WRF embedded in calcium alginate granules with corn cob grits or sawdust as nutritive amendments (alginate pellets) (Bennett et al., 1996; Childress et al., 1998).

**TABLE 1. Effect of C/N ratio of inoculum substrates on the degradation of PCP by *Phanerochaete chrysosporium*.**

| Fungus | Carbon/Nitrogen Ratio | | | | | | | | | | |
|---|---|---|---|---|---|---|---|---|---|---|---|
| | 6 | 10 | 15 | 30 | 45 | 60 | 70 | 110 | 140 | 400 | 700 |
| | Percent PCP Remaining[1] | | | | | | | | | | |
| *P. chrysosporium* | 28a | 35a | 25b | nd | 30b | nd | 18c | 8d | 4d | 3d | 1d |
| *T. versicolor* | nd | 21a | nd | 18a | nd | 24a | nd | nd | 22a | 20a | 18a |

[1] Means are average of three replicates. Means within rows followed by the same letter are not significantly different.

Pelleted inocula offer the following advantages over inocula produced using bulk substrates: (1) In contrast to inocula prepared using bulk substrates, the nutrient composition of pelleted inocula can be easily tailored (e.g., nitrogen limited) to maximize pollutant degradation by specific WRF; (2) While a significant fraction of the substrates are consumed during fungal colonization of substrates when preparing inocula using bulk materials, consumption of nutrients in pelleted inocula is initiated only when the pelleted inoculum is moistened upon mixing with the contaminated soil. This means that the amount of substrate to support fungal colonization of the contaminated soil and pollutant degradation is greater per unit weight of pelleted inoculum compared to inoculum prepared using bulk substrates; (3) The possibility of losing inoculum potential during transport of pelleted inocula from the production facility to the site is minimized because the amount of water in the pellet, which is primarily associated with the alginate, is adjusted to allow fungal survival but not growth. In contrast, "flashing", or the production of excessive amounts of metabolic heat, can occur when transporting inocula prepared from bulk substrates because these inocula are usually prepared at a moisture content of 60%. The production of excessive amounts of metabolic heat indicates that the WRF is consuming the substrate(s) and thus the inoculum potential is reduced; (4) The ability to formulate the ratio of fungal biomass to nutrient substrate allows the production of pelleted inocula with consistent inoculum potential. It is not possible to produce fungal inocula, based on bulk substrates, that possess predictable inoculum potentials because of the variability of substrate nutrition and the difficulty in controlling temperature and moisture conditions, particularly when producing inoculum on-site; (5) Operationally, the use of pelleted inocula only requires one visit to the site for mixing the contaminated soil and inoculum and construction of the biopile. This is in contrast to on-site production of inocula using bulk substrates, which requires one site visit to initiate inoculum production and a second visit to mix the inoculum and soil and construct the biopile. The additional site visits can add significantly to technology costs; (6) Finally, although the initial costs for equipment and facilities are greater, production costs for pelleted inocula will be less than for inocula produced from bulk substrates because of the significantly reduced requirements for labor.

While WRF remediation projects have been successfully completed using inocula prepared on-site with bulk substrates (White and Lamar, 1999), improvements in inoculum production techniques are necessary to make the technology less expensive and technically consistent.

Field Engineering - Initially, fungal-based remediation was practiced using a landfarming approach. A fungal inoculum was mixed with the contaminated soil and placed in a 15 cm to 40 cm deep lift that could be aerated using a rototiller or a tractor-drawn rotovator. This method was used, at the pilot-scale, with success on two occasions on pentachlorophenol (PCP)-contaminated soils. The first was treatment of a sandy gravel soil where approximately 90% decrease in the PCP concentration was obtained in 6.5 weeks (Lamar and Dietrich, 1990). The second was a field pilot-scale study, a prelude to an EPA-sponsored SITE demonstration of the technology, in which PCP/creosote-contaminated soil was treated (Lamar et al., 1993). In the latter pilot, fungal treatment resulted in a 90% decrease in the PCP concentration after two months. Using the results of that study, a large-scale demonstration of the technology was conducted with less than satisfactory results -- only 64% decrease in the PCP concentration after 20 weeks of treatment (Lamar et al., 1994). In large part, the unsatisfactory results were due to an extreme amount of rainfall during the treatment period, which saturated the soil and prevented regular physical (i.e., rotovation) aeration of the soil.

The latter experience exposed the weakness of the landfarming approach in terms of lack of control over environmental parameters like moisture and temperature that are required for consistent technology performance. The desire for more control over environmental conditions, combined with the experience and knowledge base in the commercial mushroom industry for composting of fungi, led to the use of the biopile approach in an attempt to gain control over the environmental conditions in the inoculum/contaminated soil mixture. Unfortunately, early attempts at biopile applications usually ended very quickly due to overheating of the piles (i.e., temperatures in the 60-70°C range), which resulted in cessation of growth or death of the WRF. With no or inadequate aeration, temperatures in a mixture of fungal inoculum and contaminated soil can reach almost 60°C within two days (Figure 1), as is observed in composting. Application of biopiling to fungal-based remediation is, of necessity, mesophilic composting because optimal growth temperatures for WRF typically range between 24 and 36°C (Lamar et al., 1999). Therefore, biopile temperatures should be maintained within the upper limits of the optimum for the fungus that is being used.

With the amount of metabolic heat that is generated, airflow rates of 15 to 30 L $min^{-1}$ per cubic meter of soil/substrate mixture are required to maintain pile temperatures in the range of 28°C to 35°C. These flow rates are particularly important during the first stages of treatment when fungal substrates are readily available. There are a number of ways that biopile aeration systems can be constructed. We have used two variations. In the first design, a system of perforated PVC pipes covered with 10 to 15 cm of pea gravel to act as a plenum is connected to a blower and the inoculum/soil mixture is piled on top of the gravel. Aeration is achieved by pulling air down through the pile. In the second design, an identical system of perforated PVC pipes is also placed on the top of the biopile and linked to the lower system and the blower so that the air in the pile can be recirculated. The upper aeration system may be covered by 15-20 cm of soil. In both cases, the piles are lined and covered with tarps to prevent moisture additions from precipitation. Local temperature and humidity conditions dictate which system can be used. The first design is normally only appropriate when ambient air temperatures at the beginning of remediation are significantly lower than the desired pile temperature. This approach works best where the humidity is high, so the pile does not dry out. The second design, which can allow the recirculated air to be cooled and humidified, is used when ambient air temperatures are too high at the beginning of remediation to be useful for cooling the pile. One advantage of the second design is that it can be constructed, with proper valving, to allow either direct discharge or recirculation of the air (i.e., consistent with both designs). Such a combined system was used to treat approximately 750 tons of PCP/lindane-contaminated soil at a former wood treating facility in North Carolina (White and Lamar, 1999).

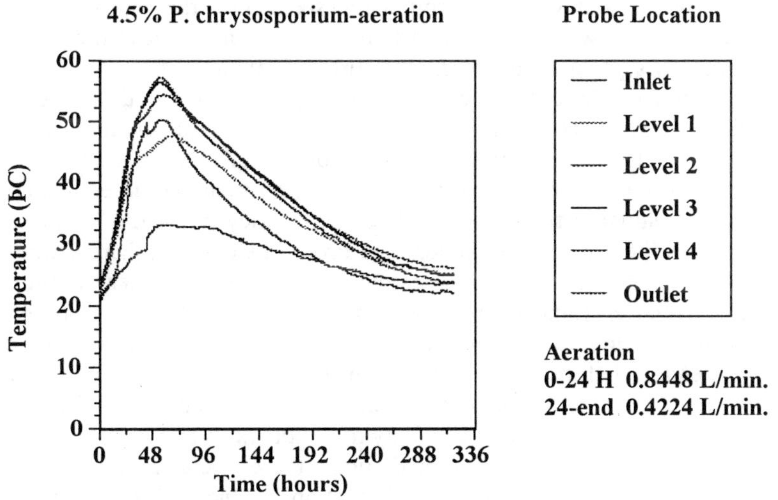


**FIGURE 1. Illustration of thermogenesis in soil inoculated with *P. chrysosporium* inoculum at a rate of 4.5%. Data were obtained from a 1 m high by 20 cm wide column insulated to mimic the interior of a biopile. Air was supplied from the bottom and temperature measured at the inlet and 15 cm (Level 1), 45 cm (Level 2), 75 cm (Level 3), 90 cm (Level 4) and in the headspace above the inoculum soil mixture (i.e., outlet).**

With appropriate aeration system design, coupled with monitoring of pile temperatures and adjustment of airflow rates, biopiles can be effectively used for fungal-based remediation.

**Lack of long-term funding.** Environmental technology endeavors follow a developmental path similar to that of other kinds of technology. Generally six stages of technology development, as they relate to capital availability, are recognized. These include the following: (1) "Idea Development" involves product/process conceptualization and initial drawings, calculations, and theoretical validation; (2) "Proof of Concept" refers to the construction of a rough, yet functioning, model of the technology which is used to test the basic operating parameters and to aid in the design of an engineering prototype; (3) "Pilot" phase consists of constructing an actual working version of the technology to test the technology's operating performance and to gauge production requirements and feasibility; (4) "Prototype" stage involves development of a full-scale model which is constructed and operated as closely as possible to final production design standards and which is used to obtain operational and economic data; (5) "Application/Demonstration" stage involves operation of the technology under market conditions, as well as third-party demonstration and evaluation of the technology (e.g., federal or state evaluation programs); and finally, (6) "Commercial Sales" is the result of the first five stages and of extensive marketing and manufacturing activities (Bridging the Valley of Death, 2000).

As a technology is developed through the six stages, the capital needs increase substantially. However, capital availability does not follow the same pattern. As with many start-up companies, the source of capital in the early stages (e.g., Idea Development through Proof of Concept) is from the developer's personal savings, investments from family and friends, and/or funding from government agencies. Unfortunately, these initial sources of funding are usually depleted before the technology developer has a final model or has commercialized the product/process, plunging the developer into the "Valley of Death." Unlike the early R&D stages (i.e., Idea Development and Proof of Concept), there is little government funding for technology demonstration activities. Further, venture capitalists and potential customers typically wait until a technology has proven itself in the demonstration, usually after the product has become a proven marketplace contender, before making an investment. Thus, if technology developers do not possess or cannot obtain the significant capital required to survive the demonstration stages, the technology does not survive.

In the case of fungal-based remediation, there were a number of government agencies (e.g., Environmental Protection Agency, Department of Defense, Department of Energy, National Institutes of Environmental Health) and private entities (e.g., Electric Power Research Institute) that considered work on the initial "Idea Development" and "Proof of Concept" stages to be appropriate and thus provided funding. This work, which took place primarily during the 1980s, demonstrated that a wide variety of pollutants could be degraded by WRF, that WRF could effectively degrade labile pollutants like PCP, TNT and some pesticides in laboratory studies in soil to very low levels, and that these results could be repeated, at a small scale, in the field.

In the developmental process, laboratory soil treatability studies and small-scale field investigations were considered proof of concept. Unfortunately, as discussed above, these investigations did not address methods for large-scale inoculum production or engineering applications for soil treatment. Thus, the technology entered the intermediate developmental stages (i.e., Pilot, Prototype and Application/Demonstration stages) without a defined "working version" for which operating performance, aside from pollutant degradation, could be adequately determined. In particular, there were no standard methods for inoculum production, quantitation of inoculum potential, and mixing of soil and inoculum, or an established engineering design with identified environmental conditions (e.g,. soil moisture content and temperature) for effective, consistent fungal growth and pollutant degradation. As a result, many different approaches were taken by various technology developers for inoculum production and field engineering when opportunities to demonstrate the technology were plentiful compared to the current situation. As would be expected from this approach, treatment success was more a result of chance then of a knowledgeable approach. Despite demonstrated potential based on positive results from laboratory treatability studies, full-scale treatment performance was often unsatisfactory due to low quality inoculum and/or inability to manage environmental conditions conducive to fungal growth and pollutant degradation. Lack of satisfactory field performance and a naive concept of the magnitude of funding required to commercialize the technology caused both public- and private-sector interest in further development of the technology to wane. For example, the Environmental Protection Agency lost interest in the technology after expending approximately $2.5 million in the period of 1984 through 1996, even though the EPA project officer, Dr. John Glaser, was not in agreement with the decision to terminate funding. In the case of private investment, at least two companies, INTECH 180 Corporation located in Logan, UT and Mycotech Corporation located in Butte, MT, found it difficult to raise continued investment because the technology was not able to achieve commercial status fast enough.

These factors have lead to a perceived "Valley of Death" status of the technology today, resulting in only a few small companies continuing the attempt to commercialize the technology by looking for opportunities to demonstrate fungal-based remediation at applicable hazardous waste sites. In our own experience, proper demonstration of the technology usually involves conducting a laboratory-scale treatability study which, if successful, is followed by on-site pilot demonstration and finally full-scale treatment. Further developments in various aspects of the technology are worked into the treatability and pilot phases and "demonstrated" during full-scale treatment. Thus, the intermediate developmental stages are overlapping the "Commercial Sales" stage, which makes marketing the technology difficult. Despite the difficulties encountered during the developmental stages, there have been several successful commercial treatments, as discussed in greater detail below, that have demonstrated the potential usefulness of the technology.

**Initial overselling of the technology.** An error that was made early in the development of fungal-based remediation that resulted in an unrealistic view of both the market and technical potential of the technology was extrapolation of results obtained from pollutant degradation work conducted in aqueous culture to predict performance in the soil environment. This extrapolation ignored the influence of factors affecting the degradation of pollutants in soil and their ramifications for technology performance. Some of the earliest papers on fungal pollutant degradation dealt with the treatment of extremely recalcitrant and hydrophobic compounds like PCBs (e.g., Arochlor 1260), DDT, and polychlorinated dioxins (e.g., 2,3,7,8-tetrachlorodibenzodioxin) (Eaton, 1985 ; Bumpus et al., 1985). Demonstrations of the ability of WRF to mineralize refractory compounds and of the *in vitro* oxidation of those compounds by extracellular fungal lignin-degrading enzymes (e.g., lignin peroxidase) led some technology developers to promote use of the technology for remediation of soils contaminated with such chemicals.

Application of fungal-based remediation to the degradation of DDT provides an excellent example of this. Initial work on the degradation of DDT by the white-rot fungus *Phanerochaete chrysosporium* demonstrated that the fungus was able to mineralize DDT in aqueous culture, under ligninolytic conditions (Bumpus and Aust, 1987), and in soil/corn cob mixtures in which the DDT had been spiked into the soil (Fernando et al., 1989). The results of this work led one of the leading bioremediation firms at the time, ECOVA Corp., to conduct a series of bench-scale treatability studies to determine the feasibility of using the technology to treat soil contaminated with DDT and metabolites from the Leetown Pesticide Site located in northeastern West Virginia (Nelson and Kinsella, 1990). Despite obtaining reportedly "confluent" growth in their soil cultures, no significant degradation of DDT or its metabolites (i.e., DDD and DDE) by *P. chrysosporium* was observed. With a water solubility of $6 \times 10^{-3}$ mg $L^{-1}$, a $K_{ow}$ of 3.98 and a $K_{oc}$ of 5.37 (Winegardner, 1996), DDT is considered to be immobile in soil and therefore not amenable to microbial degradation. This is supported by the fact that DDT is detected in soil 20 years after application (Skryabin et al., 1978). Failure to recognize early in the technology development phase that pollutant "bioavailability" exerts the same constraints on fungal-based remediation that it does on other biologically-based remediation technologies resulted in "failures" of the technology (i.e., the DDT project discussed above, other unsuccessful and unpublished bench-scale treatability studies, and unsatisfactory published performances [e.g., Davis et al., 1993; Lamar et al., 1994]). This resulted in the view held by some that the technology is ineffective. This view however, ignores the number of successful studies, projects, and full-scale applications that have addressed treatment of contaminants like PCP and TNT that have proven to be very amenable to fungal treatment (see Lamar and Dietrich, 1990; Lamar et al., 1993; White and Lamar, 1999; Plexus Scientific, 1999).

**Application by unskilled individuals.** Like many technologies, especially technologies based on biological systems, skilled application is often an art acquired through experience. On the surface, application of fungal-based remediation appears to be deceptively simple -- grow a WRF on a lignocellulosic substrate to produce a fungal inoculum, mix the inoculum with the contaminated soil, and treat the soil by landfarming or biopiling. However, successful, consistent application of fungal-based remediation requires an intimate knowledge of fungal physiology, fungal biochemistry, engineering, and other factors. Application attempts by personnel not in possession of this information have resulted in technology failures.

Successful application of the technology comes from a consistent and methodical approach. The approach we use consists of four phases: the bench-scale treatability test, the on-site pilot test, inoculum production, and full-scale treatment.

The Bench-Scale Treatability Test - The first phase involves a bench-scale study conducted using representative contaminated soil from a site. The purpose of the treatability study is to determine the most effective combination of fungal species, inoculum application rate, and necessity of amendments to achieve the desired cleanup levels. It is usually necessary to evaluate at least two fungal species because it has been demonstrated repeatedly that no single fungus is effective in all soils for treatment of all contaminants (Lamar et al., 1999). This is also the reason that spent mushroom compost from commercial mushroom growers may only be considered for use if it is evaluated during treatability testing and if the commercial fungus is effective at degrading the pollutant in the contaminated soil. Because the cost of producing the fungal inoculum is a significant part of the overall technology cost, it is important to determine the lowest amount of fungal inoculum required to achieve the desired cleanup levels by evaluating several inoculum application rates. It is also important to conduct the treatability study using the inoculum substrates that will be used to produce the inoculum for full-scale treatment. Finally, depending on the soil texture and the nature of the contaminants, the necessity of using bulking agents to provide adequate aeration and/or surfactants to enhance the bioavailability of hydrophobic pollutants also needs to be evaluated.

On-Site Pilot Test - Based on the results of the bench-scale treatability study, an on-site pilot test is usually conducted to evaluate the most effective one or two treatments identified in the treatability phase. The pilot test gives the technology applicator an opportunity to design and fine tune the engineering of the biopile system and to determine logistics to be used for soil excavation, inoculum/soil mixing, and biopile placement that will be used during full-scale treatment. Also, depending on the results, the test gives the client and regulatory agency another level of confidence in the effectivenss of the technology. These tests are usually conducted using between 1 $m^3$ and 10 $m^3$ of soil (White and Lamar, 1999).

Inoculum Production - Production of large quantities of a lignocellulosic substrate supporting an actively growing culture of a WRF is critical to successful application of the technology. As discussed above, inoculum production involves the delivery of the substrates to the site, pasteurization of the substrates using either fumigation or steam treatment, and inoculation with either a primary liquid inoculum or a solid substrate culture of the fungus. The latter is usually purchased from a commercial mushroom spawn producer. After pasteurization, the substrate(s) are inoculated with primary inoculum. If liquid primary inoculum is used, the substrates are typically placed on a tarp into rows approximately 1.8 m wide and 30 cm deep. The inoculum is sprayed on the surface and injected into the substrates. If the solid primary inoculum is used, it is mixed with the pasteurized substrates using specialized mixing equipment or a front-end loader and the inoculated substrates are formed into rows, as described. Once inoculated, the substrates are covered with polyethlene tarp. The edges of the upper tarp are secured to allow aeration but to prevent removal by wind. Use of this configuration allows release of metabolic heat and prevents the occurrence of thermogenesis in the substrate, which would prevent the growth of the fungus, particularly in the middle of the row where temperatures would be highest. The rows can be as long as is desirable. It takes from 3 to 6 weeks, depending on ambient temperatures, for the fungus to colonize the substrates. This method has been used successfully to produce large quantities of fungal inoculum. However, this method is expensive because the process is labor intensive and it requires two visits to the site -- once to prepare the inoculum and once to inoculate the contaminated soil and construct the biopile. The same techniques, at a smaller scale, are used to produce inoculum for the pilot-scale test.

Full-Scale Treatment - Once the substrate is well, but not completely, colonized it is mixed with the contaminated soil and the biopile is constructed with a ventilation system (i.e., perforated PVC piping and a blower). The ventilation system is engineered so that air can either be drawn down through the inoculum/soil mixture or re-circulated through the mixture. This ventilation allows for control of pile temperature, through removal of metabolic heat produced by the fungi, and aeration (i.e., provision of oxygen). The pile is generally instrumented with moisture and temperature probes and air flow monitors. The probes are connected to a data logger that allows collection of data and response to conditions in real time.

## CURRENT TECHNOLOGY STATUS

Despite the problems encountered in development of a fungal-based soil remediation technology as discussed above, several successful commercial projects have been completed. Most notably, approximately 700 $m^3$ of PCP- and lindane-contaminated soil at a former wood-treating operation in North Carolina was remediated using the technology protocol described in the preceding section (White and Lamar, 1999). Even with success, this project highlighted the difficulty and expense of preparing inoculum on-site and the need for a standard engineered design for treatment of the soil. Development of inexpensive fungal inocula is critical to the commercial success of the technology as is an engineering design that can be used in a variety of climatic conditions. With these tools, the technology will be more marketable and easier to apply. There does appear to be re-kindled interest in the technology, especially in international markets. In the United States, convincing potential clients to even consider a laboratory treatability test often remains difficult.

**REFERENCES**

Barr, D.P. and S.D Aust. 1994. "Mechanisms White Rot Fungi Use to Degrade Pollutants." *Environmental Science and Technology*. 28(2):78A-87A.

Bennett, J. W., A. J. Turner, A. K. Loomis, and W. J. Connick, Jr. 1996. "Comparison of alginate and "pesta" for formulation of *Phanerochaete chrysosporium*." *Biotechnology Techniques* 10(1):7-12.

Bradley, C. A., R. D. Kearns, P. P. Wood, W. E. Black. 1996. "Bioremediation method using a high nitrogen-containing culture of white rot fungi on sugar beet pulp." United States Patent # 5,486,474

Bridging the Valley of Death. Chapter 2 *in* "Financing Technology for a Sustainable Future." http://web.dandp.com/enviroweb/asnbr2.htm

Bumpus, J. A., M., Tien, D. Wright, and S. D. Aust. 1985. "Oxidation of persistent environmental pollutants by a white-rot fungus." *Science* 228(5):1434-1436.

Bumpus, J. A. and S. D. Aust. 1987. "Biodegradation of DDT [1,1,1-trichloro-2,2-bis(4-chlorophenyl)ethane] by the white rot fungus *Phanerochaete chrysosporium*." *Appl. Environ. Microbiol.* 53:2001-2008.

Bumpus, J. A., T. Fernando, M. Jurek, G. J. Mileski. and S. D. Aust. 1988. "Biological treatment of hazardous waste by *Phanerochaete chrysosporium*." *Proceedings of the Engineering Foundation Conference*, April 1988, Longboat Key, Fl.

Childress, A. M., J. W. Bennett, W. J. Connick, Jr. and D. J. Daigle. 1998. "Formulation of filamentous fungi for bioremediation." *Biotechnology Techniques* 12(3):211-214.

Davis, M. W., J. A. Glaser, J. W. Evans, and R. T. Lamar. 1993. "Field evaluation of the lignin-degrading fungus *Phanerochaete sordida* to treat creosote-contaminated soil." *Env. Sci. Technol.* 27:2572-2576.

Chang, H. M., T. W. Joyce and T. K. Kirk. 1987. "Process for treating effluent from a pulp or papermaking operation." US Patent No. 4, 554, 075.

Eaton, D., H. M. Chang, and T. K. Kirk. 1980. "Decolorization of kraft bleach effluents." *Tappi J.* 63:103-106.

Eaton, D. 1985. "Mineralization of polychlorinated biphenyls by *Phanerochaete chrysosporium*:a ligninolytic fungus." *Enzyme Microb. Technol.* 7:194-196.

Eggen, T. 1999. "Application of fungal substrate from commercial mushroom production-*Pleurotus ostreatus*-for bioremediatrion of creosote contaminated soil." *Int. Biodeter. Biodegrad.* 44:117-126.

Fernando, T. J. A. Bumpus, and S. D. Aust. 1990. "Biodegradation of TNT (2,4,6-trinitrotoluene) by *Phanerochaete chrysosporium*." *Appl. Environ. Microbiol.* 56:1666-1671.

Fukuzumi, T., A. Nishida, K. Aoshima, and K. Minami. 1977. "Decolourization of kraft waste liquor white rot fungi. I. Screening of the fungi and culturing condition for decoulorization kraft waste liquor." *J. Jap. Wood Soc.* 23:290-298.

Huynh, V. B., H. M. Chang, T. W. Joyce and T. K. Kirk. 1985. "Dechlorination of chloro-organics by white-rot fungus." *Tappi J.* 68:98-101.

Kirk, T. K., W. J. Conners, R. D. Bleam, W. F. Hackett and J. G. Zeikus. 1975. "eparation and microbial decomposition of synthetic lignins." *Proc. Nat. Acad. Sci.* 72:2515-2519.

Lamar, R. T., D. Lestan, C. E. Smith, D. Dietrich. "Fungal inoculum preparation." U. S. Patent 5,786,188.

Lamar, R. T. and D. M. Dietrich. 1990. "In situ depletion of pentachlorophenol from contaminated soil by *Phanerochaete* spp." *Appl. Environ. Microbiol.* 56:3093-3100.

Lamar, R. T., J. W. Evans and J. A. Glaser. 1993. "Solid-phase treatment of a pentachlorophenol-contaminated soil using lignin-degrading fungi." *Environ. Sci. Technol.* 27:2566-2571.

Lamar, R. T., M. Davis, D. M Dietrich, and J. A. Glaser. 1994. "Treatment of a pentachlorophenol- and creosote-contaminated soil using the lignin-degrading fungus *Phanerochaete sordida*: A field demonstration study." *Soil Biol. Biochem.* 26:1603-1611.

Lamar, R. T., L. M. Main, D. M. Dietrich and J. A. Glaser. 1999. "Screening of fungi for soil remediation potential." pp. 437-456 *in* (Adriano, D. C., Bollag, J. M., Frankenberger, W. T., Jr., and R. C., Sims) *Bioremediation of Contaminated Soils*. Agronomy No. 37.

Lestan, D. and R. T. Lamar. 1996. "Development of fungal inocula for bioaugmentation of contaminated soils." *Appl. Environ. Microbiol.* 62:2045-2052.

Lestan, D., M. Lestan, J. A. Chapelle and R. T. Lamar. 1996. "Biological potential of fungal inocula for bioaugmentation of contaminated soils." *J. Ind. Microbiol.* 16:286-294.

Lestan, D., M. Lestan, and R. T. Lamar. 1998. "Growth and viability of mycelial fragments of white-rot fungi on some hydrogels." *J. Ind. Microbiol Biotechnol.* 20:244-250.

Lestan, D. and R. T. Lamar. 1999. "Influence of humidity on production of pelleted fungal inoculum." *World J. Microbiol. Biotechnol.*

Loske, D., A. Hutterman, A. Majcherczyk, F. Zadrazil, H. Lorsen and P. Waldinger. 1990. "Use of white-rot fungi for the clean-up of contaminated sites." Pp. 311-322 *in Advances in Biological Treatment of Lignocellulosic Materials* (Coughlan, M. P. and M. Y. A. Collaco, eds.) Elsevier, London and New York.

Lundquist, K., T. K. Kirk and W. J. Conners. 1977. "Fungal degradation of kraft lignin and lignin sulfonates prepared from synthetic 14C-lignins." *Arch. Microbiol.* 112:291.

Nelson, M. J. and J. V. Kinsella. 1990. "Bench-scale treatability study." Leetown Pesticide Site NUSP/89-0526. *REM III Program-EPA Contract 68-01-7250.*

Okeke, B. C., J. E. Smith, A. Paterson, I. A. Watson-Craik. 1993. "Aerobic metabolism of pentachlorophenol by spent mushroom culture of 'shitake' mushroom (*Lentinus edodes*) in soil." *Biotechnol. Lett.* 15:1077-1080.

Paszcynski, A. and R. L. Crawford. 1995. "Potential for bioremediation of xenobiotic compounds by the white-rot fungus *Phanerochaete chrysosporium.*" *Biotechnol. Prog.* 11:368-379.

Plexus Scientific Corporation. 1999. "Biotechnology Demonstration Final Report." U. S. Army Environmental Center. *DAAMo1-94-D-0020.*

Ryan, T. P. and J. A. Bumpus. 1989. "Biodegradation of 2,4,5-trichlorophenoxyacetic acid in liquid cluture and in soil by the white-rot fungus *Phaerochaete chrysosporium.*" *Appl. Env. Microbiol. Biotechnol.* 31:302-307.

Skryabin, G. K., L. A. Golovelva, E. L. Golovlev, R. N. Pertsova, A. M. Zyakun, Yu. V. Shurukhin and B. P. Strekozov. 1978. "Decomposition of DDT and its analogs by soil microflora." *Izvestiya Akademii Nauk SSR, Seriya Biologicheskaya* 3:352-365.

White, R. B. and R. T. Lamar. 1999. "Degrading ability of white rot fungi." *Soil and Groundwater Cleanup* Dec./Jan.:11-16.

Winegardner, D. L. 1996. *An introduction to soils for environmental professionals*. CRC Press, Inc. 270 p.

# FLOCCULANT SELECTIVITY FOR ORGANIC COMPOUNDS AND METALS REMOVAL FROM PROCESS WATER.

***Kei Shiohara***, Abdolhamid Borazjani and Mary Hannigan
(Mississippi State University, MS State, MS)

**ABSTRACT:** Wood treating operations produce a large quantity of process water containing organic and inorganic chemicals during their treatment cycle. Various flocculants are used for the treatment of contaminated wastewater before the water is discharged or sent for further treatment. In this study, five commercial polymeric flocculants were evaluated for the treatment of wood-preservative contaminated process water. The wastewater contained high levels of oil and grease, suspended solids, petroleum hydrocarbons, polycyclic aromatic hydrocarbons (PAHs) associated with creosote, pentachlorophenol (PCP), and chromated copper arsenate (CCA). Analyses of selected PAHs revealed that all five flocculants removed more than 85% of the total selected PAHs with similar results. The PCP results varied from 89% to 98% removal. Significant reduction of total petroleum hydrocarbons (TPH), total suspended solids (TSS), oil and grease, copper, chromium and arsenic were also observed for all treatments.

## INTRODUCTION

A large amount and variety of flocculants and coagulants are consumed by industries such as pulp and paper, and mining for wastewater treatment every year. Coagulants/flocculants also remove suspended substances from water at drinking water and municipal wastewater treatment facilities. Coagulants destabilize fine and stable colloidal suspensions and induce precipitation. Flocculants help individual particles combine into larger sizes and speed up the settling process. Inorganic materials like alum (aluminum sulfate), aluminum chloride, ferric sulfate and ferric chloride, and organic polymers such as polyamines and polyacrylamides have been widely utilized for such purposes (Ouellette 1996). While alum is still the most widely used coagulant/flocculant, the application of organic polymeric materials such as polyacrylamides is increasing. The main reasons for increasing application of organic polymeric flocculants, are treatment efficiency, the reduced amount of sludge produced from the treatment process, and less cost associated for further processing such as dehydration and disposal of the slugs.

Many studies have been done on the application, cost and effectiveness of polymeric coagulants/flocculants in the past decade. Cathalifaud et al. (1993) reported the usage of a suspension of kaolinite and ferric chloride could improve coagulation-flocculation process, and de Velasquez et al. (1998) demonstrated 87% turbidity removal and 87% total suspended solid elimination by treating wastewater effluent from biological and chemical treatment systems with anionic and zwitterionic polymers in combination with metallic hydroxide. Boneva et al. (1994) and Bolto (1995) enhanced the treatment of wastewater with water-soluble polymers. New cationic flocculants of high charge densities and high molecular weights were introduced for the removal of oil & grease (Sang et al. 1995). To reduce the cost of treatment process, the application of synthetic organic polymers such as poly(diallyldimethylammonium chloride) was evaluated. Addition of this polymer reduced the usage of poly(aluminum chloride) in the treatment by 50% (Lee et al. 1998).

The wood-preserving industry uses a variety of chemicals such as creosote, pentachlorophenol (PCP) and chromated copper arsenate (CCA) in their treatment process. A large volume of contaminated wastewater containing these chemicals are generated that must be cleaned up. Creosote is a complex mixture of chemical constituents, including various polycyclic aromatic hydrocarbons (PAHs), phenolic compounds, and heterocyclic compounds. PCP is another widely used organic wood preservative that is used in a 5% concentration in number two diesel oil for the pressure treatment of wood. Technical grade PCP also contains lower chlorinated phenols and minute amounts of chlorobenzodioxins, chlorobenzofurans and chlorobenzenes. CCA is the most utilized water-born wood preservative for pressure-treatment of wood that resists attack by termites and fungi that cause decay. These chemical substances are introduced to the process wastewater from several different sources, including pressure treatment, steaming of wood, vapor drying or oil seasoning. Treatment of such wastewater is difficult because wastewater contains various chemical compounds having different chemical and physical properties. No major prior studies have been done on the removal efficiency of commercial flocculants/coagulants for the treatment of process water from wood treating facilities.

By removing a significant concentration of chemicals from process water so quickly, polymeric flocculants are able to reduce the time and cost for further treatments such as carbon filtration, biological treatments in bioreactors and clarification through artificial marshes.

**Objective.** This study evaluated different commercial polymeric flocculants that are current treatments of wood-preserving process wastewater containing high concentrations of oil and grease, polycyclic aromatic hydrocarbons (PAHs), pentachlorophenol (PCP), petroleum hydrocarbons, suspended solids, and chromated copper arsenate (CCA).

## MATERIALS AND METHODS

The wastewater for this research was obtained from two different wood treatment sites in Mississippi. Wastewater from one site contained PCP and CCA, while the other contained creosote. Both had high levels of oil and grease and suspended solids. The wastewater was mixed vigorously using an industrial stainless steel blender to form a 1/1 (v/v) mixture. The flocculants in this study were obtained from five different wood treatment plants, and the amounts were adjusted to the same concentration of flocculants that are currently used at each plant (Table 1). One liter amber bottles were used for the jar test of the wastewater treatments, and three replicates were made for each treatment. After the bottles were allowed to settle overnight, three 250 ml portions of the treated water were taken from each jar for continuous liquid-liquid extraction, total suspended solid analysis, and metal analysis.

**TABLE 1. Polymeric flocculants and coagulants used in study.**

| Treatment | Description of polymer* | Amount used |
|---|---|---|
| I | Acrylamide, Polymer with Acrylic Acid | 100 ppm |
| | Melamine, Polymer with Formaldehyde | 250 ppm |
| II | Tannin Based, Cationic | 100 ppm |
| | Coagulant | 75 ppm |
| III | Anionic Polyacrylamide | 5 ppm |
| | Cationic Polyacrylamide | 5 ppm |
| | Formaldehyde Resin | 250 ppm |
| IV | Aminomethylated Tannin Polymer | 70 ppm |
| V | Acrylamide, Polymer with Acrylic Acid | 100 ppm |

* From the material safety data sheets (MSDS), provided by the manufacturer.

Continuous liquid-to-liquid extraction with dichloromethane (EPA Method 3520; U.S. EPA 1992) was accomplished with 250 ml of the treated water, concentrated to 5 ml by Kuderna-Danish apparatus, and analyzed by gas chromatography (GC) for selected PAHs (TABLE 2), PCP (EPA methods 8040 and 8100; U.S. EPA 1992) and total petroleum hydrocarbons (MS method). The analysis of total suspended solids (TSS) and oil and grease were performed according to methods 2540D and 5520B, respectively, of "Standard methods for the examination of water and wastewater" (Clesceri et al. 1989). The content of arsenic, chromium and copper were analyzed by inductively coupled plasma (ICP) atomic absorption according to methods 200.9, 218.1 and 220.1, respectively, of "Methods for Chemical Analysis of Water and Wastes" (U.S. EPA 1983).

**TABLE 2. Selected polycyclic aromatic hydrocarbons (PAHs) in this study.**

| | | |
|---|---|---|
| naphthalene | acenaphthene | 1,2-benzanthracene |
| 1-methylnaphthalene | acenaphtylene | chrysene |
| 2-methylnaphthalene | fluorene | pyrene |
| biphenyl | carbazole | benzo[a]pyrene |
| anthracene | dibenzofuran | benzo[ghi]perylene |
| phenanthrene | fluoranthane | |

The data were analyzed using the completely random design (CRD) method ($\alpha = 0.05$), using the S.A.S. software (Release 6.12, Statistical Analysis System Institute, Cary, North Carolina).

## RESULTS AND DISCUSSION

ICP atomic absorption analysis of metal elements (Figure 1) revealed that the treatments reduced copper, chromium and arsenic from the initial concentrations of 150, 50, and 13 ppb to less than 30, 10, and 5 ppb respectively. Although the treatments reduced the concentration of metals, statistical analysis of the data revealed that there were no significant differences between the five treatments, due to a large deviation in the data obtained. The concentration of the metals after treatment were very close to the detection limit of the analyses, and in some cases one or more of the replicates for each metal at each treatment were under the detection limit. The initial concentration level of copper, chromium and arsenic in the wastewater used in this study were very low and this could have contributed to this situation.

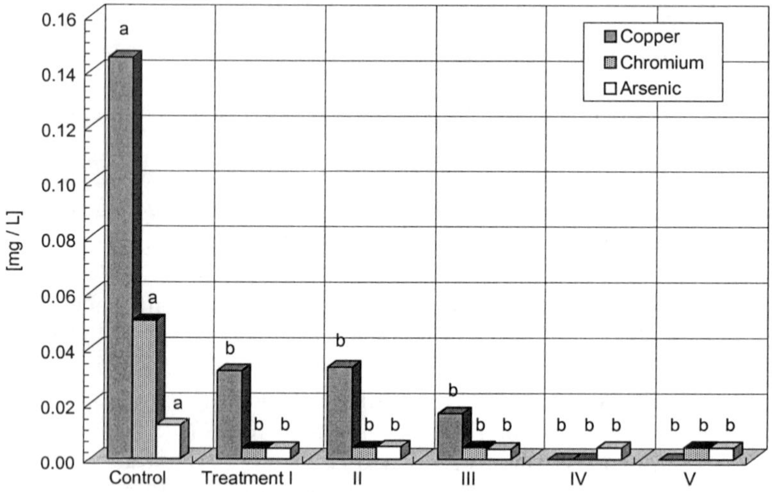


**FIGURE 1. Inductively coupled plasma atomic absorption analysis of copper chromium and arsenic in process water after treatment with flocculants. Each column represents an average of three replicates, and columns with the same letter are not significantly different ($\alpha = 0.05$).**

The data for PAH concentration obtained by GC analysis are shown in Figure 2. These data show that total selected PAHs significantly decreased from an initial concentration of approximately 760 mg/L to under 100 mg/L for each treatment. These results show treatments III, I and IV were the most effective on the removal of selected PAHs. The concentrations were reduced to 48, 37 and 30 mg/L, or 93, 95 and 96% reduction respectively.

Figure 3 depicts the decrease of PCP concentration for the five treatments. The average initial concentration of 45 mg/L was reduced to 1-5 mg/L after treatments. Treatments III, I and IV were the most effective on the removal of PCP, and concentrations were reduced to 2.5, 1.3 and 0.8 mg/L respectively, or 95, 97 and 98% reduction respectively.

Total petroleum hydrocarbons (unbranched C10~C25) data are

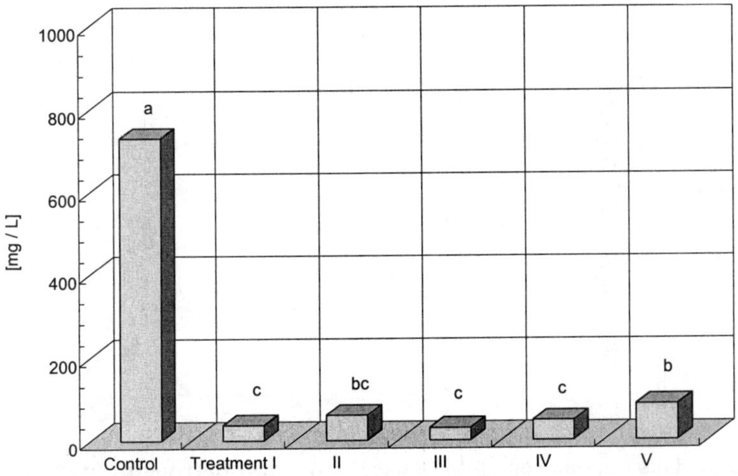


**FIGURE 2. Gas chromatographic analysis of total selected PAHs in process water after treatment with flocculants. Each column represents an average of three replicates, and columns with the same letter are not significantly different ($\alpha = 0.05$).**

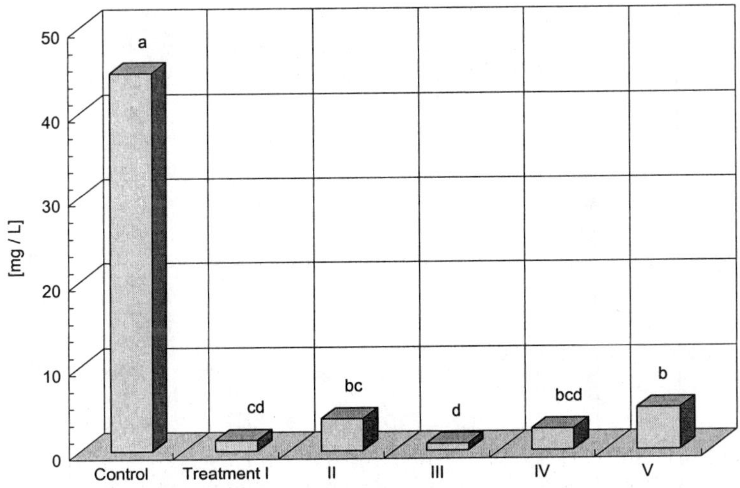


**FIGURE 3. Gas chromatographic analysis of PCP in process water after treatment with flocculants. Each column represents an average of three replicates, and columns with the same letter are not significantly different ($\alpha = 0.05$).**

summarized in Figure 4. For treatments I through IV, 97, 95, 98, and 95% of total petroleum hydrocarbons were removed respectively, while treatment V only removed 85% of the total petroleum hydrocarbons.

The data obtained from oil and grease analysis are shown in Figure 5. The oil and grease analysis reflects not only the PAHs, PCP, and TPH, but also many

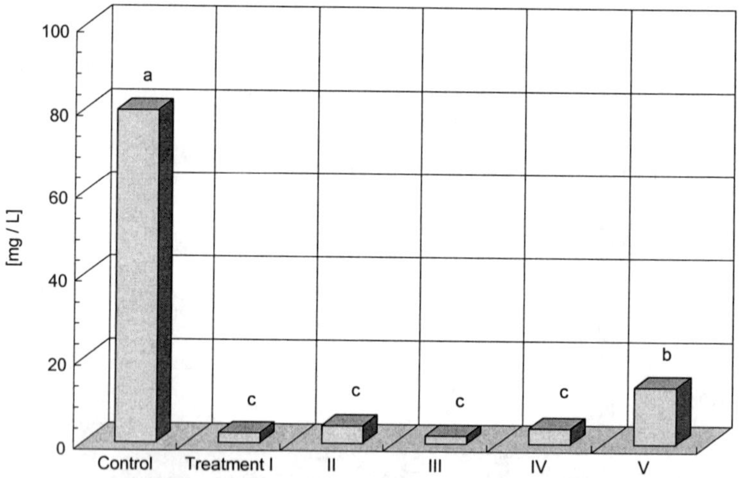


**FIGURE 4. Gas chromatographic analysis of total petroleum hydrocarbons (unbranched C10~C25) in process water after treatment with flocculants. Each column represents an average of three replicates, and columns with the same letter are not significantly different ($\alpha = 0.05$).**

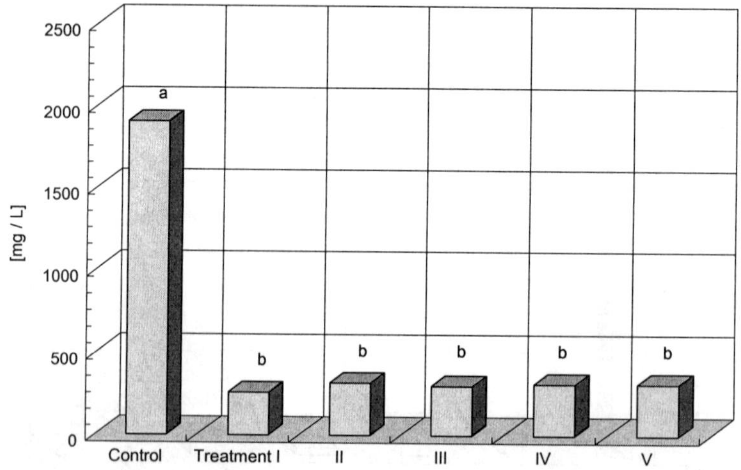


**FIGURE 5. Oil and grease analysis of process water after treatment with flocculants. Each column represents an average of three replicates, and columns with the same letter are not significantly different ($\alpha = 0.05$).**

other organic compounds that are not individually identified. The initial concentration of 1900 mg/L was reduced to approximately 300 mg/L, which resulted in the removal of 86% of oil and grease in treatment I, 84% in treatments III, IV and V, and 83% in treatment II. No significant differences among the five treatments was observed.

Figure 6 summarizes the analysis of total suspended solids (TSS). Treatments I and III reduced the initial concentration of 590 mg/L TSS to less

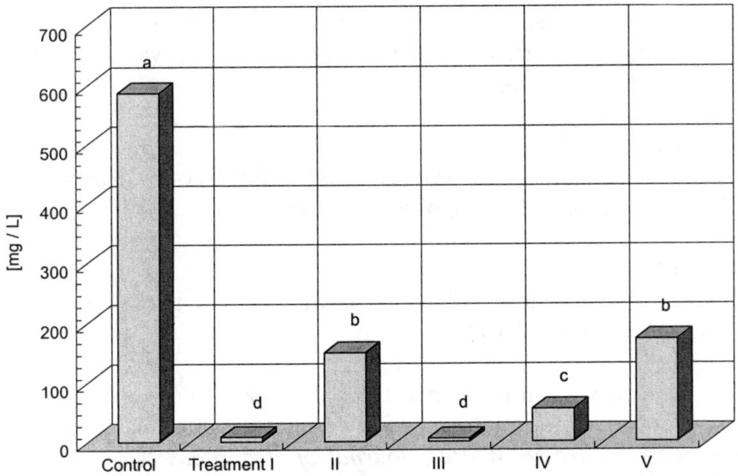


**FIGURE 6. Total suspended solids (TSS) analysis of process water after treatment with flocculants. Each column represents an average of three replicates, and columns with the same letter are not significantly different ($\alpha$ = 0.05).**

than 10 mg/L (more than 98% reduction of TSS), while treatments II and V removed 75% and 70% respectively. Unlike oil and grease that did not show significant differences between the treatments, results for TSS show significant differences between the treatments, and removal efficiencies of TSS resembles the results for selected PAHs, PCP and TPH in which treatments I and III showed the best performance. This suggests that the compounds analyzed here form a suspended solid or are being absorbed on suspended solid particles that can be filtered. Therefore, in the treatment of wastewater generated from wood treatment plants, the reduction of total suspended solids seems to be essential in the removal of PAHs, PCP and TPH.

In all treatments, significant decrease of the concentration of chemicals in the treated water was observed. Compared to other treatments, treatments I and III seemed to be more effective in the removal of TTS and organic chemicals included in process wastewater from wood treatment plants. The polymer or resins described as "Melamine, Polymer with Formaldehyde" in treatment I and "Formaldehyde Resin" in treatment III seem to be the key components in the flocculants, since the polyacrylamide in treatments I and V were the same polymer provided by the same manufacturer, and treatment V seemed to be the least effective of the five treatments. Although it is still not clear how these polymer resins function in the treatments, they appeared to enhance the elimination of total suspended solids from the treated water, which is the essential factor to be monitored resulting in the most effective removal of chemicals such as PAHs, PCP and TPH found in process wastewater.

The treatment of process wastewater using polymeric flocculants seems to be a rapid and very effective method to remove high concentrations of chemicals from process water before it is sent for further treatments such as activated carbon treatments or biological treatments such as bioreactors and artificial marsh or wetlands and finally released. However, further studies need to be conducted to measure and evaluate the efficiency of these flocculants for wastewater treatment containing high levels of CCA

**ACKNOWLEDGEMENT**
Approved for publication as Journal Article No. FP204 of the Forest and Wildlife Research Center, Mississippi State University.

**REFERENCES**

Bolto, B.A. 1995. "Soluble polymers in water purification." *Progresses in Polymer Science*, 20 (6): 987-1041.

Boneva, M., B. Boneva, and P. Pipeva. 1994. "Utilization of some water-soluble polymers as flocculants for treatment of sewage containing oil." *Symposium on the Treatment of Hazardous Wastes, Journal of Hazardous Materials*, 37: 107-115.

Cathalifaud, G., M. T. Wais Mossa, and M. Mazet. 1993. "Preformed ferric hydroxide flocs as adsorbents of humic substances." *Water Science & Technology*, 27: 55-60.

Clesceri, L.S., A.E. Greenberg, and R.R. Trussel. 1989. *Standard Methods for the Examination of Water and Wastewater.* 17th Edition. Washington, DC; American Public Health Association.

De Velasquez, M., O. Manero, J. Cardoso, and G. Martinez. 1998. "New polymeric coagulants tested in water and wastewater." *Environmental Technology*, 19: 323-330.

Lee, S. H. et al. 1998. "Improvement of flocculation efficiency of water treatment by using polymer flocculants." *Environmental Technology*, 19: 431-436.

Ouellette, J. 1996. "Coagulants and Flocculant Rise." *Chemical Marketing Reporter.* 15: 1 .

Sang, J. V., D. K. Chung, C. Emond, and R. Boudreault. 1995. "Development of efficient flocculants for oil removal in DC casting water recycling systems." *Light Met, Minerals, Metals, and Materials Soc.*, 1027-1031.

U.S. EPA. 1992. *Test Methods for Evaluating Solid Waste Physical/Chemical Methods (updated).* SW846. Washington, DC; Office of Solid Waste and Emergency Response.

U.S. EPA. 1983. *Methods for Chemical Analysis of Water and Wastes.* Cincinnati, Ohio; Environmental Monitoring and Support Laboratory.

# COMPARISON OF BIOPILE EFFECTIVENESS WITH LAND TREATMENT AT A WOOD TREATING SITE

***Gilbert M. Long, P.E.,*** C. Hunter Nolen, P.E., Camp Dresser & McKee, Houston, Texas; Darrel Stordahl, P.E., Camp Dresser & McKee, Helena, Montana; Laureen S. Kelly, Montana Department of Environmental Quality, Helena, Montana; James C. Harris, P.E., USEPA, Helena, Montana

**Abstract:** The remedial measures conducted at the Montana Pole and Treating Plant federal Superfund site in Butte, MT included construction and operation of a land treatment unit (LTU) and biopiles to biodegrade pentachlorophenol (PCP) in excavated site soils. Both the LTU and the biopiles have achieved degradation of PCP and Total Petroleum Hydrocarbons. The biopiles were more challenging to design, construct and operate than the LTU, and generally did not treat the contaminants as rapidly as the LTU.

**SITE BACKGROUND.** The Montana Pole and Treating Plant (MPTP) site is located at 220 West Greenwood Avenue in Butte, Montana. The wood treating facility began operations in 1946 and shut down in 1984. With the exception of coal tar creosote reportedly used for a short period of time in 1969, the solution used to treat timber at the facility consisted of 5 percent pentachlorophenol (PCP) dissolved in 95 percent petroleum product (similar in characteristics and composition to diesel fuel).

The Environmental Protection Agency (EPA) commenced an Emergency Removal Action (ERA) on July 10, 1985 with the U.S. Coast Guard. Removal action activities occurred during the 1985 and 1986 field seasons. EPA excavated approximately 3,000 cubic yards of highly contaminated soils, bagged them, and placed them in storage buildings constructed onsite. Two groundwater interception/oil recovery systems were installed to alleviate oil seepage into Silver Bow Creek along the northern border of the site. Oil was recovered by physical separation, and separation underflow was reinjected to site groundwater via two infiltration galleries. Contaminated areas of the site and features of the groundwater recovery system were fenced to restrict public access. The groundwater and oil recovery system was maintained and operated by the Montana Department of Environmental Quality (DEQ) until February 1993.

In June 1992, EPA instituted an additional ERA to control and recover the light nonaqueous phase liquid (LNAPL) found during the Remedial Investigation (RI). The action included the installation of an LNAPL recovery and containment system and water treatment facility.

In September 1993 DEQ and EPA issued a Record of Decision (ROD) identifying the areas and constituents of concern at the site and specifying the approach to remedying conditions at the site. The approach selected was excavation and bioremediation of site soils, offsite disposal of debris and piping materials, and containment and treatment of site groundwater and recovered LNAPL. The primary constituent of concern, PCP, was given a cleanup standard for soils of 34 ug/g established in the ROD. In addition, carcinogenic polynuclear aromatic hydrocarbons (PAH) and dioxin constituents (based on Toxicity Equivalence Factors [TEF]) had cleanup levels specified.

The Phase 1 remedial action consisted of the following major activities:

- Excavation of approximately 46,000 cubic yards of contaminated soils from the north side of the site and removal of approximately 6,000 cubic yards of highly contaminated soil from the pole barns.
- Construction of 13 biopiles, with provision for an additional five biopiles if required in the future.
- Construction of a lined 9.2-acre biological land treatment unit (LTU) with approximately 8 acres of active treatment area.
- Construction of two groundwater recovery trenches (the near creek recovery trench and the near highway recovery trench).
- Construction of *in situ* infiltration basins on the site's north side.
- Construction of significant modifications to the water treatment system.

The Phase 1 remedy was constructed by September 1997. The Montana Bureau of Mines and Geology (MBMG) has been operating the site since January 1997.

During early 1999, the Phase 2 remedial action was begun, consisting of the removal and disposal of remaining debris at the site, along with other minor miscellaneous site work. The Phase 3 remedial action, consisting of removal of treated soils from the LTU, excavation of south-side soils, installation of *in situ* treatment systems on the north and south sides of the site, and other miscellaneous site work, was conducted from Summer 1999 through Winter 2000.

**ENGINEERING SELECTION PROCESS.** The remedy selection process was strongly influenced by the need to excavate a large quantity of soil during one season, due to additional remedial response activities being performed at other NPL sites along Silver Bow Creek. All the accessible impacted soils north of I-15/90 were excavated in one operation. Since the LTU was designed based on the natural site contours and had a capacity of approximately 25,000 cubic yards (CY) [19,100 cubic meters (CM)] of fluffed soil at a depth of 18 inches (0.46 m), but 46,000 CY (35,200 CM) of soil were excavated, the remaining soil required proper covered storage to minimize its impact on the environment after excavation. This led to the concept of using soil staging/pretreatment units (SSPs), using a biopile approach, to provide the necessary containment for the soils while including some capacity for hydrocarbon degradation. These units were sized to contain approximately 2,000 CY (1,520 CM) of soil each, so 13 biopiles held the remaining excavated soil and provided the opportunity to store and pre-treat the highly impacted soil stockpiled in containers in the Pole Barns. The LTU was designed with a HDPE bottom liner, a large external retention pond to provide runoff water management and a water supply for irrigation. The biopiles were designed with a HDPE bottom liner; air piping connected to a series of vacuum pumps to provide oxygen to the soil; an irrigation drip line system across the top of the piles to provide water and nutrient addition; and a top liner to prevent precipitation from flooding the piles and leaching contaminants from the soil.

Given the need to store materials in biopiles, a comparison of advantages and disadvantages of the two bioremediation methods was conducted. These are summarized in Table 1.

**TABLE 1. Design Considerations- LTU vs. Biopiles**

| *Design Consideration* | *Land Treatment* | *Biopile* |
|---|---|---|
| Technology Status for PCP | Well Proven in similar climate | Relatively Unproven for PCP |
| Space requirements | Large | Reduced |
| Treatment Time | Rapid | Slower |
| Capital Construction Cost | Medium | Medium |
| O&M Costs | Low | Medium |
| Peripheral Equipment | Irrigation, Tractor, Nutrient System | Vacuum System, Water/Nutrient System |
| Soil Preparation | Simple | Difficult |
| Dust Control | Irrigation | Cover |
| Sampling | Simple | Labor Intensive |
| Weather Effects | Flooding, Temperature | Covered, Warmer Longer |
| Water Usage/ Leachate Generation | High from Precipitation | Low |
| Soil Mixing | Simple | Impractical |

**LESSONS LEARNED DURING CONSTRUCTION AND OPERATION OF LTU.** The construction of the LTU proceeded uneventfully, as this technology has been well documented historically, and construction techniques are rather simple. The construction included preparation of the area, placement of an HDPE liner under the entire LTU, construction of earthen berms to contain runoff, placement of leachate collection piping and a sand drainage layer, and loading the impacted soils. Some construction considerations to be kept in mind include:

- Preparation of the area beneath the plastic liner included removal of large rocks and debris to prevent liner puncture.
- Light ground pressure vehicles were used to spread the sand layer over the liner.
- A ramp was required to load the impacted soil onto the LTU, and a low ground pressure tracked vehicle was used to spread the soil to the desired thickness. The first lift was spread to approximately 18-24 inches (0.46-0.6 m) deep, but it was found later that much deeper lifts could be applied.
- The berm was constructed with at least 12 inches (0.3 m) of freeboard above the top of the soil if possible.
- A triangle shape for the LTU permitted the efficient use of a commercial center pivot irrigation unit, so no special equipment or labor-intensive irrigation was required.
- The effective treatment depth extended at least 1 foot (0.3 m) beneath the tilled zone.

- Application of solid fertilizer on the surface of the LTU after soil application proved effective, and liquid nutrient addition via the irrigation system was not necessary.
- Soil screening during soil application can be a severe bottleneck if not set up properly, particularly during the wet season.
- Very little makeup water for irrigation besides captured leachate was required for moisture control. Most of the water loss was evaporative from the pond.

Although a liquid nutrient injection system was designed into the center pivot irrigation unit, solid fertilizer was added to the impacted soil after its initial placement to accelerate the biodegradation process. This first dose of nutrients was sufficient to biodegrade the initial lift of soil to below the cleanup standard with no further nutrient additions.

The first lift was removed from the LTU in the summer of 2000 and a second, larger excavation resulted in placement of 132,000 CY (101,000 CM) of soil to an average depth of 8 feet (2.4 m). After this lift was spread, it also was treated with solid fertilizer and tilled. After approximately 3 months of treatment, the top 30-inch (0.76 m) lift was sampled and found to be below the cleanup standard. This soil was then backfilled into open excavation areas at the site, and the remaining 5.5 feet (1.7 m) of soil is currently being treated.

**LESSONS LEARNED DURING CONSTRUCTION AND OPERATION OF BIOPILES.** The construction of the biopiles was not highly complex, but required more specialized and carefully orchestrated construction techniques than did the LTU. The basic components of construction were as follows:

1. Subgrade preparation and leachate collection/storage appurtenances construction
2. Liner placement
3. Shallow (1 ft) soil layer placement
4. Air piping placement
5. Remaining soil placement (and monitoring probe placement)
6. Irrigation piping placement
7. Outer liner placement and tie down
8. Connection to vacuum blower systems

The major lessons learned during the construction and operation of the biopiles are as follows:

- As with the LTU, proper planning and the use of properly sized equipment for soil preparation and handling was a critical component of successful construction. It was important to have high rate screening equipment and pug milling equipment in order to prevent bottlenecking at these points during construction.
- Moisture/oxygen/carbon dioxide probes installed within the piles did not work effectively after completion of construction. Monitoring of pile exhaust air proved adequate for evaluating pile air conditions.
- Outer liners are subject to very large wind loads and had to be tied down extremely well.

- Access to interior soils for performance sampling was difficult and labor intensive. Sampling also occurred through holes made in the liner, which required patching.
- The piles stayed warm later in the fall than the LTU, and therefore offered an apparently longer treatment season than the LTU.
- The piles were warmed in the Spring by running the blowers to pull warm air through the piles on warm days, thereby jump-starting the treatment season.
- Very little moisture addition was required over the treatment season, due to the presence of the liners and the creation of metabolic moisture during the biodegradation process.
- Care must be taken when irrigating the piles to prevent over-watering and pile slumping.
- In general, the piles did not appear to treat the site contaminants as rapidly as the LTU.

**PCP DEGRADATION RESULTS.** Based on a four-month operating season in the relatively cold climate of Butte, Figure 1 presents the degradation results for the top and lower layers of the first lift of soil on the LTU, compared with the degradation results for an average of 10 biopiles constructed from the same source of contaminated soil over a three-year period.

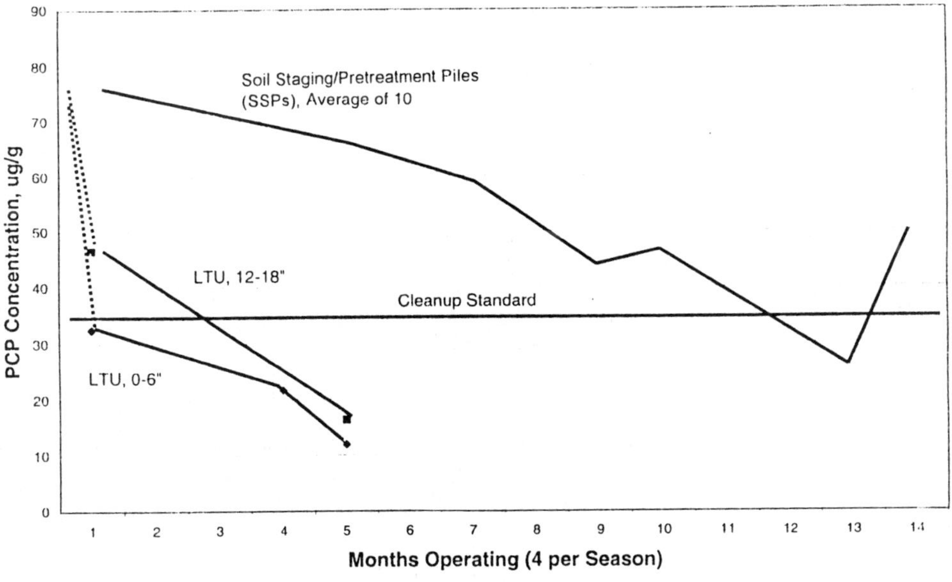


**FIGURE 1. PCP Degradation Results**

The initial sampling was conducted approximately one month after the LTU and biopiles were loaded; the biopiles exhibited a higher initial average contaminant concentration, which may be caused by the apparently rapid degradation in the LTU;. The top layer of soil on the LTU was below the cleanup standard by the time the initial sampling occurred. The average concentration of PCP in both layers of the LTU was reduced approximately 65% after one treatment season (July 1997 to June 1998), and was well below the cleanup standard of 34 ug/g.

The biopiles that were constructed from the same soil as the LTU exhibited considerably slower PCP degradation. Three operating seasons appeared to have reduced the PCP concentrations below the cleanup standard; however, confirmation sampling at the beginning of the fourth season indicated that the concentrations were still above 34 ug/g.

Although Total Petroleum Hydrocarbons (TPH) were not specified in the ROD as a cleanup requirement, TPH was measured in many of the same samples as PCP. These results are presented in Figure 2.

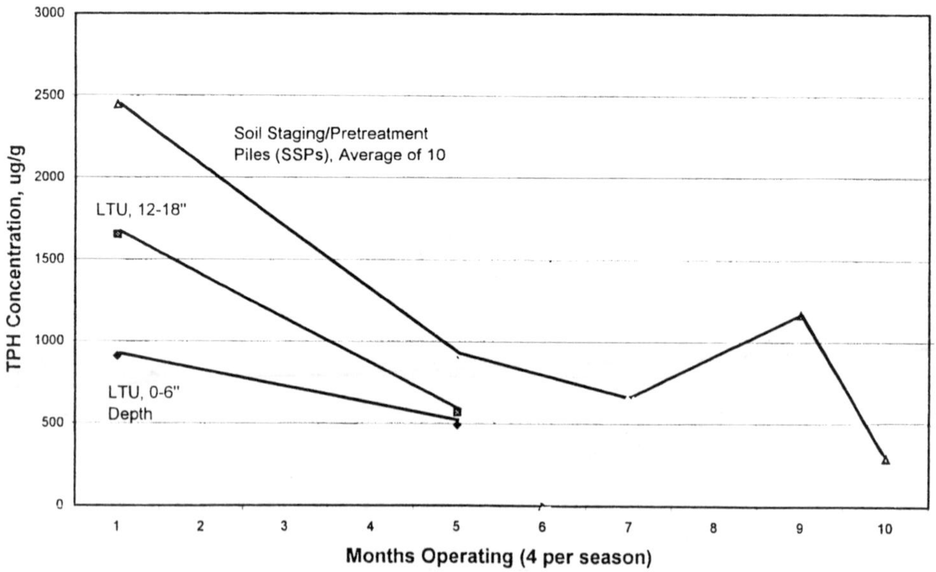


**FIGURE 2. TPH Degradation Results**

As with the PCP, the TPH in the bottom layer of the LTU degraded approximately 65% in one season. The TPH results for the top layer indicated a lesser rate of remediation, but both layers achieved concentrations of approximately 500 ug/g. The TPH concentrations at the time of the initial sampling were also higher in the biopile samples than for the LTU, probably for the same reasons indicated above for PCP. The TPH concentration average in the biopiles reached the 500 ug/g range in the third operating season, and again there is considerable variability in the data.

The highly contaminated soil that was stored in the storage buildings prior to construction of the biopiles was the subject of an additional test of a soil amendment demonstrated in laboratory studies to enhance the biodegradation of organic compounds. Two piles were amended with the additive and one pile was constructed identically except with no addition of the amendment. The piles were constructed using the same design as the biopiles described above.

The amended piles were observed to remain above the required minimum operating temperature of 50°F. (9°C.) for approximately two months longer than the untreated pile. However, the PCP degradation curves shown in Figure 3 do not seem to indicate improved degradation of PCP. These three biopiles achieved a higher percentage reduction of PCP than the less impacted soil piles, with greater than 50% reduction in PCP concentration compared with approximately 40% in the other biopiles.

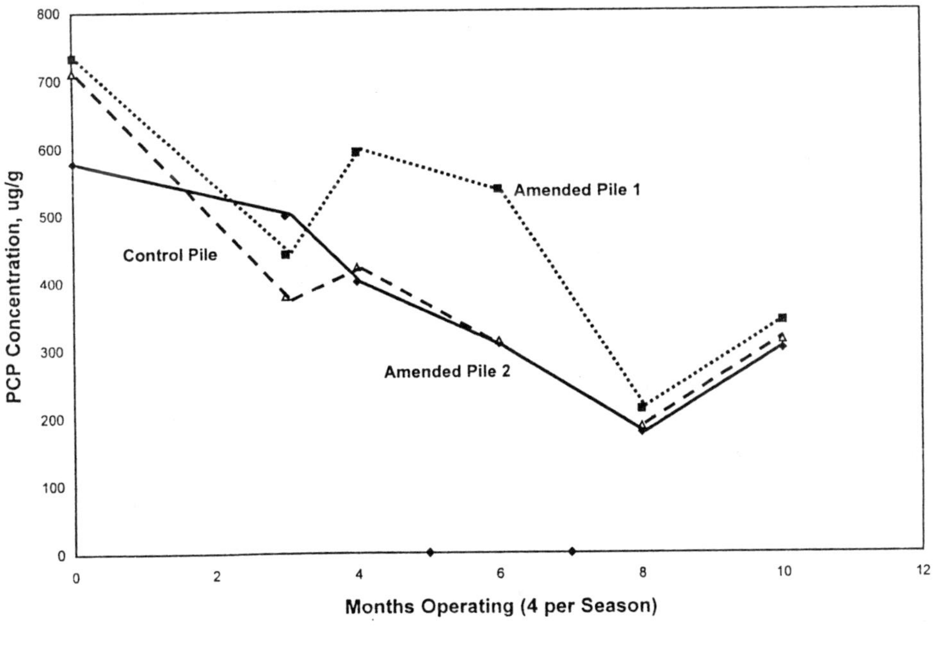


**FIGURE 3. PCP Degradation Results- High Concentrations**

**CONCLUSIONS.** Based on the results presented above, the LTU appears to be a more rapid soil treatment tool than biopiles. This has been hypothesized to result at least partially from the lack of mixing available in a biopile configuration. It also appears that significant degradation of hydrocarbons may occur during the excavation, soil preparation, and loading stages for both the LTU and biopiles.

It is not apparent from the available data whether the soil in the three highly contaminated biopiles will achieve the cleanup standard using biopiles, although significant reductions in PCP concentration occurred. These soils may eventually require placement in the LTU for further treatment. The amendment used in two of the biopiles did not exhibit increased degradation of PCP compared to the control biopile, even though the temperature of the amended piles was above 50°F. (9°C.) for approximately two months longer than the control pile.

## REFERENCES

*USEPA. 1993. Record of Decision, Montana Pole and Treating Plant Site, Butte, Montana. September 1993.*

Camp Dresser & McKee, 1995. *Preliminary Design Submittal and Revised Design Basis Report.* August 1995.

# LAND TREATMENT OF CONTAMINATED SOIL FROM WOOD-TREATING SITES

***James C. Harris, P.E.,*** (U.S. EPA, Region 8, Helena, Montana)

**Abstract:** Four former wood-treating Superfund sites located in the State of Montana are currently employing the use of biodegradation in engineered Land Treatment Units (LTUs) to achieve site remediation levels in contaminated soils for pentachlorophenol (PCP) and polynuclear aromatic hydrocarbon (PAH) compounds. The Superfund sites are the Libby Ground Water site, Libby, Montana, the BN Somers site, Somers, Montana, the Idaho Pole Company site, Bozeman, Montana and the Montana Pole and Treating Plant site, Butte, Montana. Contaminated soil from source areas at each site was excavated and placed in an LTU where the soil was irrigated, tilled and fertilized during the growing season from June through September. Remediation levels for soil, as required by the respective Records of Decision (RODs), have been achieved at all four of the sites. The remediation levels in soil for PCP range from 34 mg/kg to 48 mg/kg and the total carcinogenic PAH (cPAH or B2 PAH) levels range from 4.2 mg/kg to 59 mg/kg expressed as benzo(a)pyrene equivalents.

## INTRODUCTION

The four Montana wood-treating facilities that are the subject of this case study operated during the period from the early 1900s until as recently as 1997. Spills, tank overflows and leaks, piping breaks and drippage from treated products at each site resulted in soil and ground water contaminated with wood-treating fluids. Although a variety of wood-treating chemicals were used at each site, the primary treating fluids contained either PCP or creosote. Creosote is a compound that consists of numerous polycyclic compounds several of which are considered to be carcinogens. The PCP and creosote were typically dissolved in a carrier oil that varied in specific gravity from diesel fuel (light) to mineral spirits (heavy).

The study sites were listed on the National Priority List (NPL) of Superfund sites in the mid to late 1980s and RODs were subsequently issued for each site. The soil component of each remedy selected in the RODs included excavation of contaminated soil and treatment on-site of the contaminated soil in engineered Land Treatment Units (LTUs). The selection of bioremediation as the primary soil treatment technology for the four sites was based on the results of both field treatability test results and full scale performance of land treatment units used in the petroleum industry.

Design and construction of the LTUs began after ROD issuance with the first LTU being completed in 1990 and the most recent being completed in 1996. Each LTU featured an irrigation system to optimize soil moisture and to control dust, and each LTU was tilled and nutrients were added as required. The LTUs varied in size from 1 to 14 ac (0.4 to 5.7 ha) and were either rectangular or more typically pie-shaped to adapt to a center-pivot sprinkler configuration.

A total of nearly 25 years of operational and performance data for the LTUs at the four wood-treating sites is available and these data are presented in this case study.

**Objective.** The objective of this case study is to present data describing the performance of LTUs at 4 former wood-treating sites designed and constructed to reduce soil contaminants to levels specified in the ROD for each site. Design, construction and operational considerations based on experience with the 4 LTUs are also presented.

**Description of Sites:** A brief description of each site addressed in this study is presented below:

The **Libby Ground Water** site, located in Libby, Montana, is the location of a former wood-treating operation owned at one time by the J. Neils Lumber Company, the St. Regis Corporation, Champion International Corporation, now International Paper (IPC), (who continues to manage site remediation activities) and is currently owned by the Stimson Lumber Company. Soil contaminated with treating fluid (PCP and creosote formulations were used) was excavated from several source areas and was placed in an on-site storage/staging unit for placement on either of two rectangular 1-acre (0.4 ha) LTUs also located on-site (Figure 1). A total of approximately 45,000 loose cu yd (34,400 cu m) of contaminated soil was excavated at the site in 1988-89, and by the end of the 1998 treating season 13,758 cu yd (10,520 cu m) of soil in 6 to 8 in (15 to 20 cm) lifts had been treated to remediation levels. An additional LTU was constructed on-site in 1998 with a surface area of 12 ac (4.8 ha) and the remaining untreated soil was placed on the unit to a depth of approximately 1 ft (0.3 m). The additional LTU should allow IPC to complete the soil treatment process in 3 to 5 years instead of 21 years as predicted for the two 1 ac (0.4 ha) plots: a time savings of from 16 to 18 years.

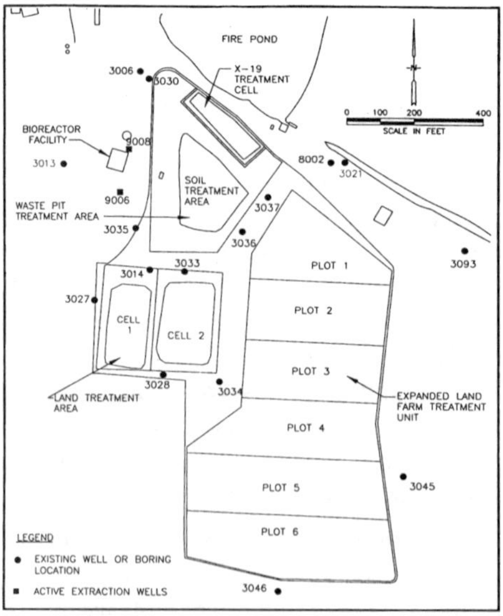


**FIGURE 1. Libby LTU.** (CIC, 1998)

The **BN Somers** site is a former railroad tie treating plant located in the Town of Somers, Montana. Site remediation activities have been and are currently being managed by the Burlington Northern Santa Fe Railway Company. Approximately

67,000 cu yd ( 51,255 cu m) of creosote (PAH) contaminated soil was excavated from several source areas at the site in 1993-94 and was placed in a 13.3 ac (5.4 ha) LTU. An 8 in (20 cm) lift of soil was applied to 10.4 ac (4.2 ha) of the LTU and the remaining volume was staged in a stockpile on the unused portion of the LTU. Three lifts of soil have been applied to the pie-shaped LTU (Figure 2). A new lift was applied once the lift being treated reached the remediation levels. As of May 1999, the final lift of contaminated soil on the LTU reached the required remediation levels for PAHs and one additional year of treatment required by the ROD was completed.

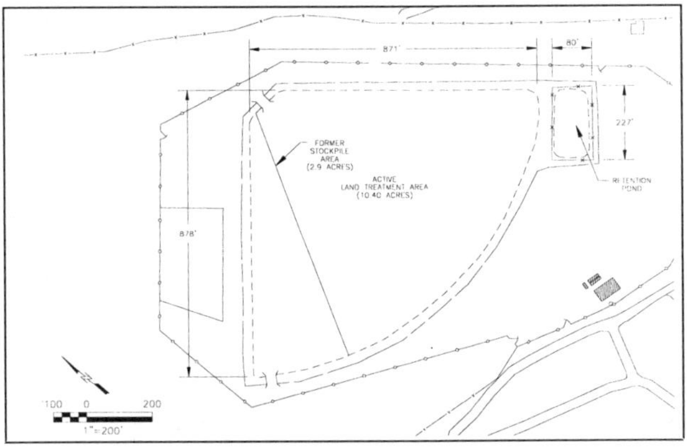


**FIGURE 2. BN Somers LTU.** (ThermoRetec, 2000)

The **Idaho Pole Company** site is located in Bozeman, Montana, near the East Gallatin River. Wood treating activities at the site ceased in 1997 when the Idaho Pole Company closed operations. Utility poles had been treated at the site in a pressure cylinder and in full-length and butt vats since 1952 with a 5% PCP in diesel oil solution. From 1945 until 1952, creosote was used as the treating fluid. The ROD issued in 1992 and a subsequent Explanation of Significant Differences (ESD) required the excavation of 16,000 cu yd (12,240 cu m) of contaminated soil from the site and placement of the soil on a 5.25 ac (2.1 ha) pie-shaped LTU. However because the plant was operating at the time (1995), soil contamination associated with site structures was not addressed. During the 1999 construction season, after plant closing, an additional 3000 cu yd (2300 cu m) of contaminated soil was placed on the LTU and a lift of treated soil was removed and used as backfill at the site. Remediation levels for the soil remaining on the LTU were reached in 2000.

The **Montana Pole and Treating Plant** site, located in Butte, Montana, formerly treated posts, poles, and railroad ties with creosote and PCP. The plant ceased operations in 1984 and a Record of Decision was issued for site cleanup in 1993. The ROD required excavation and bioremediation of an estimated 208,000 cu yd (159,000 cu m) of contaminated soil. EPA and the State of Montana are managing the cleanup project as the result of a settlement agreement with the responsible parties for the site. An engineered 8ac (3.2 ha) triangular-shaped LTU (Figure 3), designed by CDM, Inc., was constructed in 1997 and 30,000 cu yd (23,000 cu m) of contaminated soil excavated from the north side of the site was placed on the LTU to a depth of 2 ft (0.6 m). The soil remediation levels were reached in 1998 and all of the soil was removed from the LTU in 1999 and used as backfill in previously excavated areas. The remaining 70,000 cu yd (53,500 cu m) of contaminated soil at the site was placed on the LTU in July 2000 to a depth of approximately 7 ft (2.1 m). The upper 2.5 ft (0.8 m) of soil reached cleanup levels and was removed from the LTU in November 2000. Approximately 52,000 cy (39,500 cu m) of soil has been treated on the LTU.

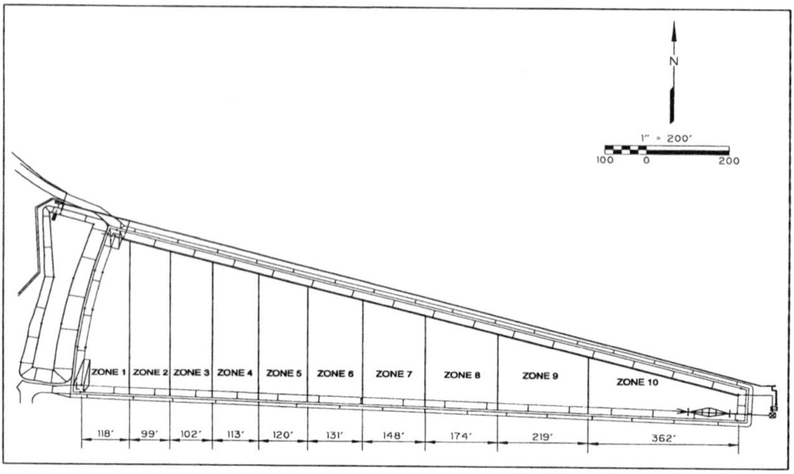


**FIGURE 3. Montana Pole and Treating Plant LTU.** (CDM Inc., 1999)

Table1 provides a summary description of the four wood-treating sites included in this case study.

**TABLE 1. Description and Status of Land Treatment Units (LTUs) at Wood-Treating Sites.**

| Site Name | LTU Area ac (ha) | LTU Shape | Soil Volume cu yd (cu m) | LTU Status |
|---|---|---|---|---|
| Libby Groundwater | 14 (5.6) | Rectangle | 45,000 (34,400) | 3-5 years of operation remaining |
| BN Somers | 10.4 (4.2) | Pie | 67,000 (51,255) | Treatment Completed 2000 |
| Idaho Pole Company | 5.25 (2.1) | Pie | 19,000 (14,540) | Treatment Completed 2000 |
| Montana Pole and Treating | 8.0 (3.2) | Pie | 208,000 (159,000) | 2 lifts Treated in 2 seasons |

## LTU OPERATIONS

**Irrigation**. Each LTU at the four wood-treating sites in this case study has irrigation capabilities for optimizing soil moisture content and for dust control. The Libby site uses berm-mounted sprinklers on the two 1 ac (0.4 ha) plots and two moving hydraulically controlled sprinklers on the new 12 ac (4.8 ha) LTU. The other three sites use a center pivot irrigation system that moved across the LTU providing the required amount of moisture in a single pass. Each of the three LTUs were triangular or "pie-shaped" to accommodate the irrigation system with the pivot point located at the apex. This system has become a standard at LTUs in Montana because it requires only one water supply point at the pivot, no plumbing on the LTU, it is easily moved and can provide large amounts of water in a short period of time.

The optimum moisture content for the LTUs has been determined to be 70 to 80 percent of field capacity or 12 to 15 percent water content. Recent studies at the Libby site indicate that moisture content may be the most important limiting factor in bioremediation of soil contaminated with PAHs.

**Tilling**. Tilling or plowing of the LTU increases the rate of oxygen uptake of the naturally occurring microorganisms in the soil and provides continual mixing of the treatment zone. Depending on the type of equipment used for tilling, the treatment zone may be up to 2.5 ft (0.8 m) deep. The 2.5 ft (0.8 m) depths have been achieved using an eight-wheel drive, 200 horsepower tractor pulling a device called a "subsoiler". The subsoiler is a plow similar to a chisel-plow with tines that are up to 3ft (1 m) in length. The tractor has an electronically controlled depth adjustment that allows precise placement of the plow and can be used to till a 1 ac (0.4 ha) plot in less than 10 minutes. The amount of soil that can be treated to a remediation level in a specified time period is dependent on the size of the LTU (footprint) and the depth of soil that can be treated in each lift. A 2.5 ft (0.8 m) treatment zone will allow for four times as much contaminated soil to be treated than an 8 in (20 cm) treatment zone or the deeper treatment zone will achieve remediation in one fourth of the time on the same size LTU.

**Nutrient Addition**. Nutrients are a required component of the bioremediation process and are applied to each LTU to stimulate microorganism performance. Nutrients, nitrogen and phosphorous, are added in the form of agricultural fertilizer as needed to maintain an optimum level. The amount of nitrogen and phosphorous to be applied is calculated from analytical results of LTU soil samples and is based on the carbon content of the soil. The Montana Pole and Treating Plant site has the capability to add liquid nutrients to the LTU through the irrigation system while the other sites use granular fertilizer applied to the LTU surface with an agricultural spreader. The LTU is then tilled to mix the fertilizer into the treatment zone. Typically, nutrients are only added once to a lift of soil during the treatment period. Remediation levels are normally reached before the nutrients are depleted.

## LTU PERFORMANCE

All four of the LTUs described in this case study have been effective in reducing soil contaminant concentration to the remediation levels specified in the ROD for each site.

**TABLE 2. Performance Summary for Land Treatment Units (LTUs) at Wood-Treating Sites.**

| Site Name | Avg Initial B2PAH mg/kg | B2PAH ROD mg/kg | Time to PAH ROD level | Avg. Initial PCP mg/kg | PCP ROD mg/kg | Time to PCP ROD level |
|---|---|---|---|---|---|---|
| Libby Groundwater | 40-150 | 59 (a) | 90 days to 2 seasons (b) | 90-230 | 36 | 90 days to 2 seasons |
| BN Somers | 100-200 | 57 (a) | 2 seasons | (c) | (c) | (c) |
| Idaho Pole Company | (d) | 15 (a) | (d) | 50-220 | 48 | 1 to 2 seasons |
| Montana Pole | (d) | 4.2 (a) | (d) | 30-70 | 34 | 1 season |

(a) Benzo (a) pyrene equivalents
(b) "season" refers to treating season, June thru Sept.
(c) PCP not used at BN Somers site
(d) PAH levels do not exceed cleanup level

The amount of time required to achieve remediation levels varies with the type of treating fluid used and site-specific factors such as soil type, weather and operational characteristics.

Table 2 presents performance information for each site in summary format. The ROD specified remediation levels listed in Table 2 are the target concentrations for the contaminants of concern, PCP and /or creosote constituents, and the average contaminant concentrations were calculated for each lift and averaged for total volume of treated soil.

## CONCLUSIONS

Data collected at four former wood-treating sites that have been using LTUs for soil remediation for from 3 to 11 years, demonstrates that ROD specified cleanup levels are being achieved at each site. Remediation levels have been achieved in soil lifts in one treating season (90 to 120 days in Montana), depending on the contaminant type and initial concentration. Information provided to EPA by the operators of the wood-treating sites in this case study indicates that the average cost of treating contaminated soil to the remediation levels in an LTU varies from \$16 to \$23 per cy ( \$21 to \$30 per cu m).

## REFERENCES

Champion International Corporation. 1998. *Expanded Landfarm and Waste Pit Treatment Area Design and Construction Plan, Libby Groundwater Site*. January 1998.

Champion International Corporation. 2000. 1999 *Annual Operations Report for the Upper Aquifer*. February 2000.

ThermoRetec Consulting Corporation. 2000. *Phase I Groundwater Remedy Annual CERCLA Report, Somers, Montana.* June 2000.

ThermoRetec Consulting Corporation. 2000. *Technical Impracticability Evaluation for Groundwater Restoration, Former Somers Tie Treating Plant, Somers, Montana.* May 31, 2000.

ThermoRetec Consulting Corporation. 2001. *BNSF Somers Land Treatment Unit 2000 Operations Report.* January 30, 2001.

ThermoRetec Consulting Corporation. 2001. *Idaho Pole Land Treatment Unit LTU Operations Report*. January 12, 2001.

USEPA. 1988. *Record of Decision (ROD), Libby Ground Water Superfund Site, Lincoln County, Montana*. EPA Region VIII. December 1988.

USEPA. 1989. *Record of Decision Burlington Northern (Somers Plant) Superfund Site, Flathead County, Montana*. September 1989.

USEPA. 1993. *Guidance for Evaluating the Technical Impracticability of Ground Water Restoration*. Interim Final Directive 9234.2-25. September 1993.

USEPA. 1997. *Explanation of Significant Differences for the Libby Ground Water Superfund Site*. January 1997.

Woodward-Clyde International. 1999. *Technical Impracticability Evaluation Report, Libby Ground Water Site, Montana.* January 11, 1999.

# BIODEGRADATION OF PHENOL IN INDUSTRIALS EFFLUENTS: PILOT-SCALE

***Claudio Belloso*** (Centro de Investigación y Estudios Superiores, Facultad Católica de Química e Ingeniería, Rosario, Argentina)

**ABSTRACT:** The goal of the project was to degrade the phenol contained in industrial effluents of Refinery plants using phenol-degrading bacteria isolated from hydrocarbons contaminated soils. The ability to degrade phenol of 3 isolated bacterial strains of landfarming soil belonging to the Refinery San Lorenzo, San Lorenzo, Santa Fe, Argentina, was evaluated.

A pilot plant (1:200,000) simulating the lagoon belonging to the Refinery was built. Trials with chemical tracers were performed. Bacterial inoculation to inlet of the pilot plant was performed to establish the relationship between the bacterial reproduction rate and the loss of these bacteria through the discharge.

By means of a mass balance of microorganisms as a function of the lagoon dimensions and the flow rate, it was shown experimentally and mathematically that the biodegradation is not feasible in these conditions.

The loss of microorganisms due to the mechanical phenomenon of haulage in the lagoon is bigger than the bacterial reproduction rate under the conditions of the experiment.

Nevertheless, this trial is useful to show that the success of industrial effluent treatments by biological means does not depend only on the ability of microorganisms to degrade the organic compounds, either indigenous or commercial strains, but it also depends on variables as such as the dimensions of the lagoons, shapes, hydraulic residence time, and operational parameters that influence the operation of systems of this type.

## INTRODUCTION

All the industries that produce or use phenol may at some point be responsible from the release of this compound into the environment. Phenol concentrations ranging from several hundred to up to 10,000 mg/L have been reported in industrial wastewater (Fedorak and Hrudey, 1988), whereas domestic wastewater commonly contains between 0.1 and 1 mg/L (Hunter, 1971).

The U.S. Environmental Protection Agency (EPA) criterion to protect freshwater aquatic life is 600 µg/L as a 24-hour average, never to exceed 3.4 mg/L, and the drinking water limit is 1 µg/L (USEPA, 1979). Without the proper treatment, industrial wastewaters are an important source of anthropogenic phenol into the environment.

In Argentina, the presence of numerous petrochemical industries and refineries that discharge their wastewaters in rivers, cause the contamination of water with phenol in repeated occasions. The plants that produce drinking water have serious problems to neutralize the flavor and the smell that are produced by the phenol and their by-products.

There is considerable interest in the isolation of bacteria able to thrive under high concentrations of phenol, as well as other organic solvents, because such bacteria would be useful in biodegradation processes where phenolic compounds exert a toxic or inhibitory effect (Heipieper et al., 1994).

**Objective**. The objective of this study was to determine the feasibility to degrade the phenol contained in industrial effluents of Petroleum Refinery plants using phenol-degrading bacteria isolated from hydrocarbon-contaminated soils.

## MATERIALS AND METHODS

**Isolation of phenol-degrading bacteria.** The bacteria from landfarming soils were isolated (Belloso, 1999). This landfarming belongs to Petroleum Refinery San Lorenzo S.A. located in San Lorenzo City, Santa Fe Province, Argentina.

Of the isolations conducted, 3 bacterial strains were used in the study. These bacterial strains are those that were demonstrated to possess the greatest capacity of phenol biodegradation, identified as strains W1, W2 and W3. These bacterial strains were grown in Bushnell-Hass mineral broth (Bushnell and Hass, 1941) with phenol as the sole carbon source. The goal of this study was to determine which one of these three strains degrade phenol the fastest.

In what follows, we can see the different tests results for each bacterial strains carried out.

**Determination of phenol biodegradation rates:**

Bacterial strains: W1, W2 y W3
Culture media: Bushnell - Haas broth (BH)
Method: Presence – Absence test
Temperature of incubation: 30°C
Carbon source: Phenol (F) (100 mg/L).
Indicator: Resazurin (I) (1 mg/L)
Positive reaction: All those tubes in which there was change of color from blue to pink were considered positive.

**TABLE 1. Test using the W1 bacterial strain.**

| Tube N° | Combination | Results after 3 hs | Results after 12 hs | Results after 24 hs |
|---|---|---|---|---|
| 1 (Negative control) | BH + I | - | - | - |
| 2 | BH + W1 + I | - | - | - |
| 3 | BH + W1 + I + F | - | + | + |

**TABLE 2. Test using the W2 bacterial strain.**

| Tube N° | Combination | Results after 3 hs | Results after 12 hs | Results after 24 hs |
|---|---|---|---|---|
| 1 (Negative control) | BH + I | - | - | - |
| 2 | BH + W2 + I | - | - | - |
| 3 | BH + W2 + I + F | + | + | + |

**TABLE 3. Test using the W3 bacterial strain.**

| Tube N° | Combination | Results after 3 hs | Results after 12 hs | Results after 24 hs |
|---|---|---|---|---|
| 1 (Negative control) | BH + I | - | - | - |
| 2 | BH + W3 + I | - | - | - |
| 3 | BH + W3 + I + F | - | - | + |

Of the three strains studied, the W2 degraded phenol the fastest. Therefore, the next step was to determine how much phenol was degraded as a function of the time.

**Degradation of phenol vs. time**
Bacterial strain: W2
Culture media: Bushnell - Haas broth
Carbon source: phenol (100 mg/L).

At different times, the phenol concentrations were measured, (American Society for Testing and Materials, 1995), Figure 1 shows the results.

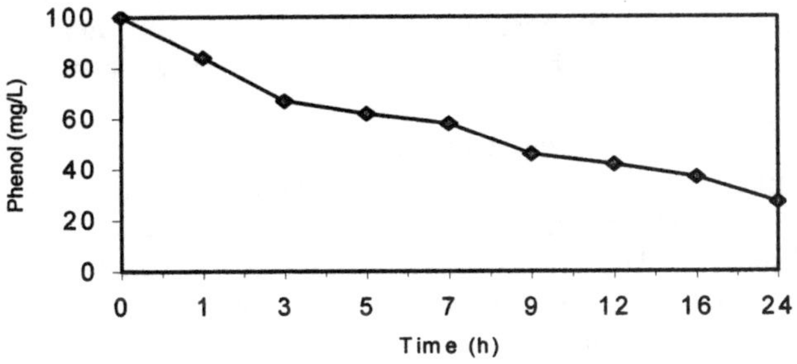


**FIGURE 1. Percentage of degraded phenol vs. time**

**Test with chemical tracers.** A trial using chemical tracers with the purpose of determining the recovery of the tracer in the discharge of the pilot plant was carried out. A pilot plant (1:200,000) simulating the lagoon belonging to the Refinery was built. Table 4 shows the design and operation parameters.

**TABLE 4. Design and operation parameters**

| | |
|---|---|
| Pilot plant, scale | 1:200,000 |
| Dimensions of the lagoon | 108:43:10 (cm) |
| Total capacity of the lagoon | 46,440 mL |
| Inlet flow | 142 mL/min |
| Fluid | Deionized water |
| Tracer | Methylene blue (0.1%) |

**Determination of the absorbance of the solution to the outlet of the pilot plant.** The tracer (5 mL of methylene blue) to the inlet of the plant was added.

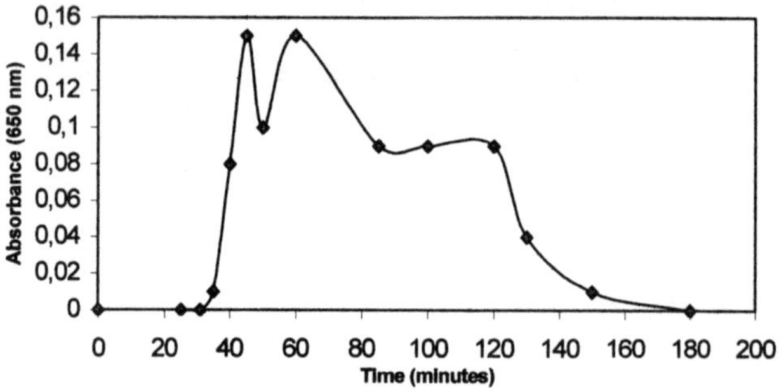


**FIGURE 2. Recovery of tracer at the outlet of the plant. Variation of the absorbance measured as a function of the time.**

We can see that the tracer reaches the maximum value at 45 minutes after the start of the test (Figure 2). After 1 hour, the concentration recovered of tracer began to decrease until its disappearance at 180 minutes. This means that bacterial reproduction rate should be sufficiently high as to prevent wash-out.

**Trial of recovery of inoculated bacteria.** $1.2 \times 10^7$ bacteria belonging to W2 strain to the plant pilot's inlet were inoculated to determine the relationship between the bacterial reproduction rate of this strain and the loss of these bacteria through the discharge. Then, total bacteria count at the outlet of the pilot plant were performed. Figure 3 shows the recovery of inoculated bacteria as a function of the time. Table 5 shows the detail of operation.

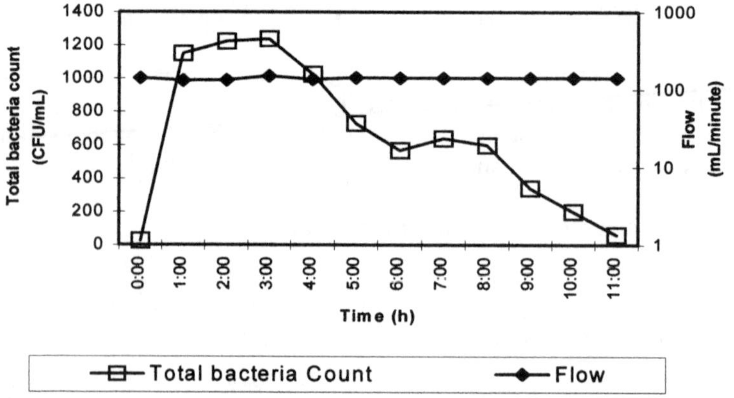


**FIGURE 3. Recovery of inoculated bacteria as a function of the time.**

**TABLE 5. Detail of operation.**

| | |
|---|---|
| Fluid | sterilized deionized water (121°C / 1 atm.)/ 30 min.) added of 1% of nutrient. |
| Nutrient | raw effluent from the refinery. |
| Flow | 142 mL/min. |
| Total bacteria count | method: pour plate, incubation: 30°C/24 h, Culture media: plate count agar (Clesceri et al., 1992). |

**Determination of the growth specific rate (μ) of W2 strain.**
Culture media: Liquid obtained from filtration (0.45μm) of industrial effluent from the Refinery before treatment.
Incubation: 35°C
Determination of the Optic Density: Spectrophotometer Spectronic 21 (520 nm).
Figure 4 shows the absorbances obtained as a function of the incubation time.

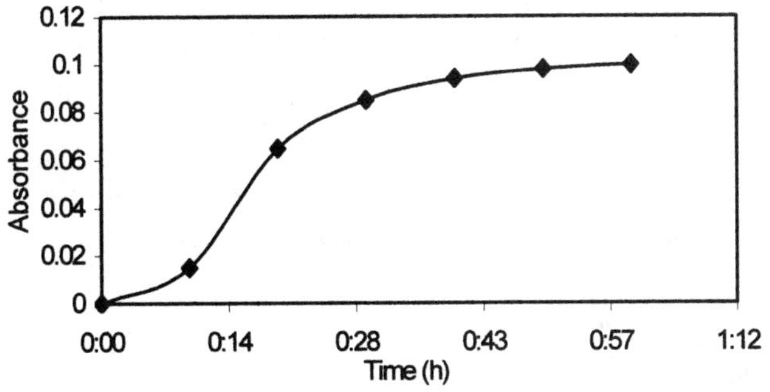


**FIGURE 4. Absorbances obtained as a function of the incubation time.**

The growth specific rate of W2 strain calculated (Malek et al. 1964) was $\mu = 0.076\ h^{-1}$.

## RESULTS AND DISCUSSION

**Results summary.** For a biological treatment of this type be successful, that is, to achieve the degradation of the compound present in the effluent, the population of microorganisms in the liquid should be constant, in spite of the fluid flow. In the trial of recovery of inoculated bacteria carried out, we can see the great loss of microorganisms that takes place in approximately 11 hours (Figure 3).

**Theoretical matters.** In what follows we develop a group of equations that describe the system behavior. The fluid enters to the plant pilot at a defined and constant rate and it leaves at the same rate by overflowing. In this system the most important factor that controls the development of the microorganisms is the untreated liquid inlet rate. The relationship between the volume of the lagoon and the liquid inlet rate is denominated hydraulic residence time.

$$\theta = V / F \tag{1}$$

Where: $\theta$: hydraulic residence time (T)
V: volume of the lagoon ($L^3$)
F: liquid inlet rate ($L^3/T$)

In this case in particular we have:

$$\theta = \frac{V}{F} = \frac{46{,}440\ \text{mL}}{142\ \frac{\text{mL}}{\text{min}}} = \frac{46{,}440\ \text{mL}}{8{,}520\ \frac{\text{mL}}{\text{h}}} = 5.45h \tag{2}$$

In the lagoon, the microorganisms are multiplying but they are also being removed of the system through the discharge. Consequently the net change in the number of microorganisms in the time will depend on the cellular balance represented by the equation 3 (Malek et al., 1964).

$$\frac{F_0}{V}X_0 - \frac{F}{V}X + \mu X - \alpha X = \frac{dX}{dt} \tag{3}$$

Where:
$X_0$ = Number of cells for unit of volume that enters continually to the system.
X = Number of cells in the lagoon.
$F_0 = F$ = inlet and outlet fluid flow rate (mL/h)
V = volume (mL)
$\mu$ = growth specific rate ($h^{-1}$)
$\mu_e$ = growth specific rate in the equilibrium point ($h^{-1}$)
$\alpha$ = death specific rate ($h^{-1}$)
t = time (h)
$(F_0 / V) X_0$ = cells inlet
$- (F / V) X$ = cells outlet
$\mu X$ = cellular growth
$- \alpha X$ = cellular death
$dX / dt$ = accumulation

Since the inoculation is carried out only once at the beginning of the trial, there is not additional entrance of microorganisms, that means that we can assign a value of 0 to the first term of the equation 3. Since the inoculated strain is in phase of exponential growth, we assume that $\mu$ is much greater than $\alpha$ ($\mu >> \alpha$), and then equation 3 may be re-written as:

$$-\frac{F}{V}X + \mu X = \frac{dX}{dt} \tag{4}$$

When the system reaches the equilibrium point, the concentration of microorganisms in the lagoon stays constant, and therefore $dX/dt = 0$ consequently, equation 4 is written as:

$$\mu_e = \frac{F}{V} = \frac{1}{\theta} = 0.183\ h^{-1} \qquad (5)$$

This means that the growth specific rate is given by the feeding flow when the system is in dynamic equilibrium (Herbert et al. 1956). To know the actual state of our system, we should compare the theoretical value of $\mu$ in the dynamic equilibrium ($\mu_e$) with the actual growth specific rate of the W2 strain calculated above.

**Possibilities:**

if: $\mu > \mu_e$ , then: $dX/dt > 0$ , there is accumulation of microorganisms in the lagoon; if: $\mu = \mu_e$, then $dX/dt = 0$, there is a dynamic equilibrium in which there are not gain neither loss of microorganisms. When we say that "gain" of microorganisms in the system we refer to the increase of the number of microorganisms by cellular reproduction. On the other hand, the "loss" of microorganisms of the system is related to the decrease of the number of them, either due to the loss of microorganisms through the discharge of the system or by microorganisms death; if: $\mu < \mu_e$ , then $dX/dt < 0$ , there is decrease of the number of microorganisms in the lagoon.

In our case, $\mu = 0.076\ h^{-1}$, that is, smaller than $\mu_e$ , therefore in our system there will be decrease of the number of microorganisms.

In this system there will be a decrease on the number of microorganisms as a function of the time due to the loss of microorganisms through the discharge (insufficient retention time) and for the phenomenon of death.

This is very complicated for the biodegradation process, since without microorganisms there is not possible biodegradation. Mathematically, the only possibility to revert this result consists on increasing the growth rate of the inoculated microorganism ($\mu$), either, inoculating a microbial strain with a greater cellular growth rate or to diminish the term F/V increasing the volume of the lagoon or diminishing the flow of the influent.

As we can see, the success or failure of a biological treatment will depend on the interrelation among the characteristics of the present microorganisms or inoculated in the system and the design of the treatment units.

Nowadays in Argentina, mixtures of microorganisms are being marketed for various treatment types, with more failures than successes, not only due to the microorganism "type" used but also to the lack of physical, chemists and microbiological studies that demonstrate the advantages or disadvantages of implementing such treatment especially when the treatment units have already been built without previous microbial kinetics studies.

This simple work is useful to meditate about the advantages that are obtained when in the design, operation and start-up of a biological treatment plants of industrial effluents, the expertise knowledge of specialists in environmental engineering is taken in account along with the engineering design.

**ACKNOWLEDGMENTS**

I wish to thank Francisco A. Casiello Ph.D. for their assistance with this project.

**REFERENCES**

American Society for Testing and Materials, 1995. *Standard Test Methods for Phenolic Compounds in Water* D 1783-91 A, Philadelphia, PA..

Belloso, C. O. 1999. "Influences of Hydrogen Peroxide and *Azospirillum* sp. Hydrocarbons Biodegradation in Soils" In B. C. Alleman and A. Leeson (Eds.), *In Situ Bioremediation of Petroleum Hydrocarbon and Other Organic Compounds*. Battelle Press. (3) 331-336, Ohio, USA.

Bushnell L. D., and H. F. Hass. 1941. "The utilization of certain hydrocarbons by microorganisms". *J. Bacteriol. 41*: 653-673.

Clesceri, L. S., A. E. Greenberg, and R. Trussel Rhodes. 1992. "*Standards Methods for the Examination of Water and Wastewater*". 18th. ed., American Public Health Association, N.Y.

Fedorak, P. M. and S. E. Hrudey. 1988. Anaerobic Degradation of Phenolic Compounds with Application to Treatment of Industrial Waste Waters. In: *Biotreatment Systems*, Vol. I., pp. 170-212 (D.L. Wise, Ed.). CRC Press: Boca Raton, FL.

Heipieper, H. J., F. J. Weber, J. Sikkema, H. Keweloh, and J.A.M. De Bont. 1994. Mechanisms of Resistance of Whole Cells to Toxic Organic Solvents. *Trends Biotechnol.* 12: 409-415.

Herbert, D. 1956. "The continuos culture bacteria. A theoretical and experimental study." *J. Gen. Microbiol.* 14: 601-622.

Hunter, J. V. 1971. Origin of Organics from Artificial Contamination In: *Organic Compounds in the Aquatic Environment*. pp. 51-94. (S.J. Faust and J.V. Hunter, Eds.). Marcel Dekker: New York.

Malek, I., 1964. "*Continuous Cultivation of Microorganisms*". Academic Press. New York.

U.S. Environmental Protection Agency. 1979. Phenol. Ambient Water Criteria, PB 296 787. Office of Water Planning and Standards, EPA, Washington, DC.

# BIODEGRADATION OF A COMPLEX PHENOLIC WASTE-CONTAMINATED GROUNDWATER BY PRESSURIZED FLUIDIZED BED REACTOR

**Jian Xing** and Richard M. Raetz
(Global Remediation Technologies, Inc. Traverse City, MI)

**ABSTRACT**: A bench-scale pressurized fluidized bed reactor (PFBR) system was employed to treat a complex phenolic waste contaminated groundwater containing 2,4-dimethylphenol, tricresylphosphate and as many as twenty (20) other volatile and semi-volatile organic contaminants. The experimental system consists of two PFBRs operated in series, a biofilter, and a settling tank. Granular activated carbon was added to serve as the adsorbent/biomass carrier in the PFBRs. Experimental results show that the phenol waste degrading microorganisms could be cultured in the PFBR system under an operating hydraulic retention time (HRT) of 1 hour and an organic loading rate of 2.9 kg COD/m$^3$-day. The PFBR system could continuously remove organic contaminants with influent concentrations up to 7,000 $\mu$g/l (2,4-dimethylphenol) and 2,500 $\mu$g/l (tricresylphosphate) to below the GC's detection limits (<5 $\mu$g/l for 2,4-dimethylphenol and < 40 $\mu$g/l for tricresylphosphate, respectively). The PFBR system was able to resist a six fold-three hour shock load and recovered within one and a half hours when normal operating conditions were resumed.

## INTRODUCTION

Phenol and its higher homologues are aromatics containing hydroxyl, methyl, amide and sulphonic groups attached to the benzenoid molecules. Phenolic waste is one of the most frequently found organic contaminants in wastewater from fossil fuel refining processes, phenol manufacturing plants, pharmaceutical plants and a variety of other industries. Phenols are toxic to several biological reactions. However, biological transformation of phenols to non-toxic entities exists in specialized microbes, having enzymes of aromatic catabolic pathways (Kumaran and Paruchuri, 1997).

Bioreactors can be designed in a variety of configurations to accommodate the physical and chemical characteristics of the targeted pollutants. Bioreactors can contain indigenous microbes, or they may be inoculated with specially selected microorganisms capable of rapidly and extensively degrading targeted pollutants (Raetz and Xing, 1997). While various biological systems have been reported to be effective in treating phenol wastewater (Banerjee, 1996; Beltrame et al., 1980; Gonzalez and Duque-Luciano, 1992), little is known about using pressurized fluidized bed reactor for complex phenolic waste-contaminated groundwater treatment. Analytical results reveal that the target organic contaminants in the groundwater are a complex mixture of hazardous chemicals, including volatile and semivolatile compounds (Table 1). The most concentrated organic contaminants among the chemicals are 2,4-dimethylphenol and tricresylphosphate.

The goal of this study is to develop a continuously operating PFBR system to biologically degrade the complex phenolic waste and measure its performance characteristics. In this system the PFBR serves as the primary treatment device while the fixed film biofilter serves as both back-up treatment and solids retention/digestion unit.

**Table 1. Characteristics of the contaminated groundwater**

| PARAMETERS | VALUE |
|---|---|
| COD (mg/l) | 70 – 130 |
| pH | 7-8 |
| 2,4-dimethylphenol (μg/l) | 420-7500 |
| Tricresylphasphate (μg/l) | 500-2600 |
| 2 methylphenol (μg/l) | 700-1000 |
| 3/4-methylphenol (μg/l) | 550-750 |
| BTEX (μg/l) | 63-992 |

## MATERIALS AND METHODS

**Biological treatment system.** The phenolic waste biodegradation study was performed within a multistage biological treatment system as shown in Figure 1. The experimental system consisted two PFBRs operated in series each with a working volume of 0.75 liters, a biofilter with a working volume of 2 liters, and a settling tank with adjustable working volume to keep a constant hydraulic retention time of 1 hour. The PFBR contents were not directly exposed to atmosphere and its internal pressure was adjustable (from 0 to 4 atm for full-scale system). This design minimized volatilization of organic compounds and increased dissolved oxygen concentration in the reactor. To each PFBR, 200 ml of 0.75/1.3 mm granular activated carbon was added to serve as the adsorbent/biomass carrier. This activated carbon volume had a settled bed height of 25 cm while the total reactor height was 110 cm. Influent and effluent sample ports were installed at points directly prior to influent flow entering the PFBR and in the clear zone above the fluidized bed, respectively. The contents of the PFBR reactors were kept fluidized with recycling pumps. The biofilter was filled with two-liters of ceramic saddles as microbe growth attachment media. The biofilter was employed as a "polishing process" to remove volatile organic compounds not degraded in the PFBR reactors; therefore, ensuring the system effluent quality. Secondly, the biofilter contributed to the sludge retention and the sludge digestion processes.

Hydrogen peroxide was used as the oxygen source. The dissolved oxygen (DO) level in the effluent of the biological system was maintained above 2 mg/L by adjusting hydrogen peroxide supply. During the experimental period, wastewater was fed to the system via a digital peristaltic pump (Cole-Parmer, Model E-77921-00) with a flow rate varying from 7 l/d to 37.2 l/d. During the experimental period the PFBR-biofilter system was inspected and maintained daily.

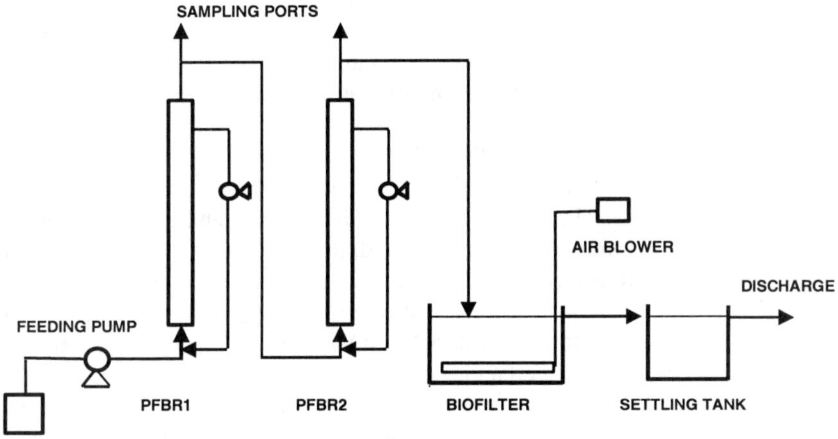


**FIGURE 1. Schematic of experimental PFBR system**

**Analytical Methods**. Liquid samples taken from the biological treatment system were transferred to clean 10-ml glass vials each with a Teflon cap and stored at 4 °C until analyzed in laboratory. Concurrently, separate 50 ml samples of influent and effluent were taken to monitor changes in DO concentration, temperature and pH.

The concentration of 2,4-dimethylphenol in liquid samples was determined by gas chromatographic analysis. Analyses were performed on a SRI Model 8610B GC (SRI Instruments, Torrance, CA) equipped with a flame ionization detector (FID). Helium was used as the carrier gas (20 ml/min) while hydrogen (150 kPa) was supplied to the FID. Phenol compounds (present in 1.0-ul injections) were separated on a Restek MXT-5 (Supelco, Bellafonte, PA) capillary column (15 m x 0.53 mm [inside diameter] with a 0.25-ml film thickness). Temperature was initially 45 °C for 1 minute, then ramped up at 12 °C /min to 120 °C, and then further ramped up at 20 °C/min to 245 °C. The concentration of 2,4-dimethylphenol present in liquid samples was calculated by comparing the peak area obtained by 1.0-ul injections with 2,4-dimethylphenol standard. The concentrations of tricresylphosphate and other volatile and semi-volatile organic compounds in liquid samples were analyzed by outside lab (Trace Analytical Laboratories, Inc., Muskegon, MI).

COD was analyzed by using Hach COD vials and a Hach COD reactor (Model 45600 Hach Company, Loveland, CO). Properly diluted liquid samples (2 ml each) were mixed with COD digestion reagent in Hach glass vials and digested at 150 °C for two hours. The COD values of the digested samples were determined using the ferrous ammonium sulfate titration method. pH was measured using a pHtestr3 Model 35624-30 pH meter (Cole-Parmer Instrument Co., Vernon Hills, IL). Before measuring the sample pH, the electrode was calibrated with standard pH solution according to the procedure supplied by the pH meter manufacturer. Reactor influent and effluent DO concentrations were analyzed using a YSI model 50B dissolved oxygen meter (YSI Incorporation, Yellow Springs, Ohio). The oxygen meter was air-calibrated 15 minutes prior to use. Reactor influent and effluent temperature was measured using a YSI model 50B dissolved oxygen meter (YSI Incorporation, Yellow Springs, Ohio).

**Experimental Conditions.** The PFBR was initiated with mixed inocula containing a known capacity to degrade phenol obtained from a batch reactor in Global Remediation Technologies, Inc. (GRT's) lab and from the site aquifer water. Before continuous feed was started, the inocula-rich mixture was recycled through the PFBR for two hours to allow sufficient contact between the GAC media and the cells. Nitrogen (N) and phosphorus (P) were added to the influent at a stoichiometric ratio of 100/5/1 (COD/N/P). A 35% commercial $H_2O_2$ solution was properly diluted and used as the oxygen source.

When steady-state operation was achieved, site waste was gradually introduced to the biological treatment system until finally replacing the laboratory prepared phenol solution. Site waste degradation test was performed under three different hydraulic retention times. The removal efficiency of the organic pollutants, especially 2,4-dimethylphenol and tricresylphosphate, in the biological treatment system was intensively studied during the treatability test through periodic sampling and analysis.

## RESULTS AND DISCUSSION

**System Start-up and Calibration Results.** The PFBR system was started with laboratory prepared phenol solution. Evidence for biodegradation of phenol was based primarily on GC analysis of influent and effluent liquid samples. In addition, changes in COD and DO concentrations were also used as a secondary or indirect indication of biological activity towards phenol biodegradation. With increased biological activity, the phenol was continuously biodegraded by the microorganisms in the PFBR. Viable biomass began to be observed on the GAC surface during the second week of PFBR reactor operations. Experimental results show that under steady-state conditions a 99% or higher phenol removal efficiency was achieved under designed operating conditions (COD loading of 3.3 $kg/m^3$-day and HRT of 2.6 hours). Of the total phenol removal, about 85% was removed in the first stage of the PFBR while the other 15% was removed in the second stage.

When steady-state operation was achieved in the PFBR system, site waste was gradually introduced into the system. The gradual increase in site wastewater supply allowed the microbes in the bioreactors to adapt to the toxic and complex organic contaminants. After an adaptation period of about three weeks, laboratory-prepared media was completely replaced by site wastewater.

**Site Groundwater Treatment Results.** As shown in Table 1, the target groundwater was contaminated with a complex phenolic waste. The major contaminants were 2,4-dimethylphenol and tricresylphosphate. Considerable amounts of 3/4-methylphenol, 2-methylphenol and BTEX were also present in the contaminated groundwater. More than 20 volatile and semi-volatile organic contaminants were detected from high to trace amounts in the site groundwater samples (results not shown). During the experimental period, the influent COD concentration averaged about 100 mg/l. Influent 2,4-dimethylphenol and tricresylphosphate averaged approximately 5000 $\mu$g/l and 2500 $\mu$g/l respectively.

Three experimental tests were performed in the site wastewater treatability study. In the study, the PFBR was first set-up at an organic loading rate of 0.6 kg/m$^3$-d and a hydraulic retention time of 4 hours. When enough experimental data were obtained, the applied organic loading rate was further increased to 0.8 kg/ m$^3$-d and 2.9 kg/ m$^3$-d, respectively. The influent flow rate was concurrently increased to 12.4 l/d and 37.2 l/d respectively.

The experimental results are summarized in Tables 2 to 4. As indicated by the experimental results, the PFBR system was capable of being used for site wastewater remediation. The PFBR system worked under a wide range of organic loading rates and operating HRTs while reducing subject organic contaminants to levels lower than GC detection limits (5 $\mu$g/l for 2,4-dimethylphenol, 40 $\mu$g/l for tricresylphosphate and 1 $\mu$g/l for each BTEX compound). The majority of the toxic organic compounds were degraded in the first stage of the PFBR. The second stage of the PFBR and the biofilter served mainly as a scavenger reactors to remove daughter decomposition products and trace amounts of the original organic contaminants that escaped from prior stages. The biofilter also partially acted as a backup system should severe operational upset have occured to the PFBRs.

Under the experimental conditions (100 mg/l COD and 5000 $\mu$g/l 2,4-dimethylphenol in reactor influent) the PFBR-biofilter system had relatively low biomass production. The volatile suspended solids (VSS) concentration in the PFBR stage two effluent varied from <1 mg/l to 8 mg/l, while the VSS concentration in the biofilter effluent varied from 2 to 4 mg/l. The total suspended solids (TSS) concentration in PFBR2 and the biofilter effluent varied from < 10 mg/l to about 45 mg/l. This low VSS and TSS concentration observed in the PFBR-biofilter system effluent made the suspended solids separation process optional.

A pH drop was observed during biodegradation of the phenolic waste. Therefore, proper amounts of $NaHCO_3$ were added to the reactor influent to control system pH. After pH adjustment, the average influent pH was 8.0, while the average pH in PFBR stage one, stage two, and biofilter were 7.7, 7.6, and 8.1, respectively.

During the experimental period, the influent temperature was varied from 19 to 26$^o$C and had an averaged value of 22 $^o$C. The reactor effluent temperature was slightly higher (less than 1 $^o$C) than influent temperature due to mechanical recycling.

**Table 2. Summary of four-hour HRT test results**

| | Influent | PFBR1 | PFBR2 | Biofilter |
|---|---|---|---|---|
| 2,4-dimethylphenol ($\mu$g/l) | 5600 | 320 | <13* | <5 |
| Tricresylphasphate ($\mu$g/l) | 2000 | 440 | 130 | <40 |
| 2 methylphenol ($\mu$g/l) | 700 | 40 | <13 | <5 |
| 3/4-methylphenol ($\mu$g/l) | 550 | <13 | <13 | <5 |
| BTEX ($\mu$g/l) | 63 | 4 | <5 | <5 |
| COD (mg/l) | 91 | 21 | 6 | 4 |
| PH | 8.0 | 7.8 | 7.6 | 8.1 |
| DO (mg/l) | 2.8 | 11.6 | 8.1 | 6.5 |

*: Below detection limit.

**Table 3. Summary of three-hour HRT test results**

| | Influent | PFBR1 | PFBR2 | Biofilter |
|---|---|---|---|---|
| 2,4-dimethylphenol ($\mu$g/l) | 4100 | <13 | <13 | <5 |
| Tricresylphasphate ($\mu$g/l) | 2600 | 200 | <100 | <40 |
| 2 methylphenol ($\mu$g/l) | 1000 | <13 | <13 | <5 |
| 3/4-methylphenol ($\mu$g/l) | 750 | <13 | <13 | <5 |
| BTEX ($\mu$g/l) | 68 | <5 | <5 | <5 |
| COD (mg/l) | 89 | 31 | 6 | 5 |
| PH | 8.0 | 7.6 | 7.5 | 8.2 |
| DO (mg/l) | 2.8 | 11.6 | 8.1 | 6.5 |

**Table 4. Summary of one-hour HRT test results**

| | Influent | PFBR1 | PFBR2 | Biofilter |
|---|---|---|---|---|
| 2,4-dimethylphenol ($\mu$g/l) | 4500 | -* | <13 | <5 |
| Tricresylphasphate ($\mu$g/l) | 520 | - | <100 | <40 |
| 2 methylphenol ($\mu$g/l) | 800 | - | <13 | <5 |
| 3/4-methylphenol ($\mu$g/l) | 700 | - | 59 | <5 |
| BTEX ($\mu$g/l) | 48 | - | <5 | - |
| COD (mg/l) | 118 | 20 | 14 | 10 |
| PH | 8.1 | 7.6 | 7.6 | 8.1 |
| DO (mg/l) | 2.8 | 11.6 | 8.1 | 6.5 |

***: Not analyzed**

**Shock Loading Test.** In order to examine the stability of the PFBR-biofilter system under overloading conditions, a 6-fold shock loading was applied to this system. In this shock loading test, the influent organic concentration was suddenly increased by 6 times and kept at that condition for three hours. During this time, the influent COD concentration was increased from 100 mg/l to 630 mg/l, while the influent 2,4-dimethylphenol concentration was concurrently increased from 5,000 $\mu$g/l to approximately 23,000 $\mu$g/l. The system feed was returned to normal conditions after the shock loading.

The response of the PFBR-biofilter system to the shock loading was monitored by continuously sampling the PFBR and biofilter effluent. The 2,4-

dimethylphenol and COD concentrations were then analyzed to track reactor performance changes under shock loading conditions. The experimental results for 2,4-dimethylphenol and COD are presented in Figures 2 and 3, respectively. As indicated by the experiment results, the PFBR system was relatively resistant to the applied shock loading conditions. The 6-fold shock loading caused a moderate 2,4-dimethylphenol concentration increase in the PFBR stage 1 and 2 effluents. At the end of the three hour shock loading test, the PFBR stage 1 and 2 effluent 2,4-dimethylphenol concentrations increased to 5,200 and 2,400 $\mu$g/l, respectively, while the biofilter effluent 2,4-dimethylphenol concentration still remained at a low level of 200 $\mu$g/l.

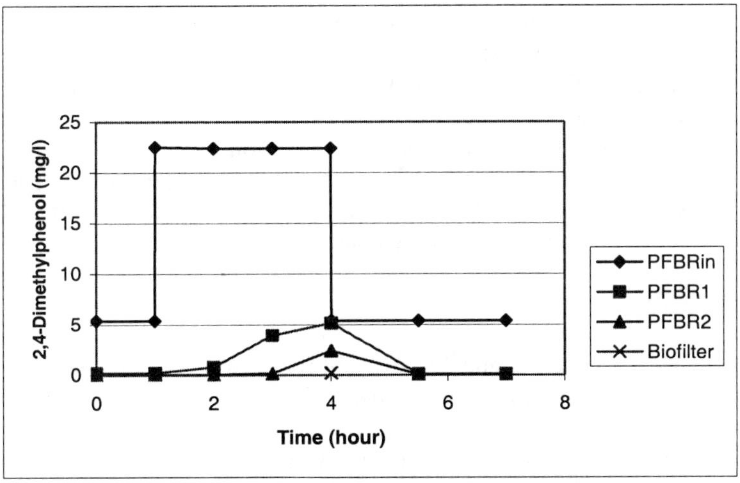


**FIGURE 2. 2,4-Dimethylphenol concentration changes during shock loading test**

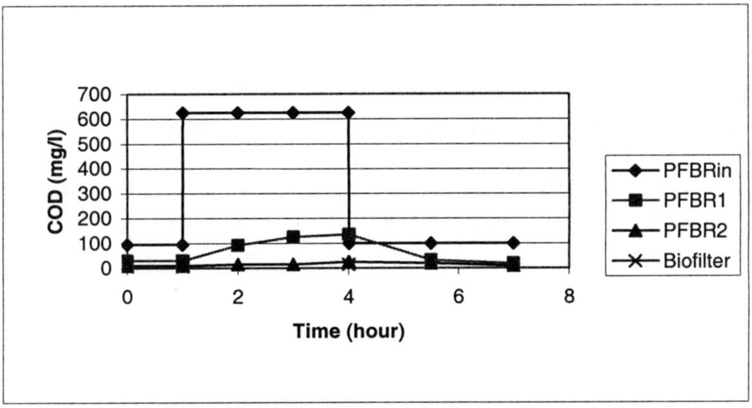


**FIGURE 3. COD changes during shock loading test**

COD measurement results show the same trend. During the shock loading test, the influent COD concentration increased from 100 mg/l to 625 mg/l. After three hours of shock loading operation, the PFBR stage 1 and stage 2 effluent concentrations increased to 135 mg/l and 25 mg/l, respectively. Concurrently, the biofilter effluent COD concentration just slightly changed from 10 mg/l to 15 mg/l.

Once a normal feeding condition resumed, the PFBR-biofilter system recovered quickly. As observed in this test; within one and a half hours the PFBR-biofilter system performance completely returned to its normal conditions.

## CONCLUSIONS

The complex phenolic waste degrading microorganisms were cultured in the PFBR system under operating HRT of 1 to 4 hours and organic loading rates of 0.6 to 2.9 kg COD/$m^3$-day. The PFBR system continuously removed 2,4-dimethylphenol and tricresylphosphate from influent concentrations up to 7,000 $\mu$g/l and 2,500 $\mu$g/l to lower than GC detection limits (<5 $\mu$g/l for 2,4-dimethylphenol and < 40 $\mu$g/l for tricresylphosphate, respectively). The PFBR system was able to resist a six fold-three hour shock load and recovers within one and a half hours when normal operating conditions were resumed.

## REFERENCES

Banerjee, G., 1996. “Phenol- and Thiocyanate-based Wastewater Treatment in RBC Reactor” *J. of Environmental Engineering*, Oct. 941-948.

Beltrame, P., et al. 1980. “Kinetics of Phenol Degradation by Activated Sludge in a Continuous-Stirred Reactor”, *Journal of Water Pollution Control Federation*, 52(1): 126 –133.

Kumaran, P. and Y. L. Paruchuri. 1997. “Kinetics of Phenol Biotransformation” *Water Research*, 31(1): 11-22.

Gonzalez, S. and J. Duque-Luciano, 1992. "Aerobic Submerged Biofilm Reactors for Wastewater Treatment", *Water Research*. 26(6): 825-833.

Raetz, R., and J. Xing. 1997. “Integrated Bioprocessing System Formidable Weapon for BTEX, PNA”, *Soil & Groundwater Cleanup*, *1997*(7):18-20.

# ENZYMATIC REMOVAL OF PHENOL FROM WASTEWATER

F.N. Sánchez, M.C.Vargas, B. Suárez, ***N.E. Ramírez*** (ECOPETROL– ICP, Colombia), and Lu-Kwang Ju (University of Akron, Akron, OH)

**ABSTRACT**: Phenol is present in wastewater from a number of industries, including coal conversion, petroleum refining, resins and plastics, dyes, textiles, mining, and pulp and paper. The removal of phenol from industrial wastewater is an important practical problem. In this study, a crude preparation of laccase was produced by the lignolytic fungi, *Pleurotus ostreatus*, by solid-state fermentation using wheat straw as the substrate. We then studied the removal of phenol from refinery effluents, at laboratory scale, by the laccase-containing broth concentrated using ultrafiltration (NMWCO = 50,000). The results showed that laccase nearly completely removed the phenol in the solutions spiked with 25 to 400 mg/L of phenol. Similar results were obtained with real samples from either refinery wastewater treatment plant or sour water effluent. The results indicated that the enzymatic treatment is effective over a wide range of pH and the phenol removal was not significantly different in systems with and without aeration. The enzyme addition strategy was also studied. Multiple additions in small quantities were more effective than a single addition at the beginning of the treatment process. The responsible mechanisms are being studied currently.

## INTRODUCTION

Phenol is present in the wastewaters from a number of industries, including coal conversion, petroleum refining, resins and plastics, dyes, textiles, mining and pulp and paper; furthermore, phenol and many of its derivatives are considered hazardous pollutants (Atlow *et al.*, 1984).

Conventional methods for dephenolization of industrial wastewaters include solvent extraction, microbial degradation, adsorption on activated carbon, and chemical oxidation. These methods, although effective and useful, suffer from such serious drawbacks as high costs, incompleteness of purification, formation of hazardous by-products, and applicability to only limited phenol concentration range (Atlow *et al.*, 1984). Therefore, alternative technologies are highly desirable.

An enzymatic approach to the treatment of phenol-containing aqueous effluents has been developed to remove phenol. Enzymes are known to be highly selective catalysts and therefore have been proposed as a means of removing targeted substances from wastewaters. A number of investigators have demonstrated previously that oxidation of phenolic pollutants can be catalyzed by several enzymes (Aitken, 1993) including horseradish peroxidase (HRP), chloroperoxidase (CPO) from the fungus *Caldaryomyces fumago*, laccase from several microbial sources, lignin peroxidase (LiP) from the white-rot fungus *Phanerochaete chrysosporium*, and tyrosinase.

Polyphenol oxidase is a monooxygenase, which catalyzes the *o*-hydroxylation of phenol and the oxidation of *o*-dihydrophenol to *o*-quinone using molecular oxygen. Most *o*-quinones are unstable and undergo a nonenzymatic polymerization yielding water-insoluble pigments. Polyphenol oxidases include two subclasses: tyrosinase and laccase. Both enzyme groups require the presence of molecular oxygen for activity instead of hydrogen peroxide (Kadhim *et al*., 1999).

The purpose of this work is to present a concise study of the laccase of *Pleurotus ostreatus* related with its production, oxidation procedure and conditions for phenol removal in spiked matrix and refinery wastewater sample.

## MATERIALS AND METHODS

**Microorganism**: *Pleurotus ostreatus* (isolated at Chapas, Mexico) maintained on malt agar plates at 4°C was used to inoculate 200 ml of Malta Broth in 1-liter Erlenmeyer flasks. Ten to twelve-day-old pellets incubated at 20°C and 200 rpm were used as the inoculum for the production culture.

**Culture conditions and enzyme production:** *Pleurotus ostreatus* was grown on Malt agar. The subculturing was realized on vinasse agar containing 1% of bacteriological agar and 0.4% of vinasse and incubated at 30 °C for seven days until complete growth. The mycelium was then smashed in a wire blender.

Maximum laccase yield was obtained from solid-state fermentation incubating *P. ostreatus* in a complex medium with vinasse as inducer. The medium contained 250 g of sterilized wheat bran and 250 mL of buffer phosphate. The fungus was cultivated in flasks at 22 °C. After 20 days, the enzyme was extracted from the culture medium (paste) by washing the paste with a phosphate buffer (pH 6.3) at 1:1 v/w ratio. The suspension was filtered through cheesecloth and centrifuged at 4 °C and 10,000 rpm for 15 minutes for solid separation. The supernatant was then concentrated by membrane ultrafiltration (Nominal Molecular Weight Cut-off – NMWCO- = 50,000). Samples were withdrawn aseptically, and protein and laccase were determined.

**Chemicals**: Laccase was prepared from solid-state fermentation using *P. ostreatus*. Phenol and all other chemicals used in this study were obtained commercially and were analytical grade. Wheat straw and vinasse were obtained from flour and yeast industry, respectively. Two different samples of phenol-containing wastewater were examined in this work. One is the sour water, which is the refinery process water that has actually come into contact with the process streams (naphtha). Another is from a refinery wastewater treatment plant. Both were provided by Colombian petroleum refinery.

**Phenol quantification**: The disappearance of phenol was monitored by high performance liquid chromatography using a HP 1090 series II with a diode-array detector and the software of HP ChemStation version A.06.03. A reverse phase column, Supelcosil $LC_{18}$ (5 µm, 4.6 mm i.d. x 25 cm), was used and the mobile phase consisted of acetonitrile and acctate buffer (50 mM, pH 4.6) at a gradient from an initial ratio of 40:60 to 100% acetonitrile in 30 min. After the prescribed time, 20 µL of reaction solution was injected. Compounds were detected by their absorbance at 270 nm. The absorption spectra of reaction solution were determined using the same spectrophotometer.

**Measurement of protein and laccase activity:** Protein concentration in the extracellular extract was determined using Bradford reagent (Bradford M. 1976). Laccase was able to oxidize ABTS [2.2'-azino-di-(3-ethylbenzo-thialozin-6-sulfonic acid)]. The assay mixture contained 0.5 mM ABTS and 40 mM citrate-buffer, pH 5.6. Oxidation of ABTS was followed by an absorbance increase at 420 nm ($\varepsilon_{420}$ = 36,000 $dm^3\ M^{-1}\ cm^{-1}$) at a temperature of 25 °C. Enzyme activity was expressed in units; 1U was defined as 1 µmol of ABTS oxidized per min.

**Laccase-catalyzed oxidation of Phenol:** Typical experiments were made in 100-ml flasks. Each contained 25 mL of the total reaction medium, including a predetermined volume of laccase concentrate (~ 25 U/mL) to achieve the desirable enzyme concentration, a specific volume of phenol-spiked solution, and the remaining volume of citrate-phosphate buffer solution (50 mM, pH 5.0). The reaction was conducted at room temperature and stirred at 100 rpm.

At regular intervals, aliquots were withdrawn, acidified to pH 2 with HCl solution (20% v/v) to stop the reaction, and assayed for phenol quantification after centrifugation to remove the precipitate formed. Each system was studied in duplicates. The controls (buffer + phenol, without enzyme) were also included for monitoring the evaporation and photo-oxidation effects.

## EXPERIMENTAL STRATEGY

**Study on spiked phenol matrix**: Four variables were selected to explore phenol oxidation performance: pH range (3.5, 5 and 7), phenol concentration (25, 100 and 400 mg/L), aeration, and enzyme addition strategies - batch (AD1: 3 mL at 0 h) versus step-addition (AD2: 1.5 mL at 0 h and 0.5 mL at 1, 2, and 4 h).

**Wastewater enzymatic oxidation:** Two types of samples were tested: sour water and refinery water. Oxidation was conducted without pH adjustment. The enzyme concentration was 5 U/mL, and the AD2 enzyme addition strategy was evaluated.

## RESULTS AND DISCUSSION

**Laccase Production by *Pleurotus ostreatus* Fermentation.** Wheat bran was selected as a low cost and readily available substrate for the growth of *Pleurotus ostreatus*. The kinetic profile for laccase production using *Pleurotus ostreatus* is shown in Figure 1. The pH of the medium began at 6.3 and dropped over the course of 31 days to pH 5.8. The laccase concentration gradually increased until leveling off at about 25 U/mL at Day 22. Protein production showed a parallel profile. Laccase and protein concentrations after 31 days were 21 (± 2) U/mL and 1530 mg/L, respectively.

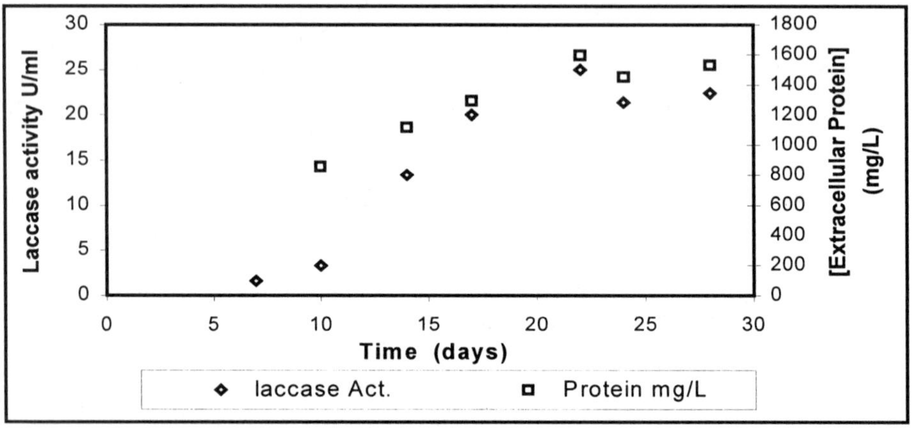


**FIGURE 1. Time course of a *Pleurotus ostreatus* solid-state fermentation**

**Phenol Removal by Crude Laccase Preparation.** When 5 U/mL of laccase were added to 15 mL of phenol solution (25-400 mg/L), the solution gradually turned black. Upon subsequent stirring, a black precipitate gradually formed. After one hour of stirring (first sampling), the precipitate was separated by centrifugation and the supernatant assayed for phenol.

***Effect of Phenol Concentration.*** It was important to establish how the efficiency of the laccase-catalyzed phenol removal depends on various parameters, such as pH, aeration, substrate concentration, and enzyme addition strategy. Figure 2 shows the phenol concentration dependence of the removal efficiency, i.e., the percentage of phenol removed. It can be seen that the enzyme was effective at all of the phenol concentrations (25, 100 and 400 mg/L) evaluated. At a fixed enzyme concentration, the maximum removal percentages varied with the initial phenol concentrations: at 400 mg/L, the phenol removal was 16% higher than that at 25 and 100 mg/L. No appreciable reduction in phenol concentration was observed in the reference systems having the same phenol concentrations but no laccase.

***Effect of Aeration.*** Laccases are copper containing enzymes, which utilize molecular oxygen as a terminal electron acceptor. Air addition strategy was evaluated to study its effect on enzymatic oxidation. Figure 3 shows the dependence of the reaction on the aeration in an experiment with solutions of 100 mg/L of phenol. While all systems were stirred, one set was under aeration and the other was not. The aeration did not cause appreciable effects, indicating that the oxygen supply by stirring was enough under the experimental conditions. The results from the controls without laccase showed that the phenol loss was 1.4% for the systems with aeration and 0.1% for those without.

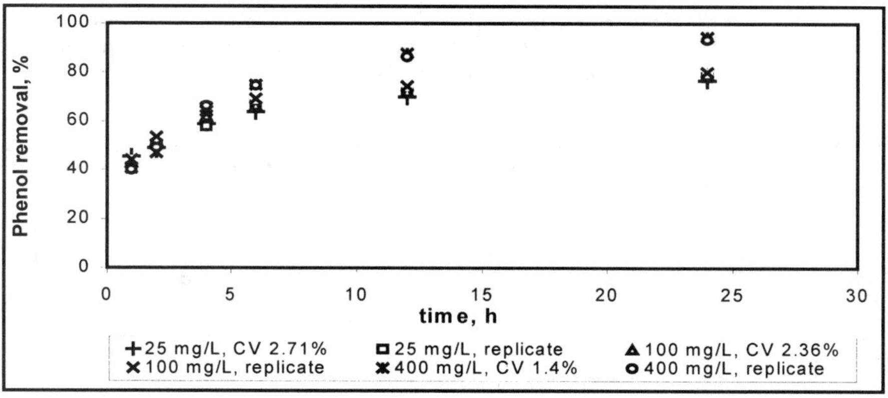


**FIGURE 2. Effect of initial phenol concentration on phenol removal; crude laccase 5 U/mL, pH 5.0, no aeration**

***Effect of pH.*** The effect of pH on phenol oxidation is shown in Figure 4. While laccase was functional in a broad range (3.5 – 7.0), the phenol removal percentage was higher at 5-7. With 400 mg/L of phenol, the highest removal was 93.9% at pH 5 while only 46.3% of removal was achieved at pH 3.5. For comparison, Dec and Bollag (1990) found that oxidation of chlorinated phenols, using laccase from *Trametes versicolor* and *R. praticola,* had optimal pH ranges of 4.5-9 and 3-8, respectively.

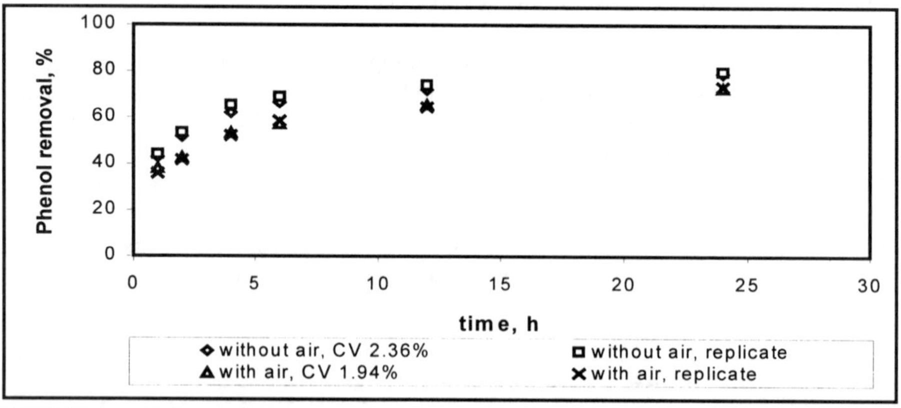


**FIGURE 3. Effect of air addition on phenol removal; 100 ppm of phenol, pH 5.0, and 5 U/mL crude enzyme**

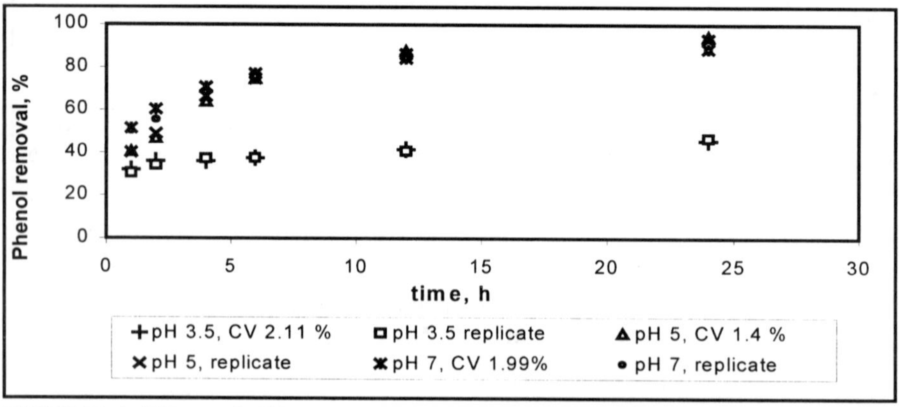


**FIGURE 4. Effect of pH on phenol removal; phenol - 400 mg/L, laccase (5 U/mL), no aeration**

***Effect of Enzyme Addition Strategy.*** As shown in Figure 5, when the entire enzyme was added at the beginning (AD1), the phenol removal efficiency was lower (77.4% and 79.5% for 25 and 100 mg/L phenol solutions, respectively) than the efficiency obtained with the AD2 strategy, i.e., 89.5% and 90.2% for 25 and 100 mg/L, respectively. (Results for 25 mg/L are not shown.) Similarly, Atlow *et al.,* 1984 found that while 93% of the substrate was removed by adding 30 U/mL of tyrosinase at the beginning, the removal was increased to 96% by adding the same enzyme concentration in three aliquots. One of the possible explanations was that the oxidation by-products deactivated the enzyme.

***Phenol Removal in Real Industrial Samples.*** Two industrial samples (wastewater treatment plant effluent and sour water) were evaluated. Removal results for both samples are summarized in Tables 1 and 2. The wastewater effluent was studied without adjustment, while the pH of sour water was adjusted to 7.3 before the experiments. Both showed phenol removal efficiency higher than 98%. It should be noted that the phenol removal percentages were higher with these real samples than with the spiked phenol solutions. The responsible mechanisms are yet to be identified. Nonetheless, the use of crude laccase preparation proved to be satisfactory for the removal of a range of phenolic compounds.

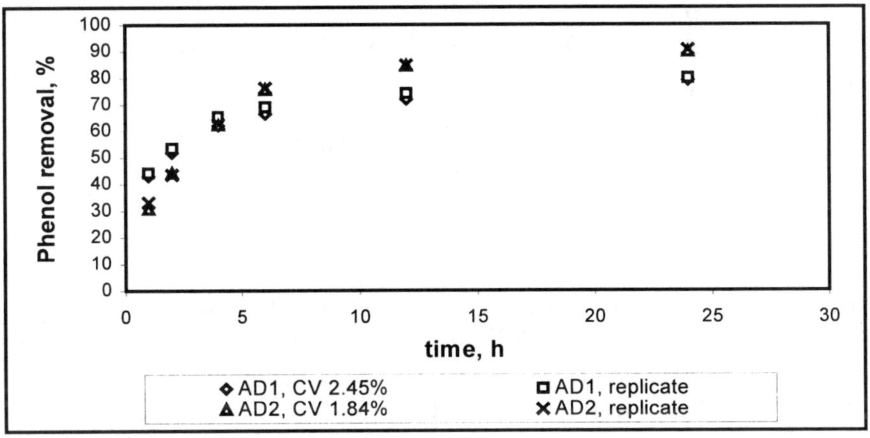


**FIGURE 5. Effect of addition strategy; laccase (5 U/mL) phenol/(100 mg/L)**

**TABLE 1. Enzymatic phenol oxidation: Effluent of wastewater treatment plant at Colombian refinery, 16.1 mg/L of phenol, pH 4.6, without aeration, laccase 5 U/mL, AD2. (Data given for the duplicate experiments)**

| Time, Hours | Concentration, mg/L | | Removal, % | | Average Removal, % | CV, % |
|---|---|---|---|---|---|---|
| 0 | 16.10 | 16.10 | 0.00 | 0.00 | 0.00 | |
| 1 | 10.80 | 11.22 | 32.92 | 30.31 | 31.61 | 5.83 |
| 2 | 8.25 | 8.40 | 48.76 | 47.14 | 47.95 | 2.39 |
| 4 | 5.25 | 5.13 | 67.39 | 67.72 | 67.55 | 0.34 |
| 6 | 3.34 | 3.13 | 79.25 | 79.67 | 79.46 | 0.37 |
| 12 | 0.22 | N.D | 98.63 | >79.67* | >98.63 | |
| 24 | N.D | N.D | >98.63 | >79.67* | >98.63 | |
| 48 | N.D | N.D | >98.63 | >79.67* | >98.63 | |

N.D: not detectable CV: coefficient of Variation

* They were calculated based on the last detectable HPLC data.

**TABLE 2. Enzymatic phenol oxidation: sour water with 248.36 mg/L of phenol, pH 7.3, without aeration, laccase 5 U/mL, AD2. (Data given for the duplicate experiments)**

| Time, Hours | Concentration, mg/L | | Removal, % | | Average Removal, % | CV, % |
|---|---|---|---|---|---|---|
| 0 | 248.35 | 248.35 | 0.00 | 0.00 | | |
| 1 | 213.60 | 207.60 | 13.99 | 16.41 | 15.20 | 11.24 |
| 2 | 188.30 | 188.0 | 24.18 | 24.30 | 24.24 | 0.35 |
| 4 | 164.60 | 155.79 | 33.72 | 37.27 | 35.50 | 7.07 |
| 6 | 117.50 | 126.10 | 52.69 | 49.22 | 50.96 | 4.81 |
| 12 | 36.48 | 39.31 | 85.31 | 84.17 | 84.74 | 0.95 |
| 24 | 0.48 | 0.51 | 99.81 | 99.79 | 99.80 | 0.01 |
| 48 | 0.39 | 0.39 | 99.84 | 99.84 | 99.84 | 0.00 |

## REFERENCES

Aitken, M.D. 1993. "Waste treatment applications of enzymes: opportunities and obstacles". *The Chemical Engineering Journal*, 52, B49-B58.

Atlow, S., Bonadonna-Aparo, L. ,and Klibanov, A.M. 1984. "Dephenolization of Industrial wastewaters catalized by polyphenol oxidase". *Biotechnology and Bioengineering*. 16: 599-603.

Bradford M. 1976. "A Rapid and Sensitive Method for the Quantitation of Microgram Quantities of Protein Utilizing the Principle of Protein-Dye Binding." *Anal. Biochem.,* 72: 248-254

Dec, J and Bollag, J. 1990."Detoxification of Substituyed Phenols by Oxidoreductive Enzymes through Polymerization Resctions". *Arch. Environ. Contam. Toxicol.*, 543-550

Kadhim, H., Graham, C., Barrat, P., Evans, C.S., and Rastall, R.A.1999." Removal of phenolic compounds in water using *Coriolis versicolor* grown on wheat bran" *Enzyme and Microbial Technology*. 24, 303-307.

# FUNGAL TREATMENT OF RECALCITRANT ORGANICS IN WASTEWATER

***Anna M. De'Ath** and* Richard M. Stuetz
(Cranfield University, Cranfield MK43 OAL, UK)

**Abstract:** Wastewaters contain an array of persistent organic compounds, which can resist breakdown during traditional biological wastewater treatment. Such recalcitrant compounds result from the synthesis of organic compounds for use in pesticide, detergent and pharmaceutical product manufacture. Local fungal isolates collected from wastewater treatment sites have been screened for their ability to grow in presence of a selection of organic compounds of varying structures and toxicity.

The presence of certain groups of compounds in trade waste has increased in recent years and has been reported to be the cause of taints and off-flavours either in recipient water-bodies or when abstracted for potable water supply. The occurrence of off-flavours with elevated levels of ammonia at an upstream treatment plant is taken to signify inhibition of the treatment process due to the toxicity of the wastes being processed.

## INTRODUCTION

The recalcitrance of industrial wastes is attributable to their low nutritional value and or the presence of inhibitory compounds. Both of these factors limit the performance of conventional biological treatment and failing treatment plants must take further steps to protect both the plant and the receiving environment.

In the UK it is common practice to treat industrial effluents together with domestic sewage in municipal wastewater treatment plants. Combining waste streams resolves the issue of nutrient deficiency but the dilution of toxic or recalcitrant compounds may not ensure their removal. The combination of wastewaters results in a greater volume of effluent requiring treatment which leads to the acceleration of costs when advanced chemical or physical treatment is required.

Conventional treatment of recalcitrant compounds is often reliant upon the physical and chemical technologies such as advanced oxidation processes (AOP) and adsorption utilising granular activated carbon (GAC). Realistically, the widespread application of AOP and GAC to the treatment of combined wastewaters is unfeasible due to the high operational costs involved. If GAC were to be used for wastewater treatment its application is likely to be restricted to final polishing of treatment plant effluent which, whilst protecting the receiving water, affords the treatment plant no protection.

The objective of this study was to obtain a suitable fungal consortium for a pre-treatment step to reduce the recalcitrance and perceived toxicity of the feed to biological trickling filters.

## MATERIALS AND METHODS

**Environmental Isolation.** Czapek Dox Modified Agar (CM97, Oxoid, UK) was used for the environmental isolation work it had the following composition (g/L): sodium nitrate 2.0; potassium chloride 0.5; ferrous sulphate 0.01; potassium sulphate 0.35; sucrose 30.0 and agar 12.0. The selective supplement Chloramphenicol (Oxoid, UK) was added to limit the development of bacterial colonies.

Czapek Dox Modified agar plates were inoculated with samples of effluent and biomass, collected from a chemical manufacturing plant and a municipal wastewater treatment plant. Inoculations were made both by pipetting sample filtrate onto the surface of the plates and by transferring the filter papers to fresh Czapek Dox Modified agar plates. Some 40 plates were obtained by routine sub-culturing in an attempt to obtain pure cultures for subsequent experimentation.

**Screening of Isolates.**

**Solid phase.** Fungal isolates were screened against 10 organic compounds, including formaldehyde, phenol and pentachlorophenol (PCP). The growth of each isolate was recorded for three concentrations of each compound with or without the inclusion of glucose in the test agar. The test concentrations were chosen by reference to published toxicity data (Verscheuren, 1996), with 'Low Rate' and 'High Rate ' being 0.5 and 2.0 times the quoted figure respectively. Formaldehyde and Hexamine are case study specific compounds and it was deemed appropriate to test the tolerance and possible degradation of these at concentrations reflecting the annual mean recorded for each (Table 1).

**TABLE 1. Organic compounds Evaluated During Study**

| Compound | Functional Group | Selection Criteria | Test Organism | Mid Rate Test concentration mg/l |
|---|---|---|---|---|
| Hexane | Alkane | I | Fish | 4[A] |
| Hexene | Alkene | I | Fish | 4 |
| Butyric acid | Carboxylic Acid | I & II | *Pseudomonas putida* | 875[A] |
| Butanone | Ketone | I | *Pseudomonas putida* | 1150[A] |
| Phenol | Aromatic | I, III, IV & V | *Pseudomonas putida* | 64 |
| Formaldehyde | Aldehyde | I & III | Site Consent | 108.4[B] |
| Hexamine | Amine | I, II, III | Site Consent | 208.6[B] |
| PCP | Substituted aromatic | I,III & IV | Microbial Mixture | 25[A] |
| Benzaldehyde | Aromatic aldehyde | I & II | *Pseudomonas putida* | 268 |
| Propyl acetate | Ester | I & II | | 228 |

[A] Value from Verschereun (1996). [B] Average values for a consented industrial discharge.
[I] Representative compound structure. [II] Associated odour.
[III] Present in industrial effluents. [IV] Subject to previous fungal research.
[V] Causal agent of taints and off-flavours in potable supply.

**Basal Media.** Media of the following composition was used; D(+)-Glucose (BDH, UK) (use in media is denoted by G) 2 $gL^{-1}$; Agar (Merck, UK)15 $gL^{-1}$; 75 ml of trace element solution was also added. The trace element solution had the following composition ($gL^{-1}$); $NH_4(SO_4)_2$ 150; $KH_2PO_4$ 100; $MgSO_4$ 40; $CaCl_2$ 10; NaCl 20; $FeCl_3$ 0.5; $ZnSO_4 \cdot 7H_2O$ 0.4; $MnSO_4 \cdot 4H_2O$ 0.4; $CuSO_4 \cdot 5H_2O$ 0.04.

**Test Compounds.** Benzaldehyde (>99%), butyric acid, 2-butanone, formaldehyde (37% w/v in $H_2O$), hexane (>99%), hexene (97%), hexamethylenetetramine (hexamine) (>99%) pentachlorophenol (98%), propyl acetate (99%) (Sigma-Aldrich, UK) and phenol (99.5%) (BDH Lab supplies, UK).

**Liquid phase.** Shake flask cultures were prepared by distributing the required volume of media (Ap-1) and distilled $H_2O$ to each flask which were sterilised by autoclaved at 121°C for 15 minutes. Once cool aliquots of the filter sterilised test compounds were added to each flask. Inoculation was via the transfer of a single 3 mm plug from maintenance Malt extract agar (Merck, UK) to the test culture.

**Analyses.**
Agar Plate Colonisation was facilitated by the use of an enclosing circle template. The diameter of the enclosing circle was recorded and these were then converted to gain categories of 5mm intervals.
Chemical Oxygen Demand (COD). HACH test vials (Camlab, UK) were used to determine COD by the reactor digestion method (Jirka and Carter, 1975) the COD mg/L was determined colorimetrically using the HACH DR/2010 spectrophotometer.
Dissolved Organic Carbon (DOC). DOC was determined by analysing filtered samples on a total organic carbon analyzer, TOC-5000A (Shimadzu, UK).
Phenol Assay. Where test isolates were incubated with phenol the residual concentration at the end of the incubation was determined via spectrophotometry exploiting the colorimetric reaction of phenols present in the sample to Folin and Ciocalteu's Phenol Reagent (Sigma, Dorset, UK) as used by Jones (1994).

## RESULTS AND DISCUSSION

### Agar plate screen

The ability of the 40 environmental isolates to grow in the presence of the selected test compounds was evaluated via the agar plate screen. An example of growth characteristics of four selected isolates in the presence and absence of glucose with a number of the target organics at mid range is given in Figure 1.
The unidentified isolates were coded in the laboratory as 7, 12-1, 27 and 42. The performance of the isolates has been assessed by mycelial development. The difference between the area colonised on the agar plate at the time of inoculation and 3 days post inoculation was calculated and termed growth gain.

The y-axis on figure one indicates the gain category of an isolates growth rather than the actual development in mm with the categories being organised into 5mm increments i.e. so that category 1 represents growth between 1 and 5mm. Growth in the presence of PCP was only evident where glucose was present in the media. In the majority of the cases where glucose was absent greater mycelium development was observed. Figure 1 clearly illustrates the differing responses of the test isolates to the recalcitrant compounds and the supplementary carbon source. The media containing glucose are identified by G in parentheses.

**Appropriate Glucose Supply.** Varying concentrations of glucose (0.2 and 2 % w/v) were initially evaluated to determine if supplementary carbon sources are likely to have an impact on the development fungal mycelium, which could affect the degradative ability of the fungal isolates.

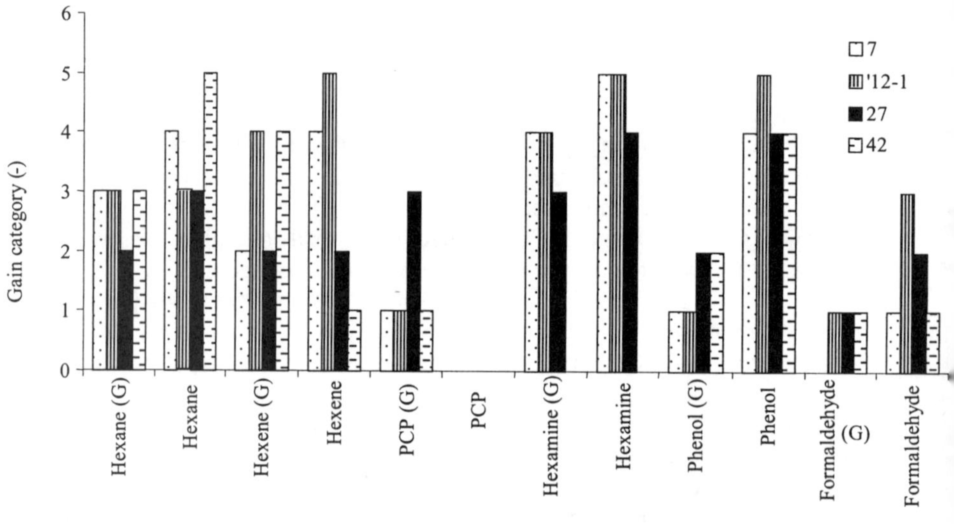


**FIGURE 1. Gain categories for isolates 7, 12-1, 27 & 42 after 3 days exposure.**

Figures 2 and 3 show the growth of four isolates on the basal media with glucose supplied at 2%w/v and 0.2%. Figure 2 illustrates the growth of isolates on 2% glucose media with isolate 37 colonising the plate more effectively than the other 3 isolates from day 7 through to day 14.

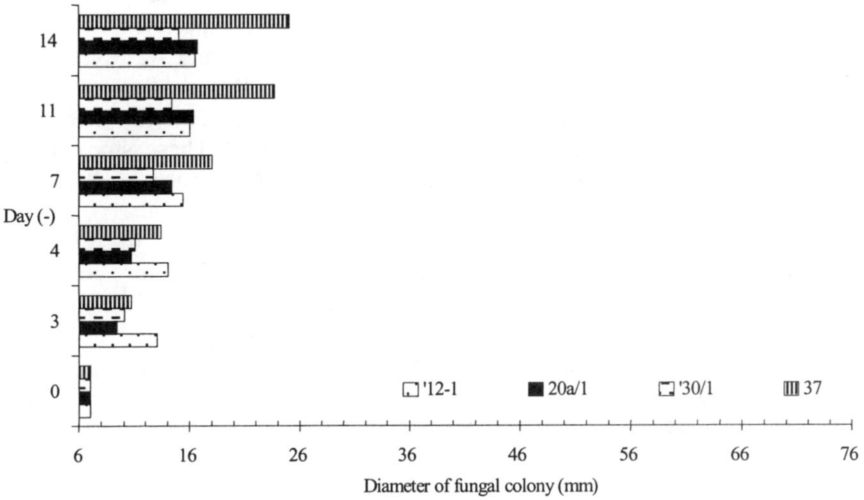


**FIGURE 2. Isolate growth on Ap-1 agar with 2% w/v glucose.**

Reduction of the carbon source to 0.2% (Figure 2a) resulted in faster mycelium development, clearly evident by day 7. On the reduced glucose media all the fungal colonies displayed increased radial growth. The greater development indicates the need for the fungus to secure further nutrients having exhausted those available in the immediate vicinity.

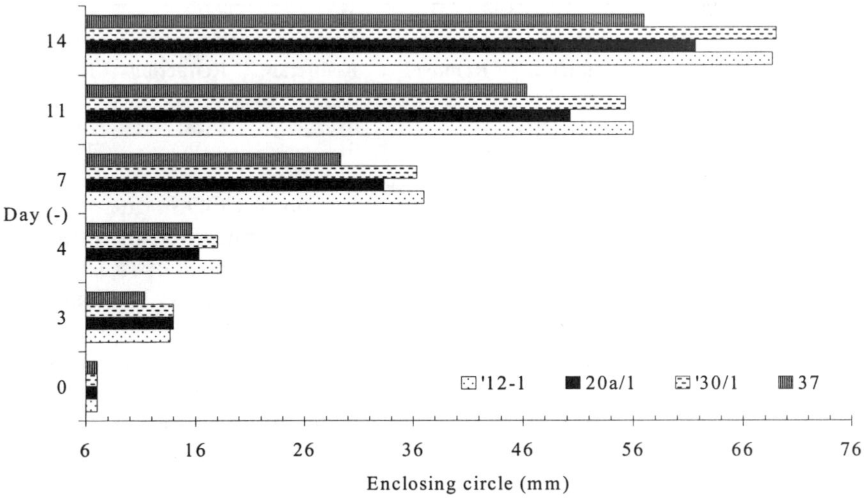


**FIGURE 3 Isolate growth on Ap-1 agar with 0.2% w/v glucose.**

**Compound Degradation in Synthetic Liquid Media.** The plate screen identified organisms suitable for preliminary batch culture experimentation using synthetic media. The 'Environmental Isolates', 7, 12-1 and 42 were tested alongside three culture strains (Cabi-Biosciences, IMI collection, UK), *Geotrichum candidum* (IMI 321760), *Trametes versicolor* (344115) *and Sporotrichum pulverulentum* (IMI 174727; ATCC 24725) These stock strains have been subject to previous waste treatment research (Alleman *et al.*,1992; Lin *et al.*, 1990; and Pallerla and Chambers, 1997).

Shake flask studies utilising a synthetic media serve to refine the selection of test isolates and establish optimal conditions for compound degradation. Comparisons of COD and DOC reduction, change in $UV_{254}$ and an assay for phenol have been used to establish degradation potential of isolates selected from the agar plate screen. The effects of readily degradable carbon sources (e.g. glucose) have also been evaluated, as such compounds are found in wastewater and are likely to influence the degradation of specific recalcitrant compounds.

Results for COD, DOC and target compound removal for media spiked with phenol (125 ppm) are given in Table 3. In both the media the presence of the selected fungi had a significant effect on the level of COD reduction. Isolate 42 removed the greatest percentage of COD in the glucose media whilst in the un-supplemented media, isolate 7 was responsible for the greatest COD reduction.

**TABLE 3. Percent COD, DOC and Phenol removal for selected isolates in AP-1 after 72 hours incubation.**

| | **Glucose (0.2% w/v)** | | | **No Glucose** | | |
|---|---|---|---|---|---|---|
| | **% COD Reduction** | **% DOC Reduction** | **% Phenol Removal** | **% COD Reduction** | **% DOC Reduction** | **% Phenol Removal** |
| **Control** | 37 | - | 13 | 2 | 8 | 6 |
| **7** | 60 | - | 12 | 81 | 12 | 97 |
| **42** | 94 | 41 | 96 | 41 | 6 | 40 |
| **12-1** | 78 | 0.7 | 8.9 | 71 | - | 92 |

Over a five day period (Figure 4) the reduction of COD in glucose media was significant and considered in isolation it could be dismissed as the utilization of the readily degradable co-substrate. However, in the instance of isolate 42 the reduction corresponds to a significant removal of phenol within 72 hours (Table 3).

Figures 4 and 5 illustrate the reduction of COD and DOC facilitated by selected fungal isolates over a period of five days. In Figure 3 the isolates demonstrates different degradative ability depending upon the provision of supplementary carbon in the test media.

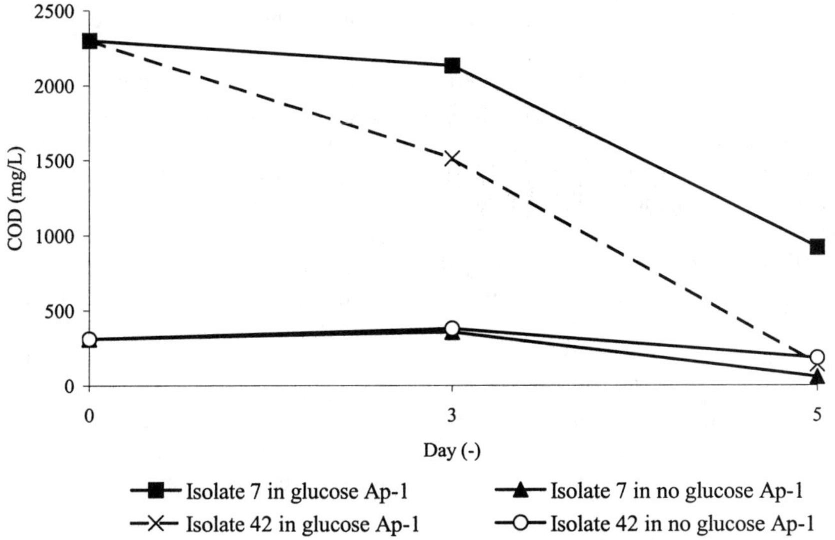


**FIGURE 4. Reduction of COD in AP-1 media containing phenol inoculated with isolate 7 or isolate 42 over 5 days.**

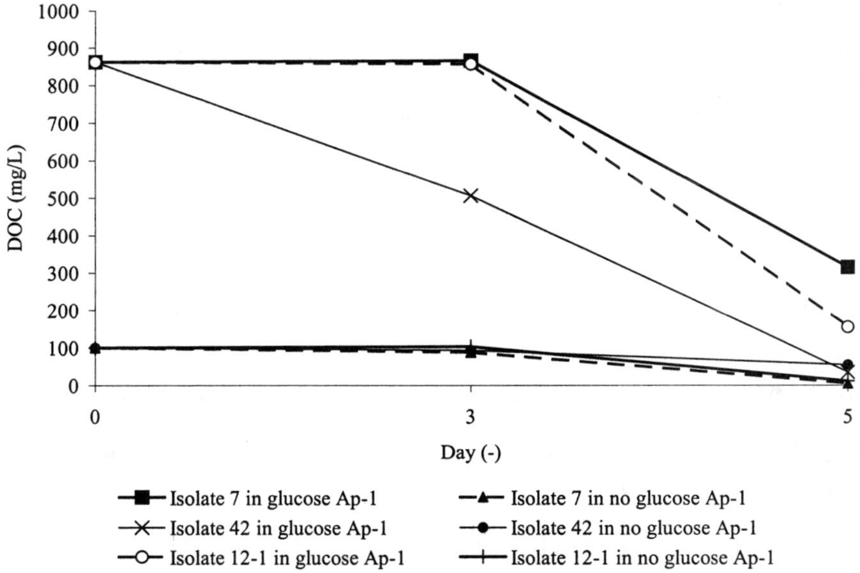


**FIGURE 5. Reduction of DOC in Ap-1 media containing phenol (125ppm) inoculated with isolate 7, 12-1 or 42 over 5 days.**

## CONCLUSIONS

Until recently, attention has focused upon the development of mycological remediation processes centred around the ligninolytic white rot fungi. This study shows that locally sourced fungi have the potential application to treat industrial wastewaters with phenol representing a common contaminant of such wastes (Davi and Gnudi, 1999).

The issue of supplementary carbon sources is also very important for the subsequent treatment of municipal wastewaters, as adequate nutrients are required for conventional bacterial treatment of wastewaters. As shown different carbon concentrations may affect the degradative ability of test fungi in a numbers of ways either by enhancing (Singleton, 1994) or limiting remediation through the suppression of degradative enzymes expression (Field et al., 1993).

Current work is investigating the effects of more complex wastewater matrices on specific recalcitrant compound degradation, using both laboratory and pilot-scale fixed film reactor studies. Future studies will evaluate the application of pre-treatment of industrial wastewaters by a fungal process prior to treatment of this wastewater by a fixed film reactor.

In addition to utilising locally sourced fungi to pre-treat recalcitrant industrial wastewaters, this method could be developed for the amelioration of recalcitrant groundwater contaminants via pump and treat remediation.

**ACKNOWLEDGEMENTS**

The authors wish to thank the U.K.'s Engineering and Physical Sciences Research Council and Anglian Water for their financial and technical support.

**REFERENCES**

Alleman B.C., B.E. Logan and R.L. Gilbertson 1992 "Toxicity of Pentachlorophenol to Six Species of White Rot Fungi as a Function of Chemical Dose". *Appl. Environ. Microbiol. 58*(12): 4048 – 4050.

Davi, M.L. and Gnudi, F. (1999) Technical Note: "Phenolic Compounds in Surface Water". *Water Research.* 33(14): 3213 – 3219.

Field, J.A., E. de Jong, G.Feijoo-Costa and J. A. M. de Bont. 1993. "Screening for ligninolytic fungi applicable to the biodegradation of xenobiotics". *TIBTECH 11*: 44-48.

Jirka, A M. and Carter, M. J. (1975) Analytical Chemistry, 47(8), 1397

Jones, K. 1994. "Metabolism of aromatic compounds by fungi". Ph.D. Thesis, Aberystwyth, Wales, UK.

Lin J-E., H. Y. Wang, and R. F. Hickey 1990. "Degradation Kinetics of Pentachlorophenol by *Phanerochaete chrysosporium*". *Biotechnology and Bioengineering 35*: 1125 – 1134.

Pallerla, S. and R. P. Chambers, R. P. 1998. "Reactor development for biodegradation of Pentachlorophenol". *Catalysis Today 40*: 103 - 111

Singleton, I. 1994 "Microbial Metabolism of Xenobiotics: Fundamental and Applied Research." *J. Chem. Tech. Biotechnol. 59*:9-23.

Verscheuren, K. 1996. *Handbook of Environmental Data on Organic Chemicals.* 3rd ed. Van Nostrand Rheinhold, New York.

# A KINETIC STUDY ON ANAEROBIC SULPHATE REDUCTION – THE EFFECT OF TEMPERATURE

***Shehnaaz Moosa***, Mehdi Nemati and Susan T L Harrison
Department of Chemical Engineering, University of Cape Town, South Africa

**ABSTRACT:** Kinetics of anaerobic reduction of sulphate was studied in a continuous bioreactor. The effects of temperature and volumetric loading of sulphate were investigated. Acetate was used as an electron donor. For the whole range of applied temperatures, 20 to 35 °C, the increasing of volumetric loading rate up to 0.08 to 0.1 kg $m^{-3}$ $h^{-1}$ resulted in a linear increase in reduction rate. Beyond this level the reduction rate became independent of volumetric loading. Despite the similar trends with volumetric loading, the reduction rate was influenced by temperature, with a maximum rate of 0.075 kg$m^{-3}h^{-1}$ observed at 35 °C. The effect of temperature on maximum specific growth rate ($\mu_{max}$) and bacterial yield was insignificant. The values of maximum specific growth rate and yield were 0.06 $h^{-1}$ and 0.56-0.60 kg bacteria (kg sulphate)$^{-1}$ respectively. Saturation constant ($K_s$) and decay coefficient ($K_d$) were both temperature dependent. The increase of temperature while decreasing the value of $K_s$, resulted in higher values for $K_d$. Using the experimental data a kinetic model incorporating terms for sulphate and bacterial concentrations, as well as temperature was developed for anaerobic reduction of sulphate.

## INTRODUCTION

Sulphate-reducing bacteria (SRBs) are an important group of microorganisms from an environmental and industrial point of view. Their potential application in wastewater treatment has been studied extensively (5, 6, 13, 17, 18, 27, 29). Anaerobic reduction of sulphate has been shown to be an important step in the treatment of process waters and mining wastes (3, 22). While playing a crucial role in the treatment of various effluents, production of $H_2S$ by SRBs is known to be responsible for serious corrosion problems in petroleum industries. As such an understanding of the principles governing the activities of SRBs not only provide basic information for their application in large scale treatment systems but also can be used as a tool to control and prevent their undesirable activities.

SRBs are obligate anaerobes that utilize sulphate or other sulphur oxoanions as the terminal electron acceptor during the oxidation of organic or inorganic compounds for utilization of energy (28). The anaerobic reduction of sulphate is a complex reaction, influenced by a variety of parameters such as availability of electron donor, temperature, pH, as well as inhibitory effects of heavy metals and sulphide. Extensive work has been done regarding the choice and efficient utilization of the electron donor. Electron donors that have been studied include: hydrogen; various organic compounds such as formate, acetate, propionate and butyrate; alcohols such as methanol, ethanol; fumarate, succinate and aromatic compounds (5, 9, 19, 29, 30). The type and structure of electron donor is crucial in the efficiency of sulphate reduction, but ultimately economics and availability dictate the choice of the electron donor.

The optimum pH for activity of SRBs has been shown to be around 7.0 to 7.8, though SRBs can tolerate pH values in the range 5.5 to 9.0 (7). Tuttle (24) reported the treatment of acid mine drainage at a pH of 2.5 to 3.6. This indicates the existence of acidophilic SRBs. Isolation of acidophilic SRBs have been reported recently (11). With respect the effect of sulphide on the reaction, inconsistent information have been reported in the literature. Widdel (29) observed that sulphide concentration of 0.085 $mgL^{-1}$ has an inhibitory effect on the activity of *Desulfotomaculum acetooxidans*. By contrast Visser (25) reported that a population of SRBs could tolerate sulphide concentrations as high as 1.13 $gL^{-1}$. There is, however, general agreement in the literature that at pH values above 8 sulphide has no effect on the reduction of sulphate. Above pH values of 8 sulphide in the solution is present primarily as $HS^-$.

Temperature is another important parameter which affects the kinetics of sulphate reduction and the activity of SRBs. Mesophilic strains of SRB, with growth temperatures in the range 28 to 45°C, as well as thermophilic strains with a growth temperature ranging from 54 to 70°C, have been isolated (9). Using a mixed SRBs culture in a batch experiments, Middleton and Lawrence (14) showed that the maximum specific rate of organic carbon utilization enhanced as the temperature increased from 20 to 31°C. For the same range of temperature the saturation constant ($K_S$) decreased, due to an increase in temperature. Using a continuous system Okabe and Characklis (16) found that the maximum specific growth rate ($\mu_{max}$) of *Desulfovibrio desulfuricans* was relatively constant for temperatures in the range 25 to 43°C, whereas, the saturation constant ($K_s$) increased with an increase in temperature. These authors reported that an increase in temperature led to a decrease in the yield coefficient. This contrasts with the results of Senez (23), who reported a constant yield for *Desulfovibrio desulfuricans* over a range of temperatures.

The objective of present work was to assess the effect of temperature and volumetric loading of sulphate on the activity of a mixed population of SRBs and to develop a kinetic model describing the effect of various physicochemical conditions, such as sulphate concentration, bacterial concentration and temperature, on the kinetics of anaerobic reduction of sulphate.

## MATERIALS AND METHODS

**Microorganisms.** A mixed culture of anaerobic bacteria containing acidogenic bacteria, methanogenic bacteria and SRBs was obtained from the Council for Scientific and Industrial Research (CSIR), Pretoria, South Africa. The culture had previously been used for treatment of a sulphate containing wastewater.

**Medium and culture conditions.** The medium used for growth and maintenance of the cells consisted of essential organic substrates, inorganic salts and trace metals (Table 1). This medium has been recommended for the growth of anaerobic microorganisms (8).

**TABLE 1. Composition of the medium used for growth and maintenance of the anaerobic mixed culture.**

| Components | Weight (g) |
|---|---|
| Peptone | 0.400 |
| Lab-Lemco[1] | 0.133 |
| $K_2HPO_4$ | 0.040 |
| $NaHCO_3$ | 1.250 |
| Deionised Water | 500 ml |
| **Trace Metals** | |
| $CoCl_2.6H_2O$ | 0.0119 |
| $FeCl_2.4H_2O$ | 0.0785 |
| $MnCl_2.4H_2O$ | 0.0038 |
| $NaMoO_4.2H_2O$ | 0.0038 |
| $NiCl_2.6H_2O$ | 0.0045 |
| Deionised Water | 500 ml |

[1] An extract used for general bacteriological purposes containing 12.4% N and 5.7% NaCl.

All reagents were analytical grade. Sodium sulphate was used as a source of sulphate. The stock cultures were grown in a medium containing 2.5 kg $m^{-3}$ of sulphate ions (3.7 kg $m^{-3}$ of $Na_2SO_4$). Initially 1.34 kg $m^{-3}$ of glucose was used as the major organic carbon source. This was then replaced by 17.5 kg $m^{-3}$ of acetate (24.3 kg $m^{-3}$ of sodium acetate). The substitution of glucose with acetate restricted the growth of the acidogenic bacteria. To suppress the activity of methanogenic bacteria, bromo-ethanol-sulphonic-acid (BESA) at a concentration of 3.2 kg $m^{-3}$ was added to the culture (27). The anaerobic culture was incubated in 1-litre bottles each containing 500 ml liquid medium and 50 % (v/v) inoculum at a constant temperature of 35°C on a rotary shaker at 160 rpm. Prior to incubation oxygen was stripped from the liquid using gaseous nitrogen. No pH adjustment was made.

**Experimental procedure.** The experiments were performed in four identical one litre bioreactors. Anaerobic conditions were maintained in the bioreactor by sealing the glass lids other glass fittings were with vacuum grease. The mixing in each bioreactor was achieved using overhead stirrers at a speed of 400 rpm. Fresh medium was fed into the bioreactor by a multiple speed peristaltic pump. To avoid channeling, feed was introduced near the bottom of the reactor. The effluent was discharged by gravity through an overflow tube. The reaction temperature was controlled using a circulating waterbath. The pH was maintained at 7.8 with the addition of either concentrated HCl or a saturated NaOH solution. Figure 1 shows a schematic diagram of the experimental set-up used for the kinetic studies.

In order to study the effects of temperature on the kinetics of sulfate reduction four independent experimental runs at 20, 25, 30 and 35°C were conducted. In each experiment the bioreactor was charged with 800 mL medium (Table 1, 5 kg $m^{-3}$ of sulphate, 17.5 kg $m^{-3}$ of acetate) and 200 mL of inoculum (20% v/v). The stock culture was used as an inoculum.

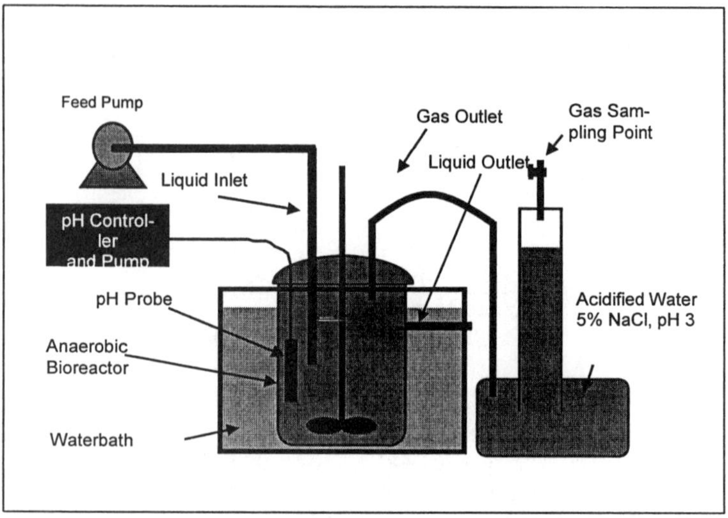


**Figure 1. Schematic representation of experimental set-up.**

The bioreactors were initially operated batchwise. Once a sufficient reduction of sulphate (90 to 95%), was achieved the bioreactors were switched to continuous mode. In each experimental run, the flow rate of the feed was varied in the range 0.004 to 0.042 L $h^{-1}$ (residence times between 10 days and 1 day). Steady state conditions were used at each flow rate to estimate the kinetics of sulphate reduction and bacterial growth. Steady state conditions were assumed to be established when both the residual sulphate concentration and bacterial concentration varied by less than 5% during a period of operation equal to three retention times. Liquid samples were taken on a daily basis and analysed for sulphate, ace tate and bacterial concentrations.

**Analytical procedures.** A turbidimetric method was used to measure the concentration of sulphate (1). Prior to sulphate determination, suspended solids were removed from the sample by centrifugation. After the addition of 0.25 ml of conditioning reagent (50 ml glycerol, 30 ml concentrated HCl, 75 g NaCl, 100 ml ethanol and 300 ml deionised water) to 5 ml of sample, an excess amount of finely ground $BaCl_2$ was added and the sample was stirred for 1 minute on a vortex mixer. The absorbance of the sample was then measured at a wavelength of 420 nm. The absorbance of the sample was used to calculate the concentration of sulphate. A calibration curve for dependency of absorption on sulphate concentration was obtained using a similar procedure. Acetate concentration was determined with a High Performance Liquid Chromatograph (HPLC, Beckmann, System Gold), with a UV detector (Detector no 168) and a glass lined Wakosil column. Phosphoric acid solution, 20 mM, with a pH of 2.5 was used as the mobile phase. The pH of the mobile phase was adjusted with sodium dihydrogen orthophosphate. The bacterial concentration was determined by dry weight measurement at 80°C.

## RESULTS

Kinetics of continuous bacterial reduction of sulphate at 20, 25, 30 and 35°C are shown in Figures 2, 3, 4 and 5 respectively. The volumetric reduction rate, calculated on the basis of initial concentration of sulphate, flow rate of the feed, working volume and conversion, was used to ascertain the kinetics of reaction.

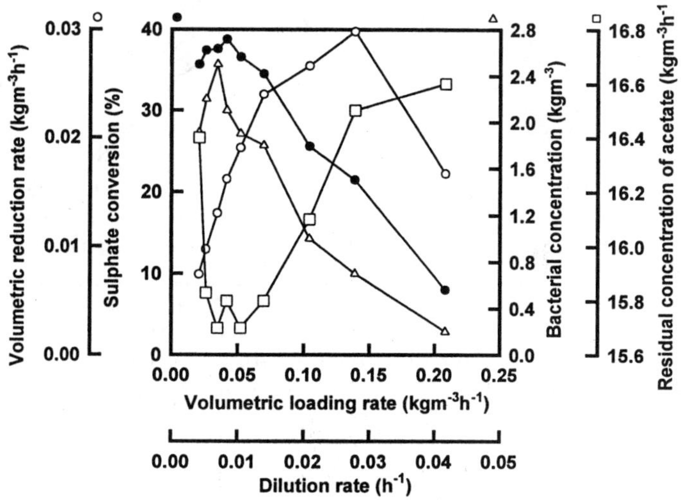


**FIGURE 2. Kinetic results of continuous reduction of sulphate at a temperature of 20°C.**

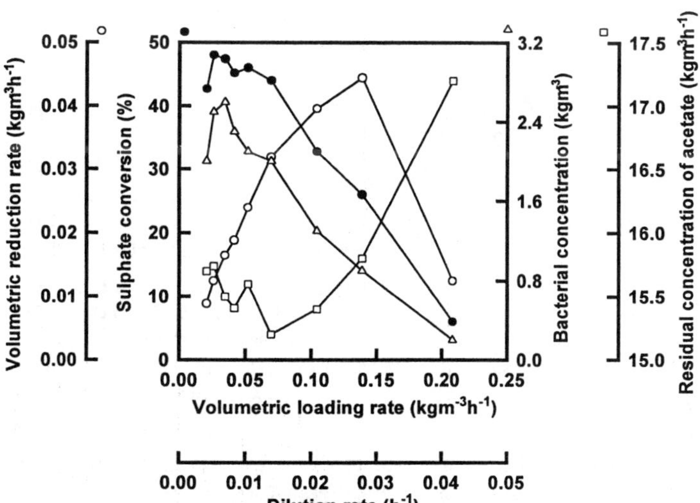


**FIGURE 3. Kinetic results of continuous reduction of sulphate at a temperature of 25°C.**

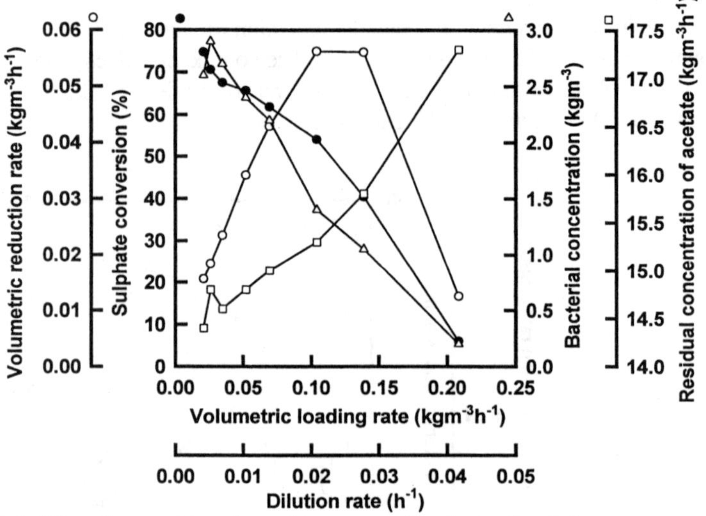


**FIGURE 4. Kinetic results of continuous reduction of sulphate at a temperature of 30°C.**

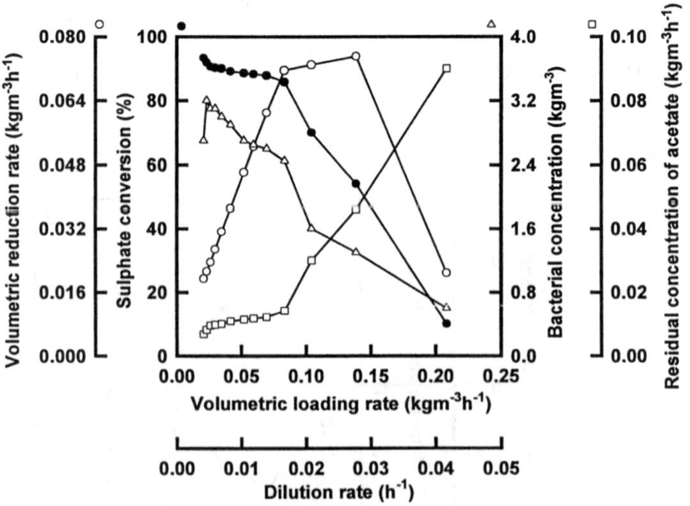


**FIGURE 5. Kinetic results of continuous reduction of sulphate at a temperature of 35°C.**

Operating the bioreactor at 20°C the maximum bacterial concentration of 2.5 $gL^{-1}$ was observed at a dilution rate of 0.007 $h^{-1}$. Increasing the dilution rate in the range 0.007 to 0.042 $h^{-1}$ resulted in a steady decrease in bacterial concentration. The Contois equation (2) with a decay term (eq 1) was used to describe the dependency of microbial growth on the concentration of sulphate. Fitting the experimental data to the Contois equation the maximum specific growth rate ($\mu_{max}$) and the saturation constant ($K_s$) were calculated to be 0.06 $h^{-1}$ and 4.75 kg $m^{-3}$ respectively. The decay coefficient ($K_d$) was 0.008 $h^{-1}$.

$$\mu = \frac{\mu_m [S]}{K_s [X] + [S]} - K_d \tag{1}$$

Where:

$\mu$ = specific growth rate of bacteria, $h^{-1}$
$\mu_m$ = maximum specific growth rate, $h^{-1}$
[S] = residual sulphate concentration, kg $m^{-3}$
$K_s$ = saturation constant, kg $m^{-3}$
[X] = bacterial concentration, kg bacteria $m^{-3}$
$K_d$ = decay coefficient, $h^{-1}$

The Pirt equation (eq. 2) was used to calculate the yield (15). The values of bacterial yield were 0.550 kg bacteria (kg sulphate)$^{-1}$ and 0.620 kg bacteria (kg acetate)$^{-1}$. As can be seen in Figure 2, the conversion of sulphate at 20°C was relatively low, with a maximum value of 39 % at a loading rate 0.042 kg $m^{-3}$ $h^{-1}$. The increasing of loading rate up to 0.139 kg $m^{-3}$ $h^{-1}$ enhanced the reduction rate. The maximum reduction rate in this set of experiments was 0.03 kg $m^{-3}$ $h^{-1}$, achieved at a retention time of 1.5 days. The corresponding conversion of sulphate was 22 %. Further increase in volumetric loading resulted in a dramatic decrease in reduction rate of sulphate. The ratio of sulphate reduced to acetate utilized was relatively constant at 0.80 moles sulphate/moles acetate for the range of loading rates applied.

$$r_s = \frac{\mu [X]}{Y} \tag{2}$$

Where:

$r_s$ = volumetric reduction rate, kg sulphate $m^{-3}$ $h^{-1}$
Y = bacterial yield, kg bacteria (kg substrate)$^{-1}$

Running the bioreactor at 25°C the bacterial concentration decreased over the range of applied dilution rates. Using the experimental data $\mu_{max}$ was found to be 0.06 $h^{-1}$, $K_s$ was 1.47 kg $m^{-3}$ and $K_d$ was 0.022 $h^{-1}$. Bacterial yield, calculated in terms of reduced sulphate was 0.560 kg bacteria (kg sulphate)$^{-1}$, and on the basis of acetate utilized was 0.600 kg bacteria (kg acetate)$^{-1}$. Applying loading rates from 0.021 to 0.070 kg $m^{-3}$ $h^{-1}$ the conversion of sulphate was relatively constant (44 to 48 %). Over this range of volumetric loadings, a linear increase in reduction rate of sulphate was observed. Further increase of volumetric loading up to 0.14 kg $m^{-3}$ $h^{-1}$, while accompanied by a decrease in conversion of sulphate, led to higher reduction rates. The maximum reduction rate of sulphate in this set of experiments was 0.036 kg $m^{-3}$ $h^{-1}$, achieved at a retention time of 1.5 days. The corresponding conversion was 26 %. The ratio of sulphate reduced to acetate utilized in this set of experiments was 0.76 moles sulphate/moles acetate.

Operating the bioreactor at 30°C, the bacterial concentration exhibited a linear decrease over the range of employed dilution rates. The values of $\mu_{max}$, $K_s$ and $K_d$ were calculated to be 0.06 $h^{-1}$, 0.61 kg $m^{-3}$ and 0.025 $h^{-1}$ respectively. The bacterial yields based on sulphate reduction and acetate utilization were 0.572 kg bacteria (kg sulphate)$^{-1}$ and 0.600 kg- bacteria (kg acetate)$^{-1}$. At 30°C the maximum conversion, 75 %, was observed at a volumetric loading of 0.021 kg $m^{-3}$ $h^{-1}$. Increasing the volumetric loadings in the range 0.021 to 0.10 kg $m^{-3}$ $h^{-1}$, while decreasing the conversion of sulphate to 62 %, led to a notable enhancement of reduction rate. The maximum reduction rate achieved was 0.056 kg $m^{-3}$ $h^{-1}$ at a retention time of 2 days. The corresponding sulphate conversion was 55 %. Increasing the volumetric loading rate beyond 0.140 kg $m^{-3}$ $h^{-1}$ resulted in a dramatic decrease in both sulphate conversion and volumetric reduction rate. For the range of applied volumetric loading rates, the ratio of reduced sulphate to acetate utilized was around 0.81.

During the experiment at 35°C the bacterial concentration decreased over the range of applied dilution rates. For this set of experiments the value of $\mu_{max}$ was calculated to be 0.06 $h^{-1}$, $K_s$ was 0.07 kg $m^{-3}$ and $K_d$ was 0.038 $h^{-1}$. Bacterial yield calculated in terms of reduced sulphate and utilized acetate were 0.610 kg bacteria (kg sulphate)$^{-1}$ and 0.580 kg bacteria (kg acetate)$^{-1}$ respectively. For volumetric loadings in the range 0.021 to 0.083 kg $m^{-3}$ $h^{-1}$ a relatively constant and high conversion of sulphate, of 87 t0 90 %, was achieved. Over this range of volumetric loadings, a linear increase in reduction rate of sulphate was observed. Further increase of volumetric loading up to 0.14 kg $m^{-3}$ $h^{-1}$, while accompanied by a sharper decrease in conversion, led to a semi-plateau region in the reduction rate curve. The maximum reduction rate of sulphate in this set of experiments was 0.075 kg $m^{-3}$ $h^{-1}$ at a retention time of 1.5 days. The corresponding sulphate conversion was 54%. The ratio of sulphate reduced to acetate utilized was relatively constant at 0.80 moles sulphate/moles acetate for the range of applied loading rates.

## DISCUSSION

The kinetic coefficients of bacterial growth and the value of yield at different temperatures are summarized in Table 2. The relatively constant value of $\mu_m$ indicates that the maximum specific growth rate of SRBs is not influenced by temperatures in the range 20 to 35°C. Growing *Desulfovibrio desulfuricans* in a continuous system, Okabe and Characklis (16) observed a similar effect. The values of $\mu_m$ obtained in this study are in good agreement with those reported in the literature. The maximum specific growth rate of *Desulfotomaculum acetoxidans* and *Desulfobacter postgatei* are reported as 0.058 $h^{-1}$ and 0.030$h^{-1}$ respectively (10, 28). Middelton and Lawrence (14) determined the value of $\mu_m$ for a mixed culture of SRBs, grown in acetate-sulphate medium, to be 0.022 $h^{-1}$.

As can be seen in Table 2, an increase in temperature results in smaller values for $K_s$. In other words the affinity of bacterial enzymatic system for the substrate (sulphate) is enhanced due to an increase in temperature. Information in the literature regarding the effect of temperature on the value of $K_s$ for SRBs is rather inconsistent. Using data from a batch system, Characklis and Marshall (4) reported an increase in the value of $K_s$ due to an increase in temperature, while analysing data from a continuous system a decrease in the value of $K_s$ was observed. Okabe and Characklis (16) noted that the value of $K_s$ for *Desulfovibrio desulfuricans* remained relatively constant as the temperature was increased from 25 to 35°C.

**TABLE 2. Kinetic coefficients of microbial growth and the values of yield at different temperatures.**

| Temperature °C | $\mu_m$ $h^{-1}$ | $K_s$ kg $m^{-3}$ | $K_d$ $h^{-1}$ | $Y_{x/acetate}$ kg bacteria (kg acetate)$^{-1}$ | $Y_{x/sulphate}$ kg bacteria (kg sulphate)$^{-1}$ |
|---|---|---|---|---|---|
| 20 | 0.06 | 4.746 | 0.008 | 0.620 | 0.554 |
| 25 | 0.06 | 1.467 | 0.022 | 0.600 | 0.561 |
| 30 | 0.06 | 0.613 | 0.025 | 0.600 | 0.570 |
| 35 | 0.06 | 0.078 | 0.038 | 0.580 | 0.607 |

The dependency of $K_s$ on temperature can be described by an Arrhenius type equation:

$$\underline{K_s} = \underline{K_o}\, e^{-\underline{E_a}/\underline{RT}} \tag{3}$$

Where:

$E_a$ = psuedo-activation energy, kJ mol$^{-1}$

$R$ = universal gas constant, kJ $K^{-1}$ mol$^{-1}$

$K_o$ = constant, kg substrate $m^{-3}$

T = absolute temperature, K

The Arrhenius dependence of the saturation constant on temperature was examined by plotting ln $K_s$ vs 1/T, for temperatures of 20, 30 and 35°C. This resulted in a straight line (Figure 6) from which $K_o$ and $E_a$ were determined. The values of $K_o$ and $E_a$ were $3.26*10^{-35}$ kg $m^{-3}$ and 198 kJ $K^{-1}$ mol$^{-1}$ respectively. (Coefficients were calculated using three sets of experimental data [20, 30 and 35°]. The experimental data was used to verify the validity of the model.)

The denaturation of proteins and enzymes in a microbial system is highly temperature dependent (21). As can be seen in Table 2, the values of the decay coefficient, $K_d$, increased due to an increase in temperature. A similar trend was observed by Lawrence and McCarty (12) for the anaerobic degradation of acetate by methanogens. The dependency of $K_d$ on temperature can also be described by the Arrhenius expression. Figure 7 shows the Arrhenius plot and the constants using the data from the experiments conducted at 20, 30 and 35°C.

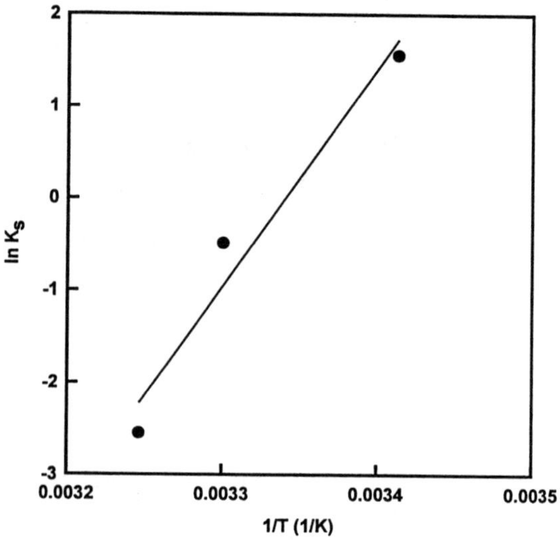


**FIGURE 6. Arrhenius plot for the dependency of $K_s$ on temperature.**

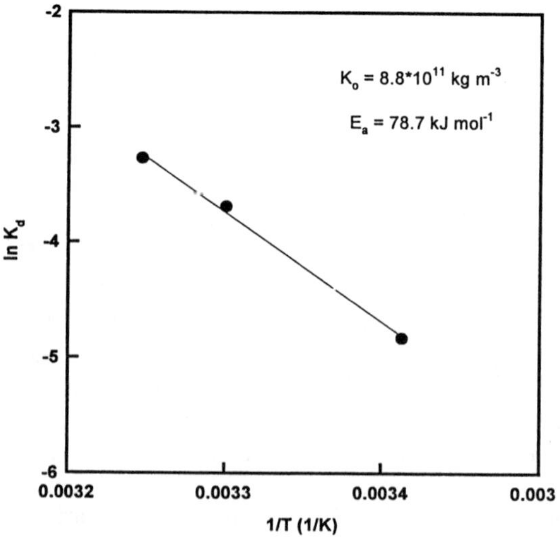


**FIGURE 7. Arrhenius plot for the dependency of $K_d$ on temperature.**

The yield coefficients calculated on the basis of sulphate reduced and acetate utilized remained relatively constant over the range of applied temperatures (variation of 10 % or less). A similar trend was observed by other researchers. Using a pure culture of *Desulfobacter desulfuricans,* and calculating yield in terms of organic substrate, Sanez (23) and Okabe and Charaklis (16) concluded that temperature had no effect on yield. There is a large variation in the reported values of $Y_{x/acetate}$. Middelton and Lawrence (14) observed that the overall yield for a mixed SRBs culture grown on acetate-sulphate medium in a batch system was 0.065 kg- bacteria (kg acetate)$^{-1}$. Widdel and Pfenning (28) reported the yield coefficients for *Desulfobacter postgatei* and *Desulfotomaculum acetoxidans* to be 0.074 and 0.095 kg- bacteria (kg acetate)$^{-1}$ respectively. Employing a mixed anaerobic culture in a continuous system Visser (27) calculated a yield coefficient of 0.043 kg bacteria (kg acetate)$^{-1}$. Values of $Y_{x/sulphate}$ are not readily available in the literature. Ingvorsen (10) determined that the yield coefficient for *Desulfobacter postgatei* grown in batch to be 0.158 kg bacteria- (kg sulphate)$^{-1}$.

The kinetics of anaerobic reduction of sulphate at 20, 25, 30 and 35°C are compared in Figure 8. As can be seen the dependency of volumetric reduction rate on volumetric loading of sulphate is similar for the whole range of applied temperatures. In all cases increasing the volumetric loading rate up to a certain level, 0.08 to 0.1 kg m$^{-3}$ h$^{-1}$, results in a linear increase in reduction rate. Beyond this level the reduction rate becomes independent of volumetric loading. Despite the similar patterns, the reaction rate is influenced by temperature. The increasing temperature results in enhanced reaction rates. At low volumetric loadings the difference in reaction rates is not pronounced. However, as the volumetric loading is increased the effect becomes significant. For instance the maximum reduction rate observed at 35°C is more than two fold higher than the maximum reduction rate at 20°C.

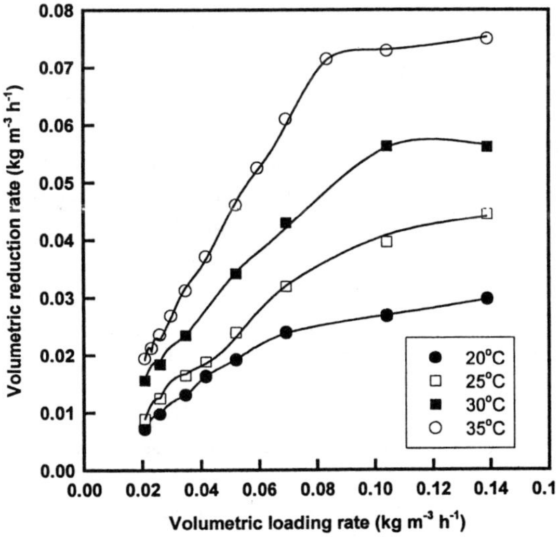


**FIGURE 8. Performance of the continuous bioreactor at 20, 25, 30 and 35°C.**

The dependency of volumetric reduction rate of sulphate on bacterial concentration can be described by the Pirt equation (eq. 2). The Contois equation (eq.

1) represents the dependency of specific growth rate, μ, on sulphate and bacterial concentration. Using the Contois expression, equation 2, can be rewritten in the following form:

$$r = \left[\frac{\mu_m [S]}{K_s [X]+[S]} - K_d\right]\frac{[X]}{Y} \qquad (4)$$

Using the Arrhenius equation (eqn. 3) to describe the dependency of $K_s$ and $K_d$ on temperature the following expression describes the reduction rate of sulphate in terms of bacterial and sulphate concentrations, as well as temperature:

$$r = \frac{[S][X](0.103 - 1.520 * 10^{12} e^{-9467/T}) - [X]^2(4.900 * 10^{-23} e^{14305/T})}{(3.260 * 10^{-35} e^{23772/T}[X] + [S])} \qquad (5)$$

Using equation 4 as the governing expression for the kinetics of anaerobic reduction of sulphate and writing a material balance, the performance of the continuous bioreactor, under different physicochemical conditions was assessed. The resulting theoretical curves are included as solid lines in Figure 8. As can be seen there is a good level of agreement between the experimental data and the prediction of the model. In other words, the rate expression is able to describe the kinetics of sulphate reduction over a range of physicochemical conditions. It should be noted that data from experiments at 20, 30 and 35°C were used to calculate the model and data at 25°C was used to verify the model.

## REFERENCES


1. American Public Health Association. 1975. Standard Methods for the Examination of Water and Wastewater 14th edition. APHA. New York.
2. Bailey, J. E. and D. F. Ollis. 1986. Biochemical Engineering Fundamentals, 2nd edition, McGraw-Hill, New York.
3. Barnes, L. J., P. J. M. Scheeren, and C. J. N. Buisman. 1992. Simultaneous microbial removal of sulphate and heavy metals from wastewater. Transactions of the Institute of Metallurgists. **101:** C183-189.
4. Characklis, W. G., and K. C. Marshall. 1989. Biofilms. Wiley-Interscience, New York.
5. Colleran, E., S. Finnegan, and P. Lens. 1995. Anaerobic treatment of sulphate-containing waste streams. Antonie van Leeuwenhoek. **67**: 29-46.
6. De Smul, A. 1998. Ph. D. thesis. University of Gent, Belgium. A combined biotechnological and physico-chemical process for the desulfurization of waste-water.
7. Fauque, G.D., L. L. Barton and J. Le Gall. 1980. Oxidation phosphorylation linked to dissimilatory sulfate reduction of elemental sulfur by *Desulfovibrio* p 71-86. *In* Sulfur in biology. Excerpta Medica, Amsterdam.
8. Grobicki, A. M. W. 1989. Ph. D. thesis. Imperial College of Science and Technology, United Kingdom. Hydrodynamic characteristics and performance of the anaerobic baffled reactor.
9. Hansen, T. A. 1988. Physiology of sulphate-reducing bacteria. Microbiological Sciences. **5:** 81-84.
10. Ingvorsen, K., A. J. B. Zehnder and B. B. Jorgensen. 1984. Kinetics of sulfate uptake by *Desulfobacter postgatei.* Applied and Environmental Microbiology. **47**: 403-408.
11. Johnson, B. W. Personal communication. 1998.

12. Lawrence, A. W., and P. L. McCarty. 1969. Kinetics of methane fermentation in anaerobic treatment. Journal of the Water Pollution Control Federation. **41**: 1659-1668.

13. Legall, J., and G. Fauque. 1988. Dissimilatory reduction of sulfur compounds, p.587-639. *In* Zehnder A. J. B. (ed.), Biology of Anaerobic Microorganisms. John Wiley and Sons, New York.

14. Middleton, A.C., and A. W. Lawrence. 1977. Kinetics of microbial sulfate reduction. Journal of the Water Pollution Control Federation. **229**: 1659-1670.

15. Nielsen, J and J. Villadsen. 1994. Bioreaction Engineering Principles. Plenum Press. New York.

16. Okabe, S. and W. G. Characklis. 1992. Effects of temperature and phosphorus concentration on microbial sulfate reduction by *Desulfovibrio desulfuricans.* Biotechnology and Bioengineering. **39:** 1031-1042.

17. Omil, F., P. Lens, A. Visser, L. W. Hulshoff Pol and G. Lettinga. 1998. Long-term competition between sulfate reducing and methanogenic bacteria in UASB reactors treating volatile fatty acids. Biotechnology & Bioengineering. 57: 676-685.

18. Oude Elferink, S. A., Visser, L. W. H. Pol and A. J. M. Stams. 1994. Sulfate reduction in methanogenic bioreactors. FEMS Microbiology Reviews. **15:** 119-136.

19. Postgate, J. R. 1984. The Sulfate Reducing Bacteria. 2 nd edition. Cambridge University Press, Cambridge UK.

20. Rintala, J. and G. Lettinga. 1992. Effects of temperature elevation from 37 to 55°C on anaerobic treatment of sulfate rich acidified wastewater. Environmental Technology. **13**: 801-812.

21. Roels, J. A. 1983. Energetics and Kinetics in Biotechnology, Elsevier Medical Press, Amsterdam.

22. Scheeren, P. J. M., R. O. Koch, C. J. N. Buisman, L. J. Barnes and J. H. Versteegh. 1991. A new biological treatment plant for heavy metal contaminated groundwater. Transactions of the Institute of Metallurgists. **101**: C188-197.

23. Senez. J. C. 1962. The considerations of the energetics of bacterial growth. Bacteriological Reviews. **29**: 95-107.

24. Tuttle J. H., P. R. Dugan and C. I. Randles. 1969. Microbial sulfate reduction and its potential utility as an acid mine water pollution abatement procedure. Applied Microbiology. **17**: 297-302.

25. Visser A., Y. Gao and G. Lettinga. 1992. The anaerobic treatment of synthetic sulfate containing wastewater under thermophilic conditions. Water Science and Technology. **25:** 193-202.

26. Visser A., I. Beeksma, F. van der Zee, A. J. M. Stams and G. Lettinga. 1993. Anaerobic degradation of volatle fatty acids at different sulfate concentrations. Applied Microbiology and Biotechnology. **40**: 549-556.

27. Visser A. 1995. Ph. D. thesis. Wegeningen Agricultural University, Wageningen, The Netherlands. The anaerobic treatment of sulfate containing wastewater.

28. Widdel F. and N. Pfennig. 1981. Studies on dissimilatory sulfate-reducing bacteria that decompose fatty acids. Archives of Microbiology. **129**: 395-400.

29. Widdel F. 1988. Microbiology and Ecology of Sulfate and Sulfur Reducing Bacteria. p. 469-585. *In* Zehnder A. J. B. (ed.), Biology of Anaerobic Microorganisms. John Wiley and Sons, New York.

30. Widdel F. and T. A. Hansen. 1992. The dissimilatory sulfate and sulfur reducing bacteria. In The Prokaryotes 2 nd edition. Springer-Verlag, New York, 583-624.

# THE EFFECTS OF NITRATE AND DO ON THE NITRATE UTILIZATION EFFICIENCY IN AEROBIC AND ANOXIC BIOREMEDIATION PROCESSES

***Chih-Jen Lu***, Chi Mei Lee, and Yan-Hung Du
National Chung Hsing University, Taichung, TAIWAN, ROC

**ABSTRACT:** This study focused on the nitrate utilization efficiency for toluene removal under different operating conditions. The nitrate utilization efficiency is defined as the ratio of the total amount of toluene removed to the total amount of nitrate consumed. The experiment was conducted with a series of batch reactors. The batch reactors were seeded with both aerobic toluene degraders and denitrifying bacteria and contained both nitrate and dissolved oxygen as electron acceptors. The experimental results indicated that toluene could be effectively removed under the aerobic or anoxic conditions, as soon as oxygen or nitrate was sufficiently supplied. The nitrate utilization efficiency in the obligate anoxic condition was only 0.22 mg-toluene/mg-nitrate, lower than that calculated from the theoretical stoichiometry equation, 0.26 mg-toluene/mg-nitrate. The results showed that toluene could be simultaneously removed by both aerobic toluene degraders and denitrifying bacteria in reactors containing both nitrate and limited oxygen. Also, nitrate utilization efficiency was relatively better in reactors containing both nitrate and limited oxygen compared to that in reactors containing nitrate alone. When the initial dissolved oxygen concentrations increased from 4.6 to 12.3 mg/L, the overall and anoxic nitrate utilization efficiencies increased from 0.31 and 0.3 to 0.51 and 0.47 mg-toluene/mg-nitrate, respectively. However, a further increase in dissolved oxygen concentration did not increase the nitrate utilization efficiency, because of the formation of a highly oxidized environment inhibiting the activity of denitrifying bacteria.

## INTRODUCTION

Aromatic compounds can be readily biodegraded in the aerobic environment. However, oxygen supply usually is the major limiting factor in the aerobic biodegradation process, primarily because of its low solubility. Generally, an alternative electron acceptor can be used to overcome the limitation of oxygen supply, such as nitrate. Nitrate helps overcome the limitation of oxygen supply, because of its high solubility and high energy yield for cell growth (Barenschee *et al.*, 1991; Hunchins and Wilson, 1994). However, it is very impracticable to keep the bioremediation process in an obligate anoxic condition by using nitrate as the alternative electron acceptor in the field. Biodegradation processes that use nitrate as the electron acceptor usually contain some dissolved oxygen (Schreiber *et al.*, 1997). Therefore, the biodegradation system may contain two electron acceptors, dissolved oxygen (DO) and nitrate, if nitrate is used as the alternative electron acceptor (Barbaro *et al.*, 1997; Schreiber *et al.*,

1997). This paper focused on the effects of the presence of both DO and nitrate on the removal of an aromatic compound, toluene. Because oxygen is usually consumed rapidly during the bioremediation process, hydrogen peroxide can be used as an alternative oxygen source to overcome the limitation of oxygen supply. Therefore, this paper also focused on the presence of DO (or hydrogen peroxide) on the nitrate utilization efficiency in the bioremediation process. Here, the nitrate utilization efficiency is defined as the ratio of the total amount of toluene removed to the total amount of nitrate consumed. A better nitrate utilization efficiency means the less nitrate requirement during the bioremediation process and the less cost in nitrate addition.

## MATERIALS AND METHOD

Microorganisms used in this study were initially collected from the effluent of a petroleum wastewater treatment plant. Microorganisms were then acclimated with toluene under aerobic and anoxic conditions with two chemostats for more than three months. The acclimated cultures were used as the inoculum enrichment using toluene as the sole carbon source. The enriched cells were concentrated and washed with buffered water and then used as the inoculum for the batch study.

The experiments were conducted with a series of batch reactors with volume of 125 mL. Each reactor contained microorganisms at an initial concentration of $10^7$ cfu/mL and nutrients-containing solution. Batch reactors were shaken at 100 rpm in the dark at room temperature ($22\pm1^{\circ}C$). During the experimental period, the reactors were sampled to analyze the remaining concentrations of toluene, nitrate, and DO by a high pressure liquid chromatography (HPLC), ion chromatography (IC), and DO meter, respectively.

In this study, the reactors were divided into three groups. The first group was an aerobic study. The carbon source, nitrogen source, and electron acceptor were toluene, ammonium ($NH_4Cl$), and oxygen, respectively. The initial toluene concentration was about 32.8 mg/L with the addition of various concentrations of hydrogen peroxide, from 0 to 101.5 mg/L. The reactors were seeded with both acclimated aerobic toluene degraders and denitrifying bacteria. The second group was an anoxic study. Nitrate was used as both the nitrogen source and electron acceptor. The reactors were seeded with only acclimated denitrifying bacteria. The third group was in the aerobic condition initially and then in the anoxic condition. The reactors contained two electron acceptors, nitrate and oxygen. Nitrate was used as both the nitrogen source and electron acceptor. The reactors were seeded with both denitrifying bacteria and aerobic toluene degraders.

## RESULTS AND DISCUSSION

The stoichiometry of toluene biodegradation using ammonium as the nitrogen source and oxygen as the electron acceptor can be expressed as equation (1):

$$C_7H_8 + 0.9\ HCO_3^- + 0.9\ NH_4^+ + 4.5\ O_2 \longrightarrow 0.9\ C_5H_7O_2N + 3.4\ CO_2 + 3.1\ H_2O \quad (1)$$

The theoretical stoichiometry equation shows that the removal of toluene at 32.8 mg/L required 51.3 mg/L oxygen. Therefore, 101.5 mg/L hydrogen peroxide was provided to supply sufficient dissolved oxygen (DO). Theoretically, the decomposition of 101.5 mg/L hydrogen peroxide could provide 47.7 mg/L DO. In addition to the original DO (7.2 mg/L), the addition of hydrogen peroxide at 101.5 mg/L provided abundant DO for toluene removal. The results showed that toluene was completely removed within 10 days with the addition of hydrogen peroxide at 101.5 mg/L (Figure 1). However, nitrate was intact during the experimental period. Because ammonium was the nitrogen source and DO was sufficiently supplied, nitrate was not used as the nitrogen source or electron acceptor. Therefore, nitrate was intact.

For the next case, nitrate was the nitrogen source. The reactors contained both nitrate and DO. Figure 1 shows that toluene removal efficiencies were 52% and 76% with the addition of hydrogen peroxide at concentrations 31.3 and 62.5 mg/L, respectively. Although, the supply of hydrogen peroxide at 31.3 and 62.5 mg/L could provide extra 14 and 28 mg/L of DO, respectively, the DO was still insufficient to have complete removal of toluene at 32.8 mg/L. The reactors contained both DO and nitrate as the electron acceptor. After the depletion of DO, nitrate at 125 mg/L was the next available electron acceptor for toluene removal. However, Figure 1 shows that nitrate was intact after the depletion of DO, even if the reactors were also seeded with denitrifying bacteria. The results suggested that the presence of hydrogen peroxide at 31.3 and 62.5 mg/L inhibited the activity of denitrifying bacteria. Therefore, nitrate was not used as the electron acceptor after the depletion of oxygen. The addition of hydrogen peroxide at concentrations of 31.3 and 62.5 mg/L resulted in partial toluene removal. Figure 1 shows that nitrate consumption was 4.2 and 7.1 mg/L for 52 and 76% toluene removals, respectively. The stoichiometry equation of toluene biodegradation using nitrate as the nitrogen source and oxygen as the electron acceptor can be expressed as following:

$$C_7H_8 + 0.446\ NO_3^- + 0.446\ H^+ + 5.82\ O_2 \longrightarrow 0.446\ C_5H_7O_2N + 4.71\ CO_2 + 2.64\ H_2O \quad (2)$$

According to equation (2) the theoretical nitrate consumption was 4.8 and 7.2 mg/L for reactors fed with hydrogen peroxide at 31.3 and 62.5 mg/L, respectively. The measured nitrate consumption was 4.2 and 7.1 mg/L for reactors fed with hydrogen peroxide at 31.3 and 62.5 mg/L, respectively. Therefore, the results of nitrate consumption indicated that nitrate was used as the nitrogen source by aerobic cells only, but not used as the electron acceptor in reactors fed with hydrogen peroxide at 31.3 and 62.5 mg/L. Because the environment was too oxidized, nitrate was used as the nitrogen source alone. Nitrate was not used as the electron acceptor simultaneously.

Figure 1 shows that toluene at the initial concentration of 31 mg/L was completely removed in the reactor without the addition of hydrogen peroxide but with sufficient nitrate. The initial DO concentration was 8.2 mg/L, which was insufficient to support 31 mg/L toluene removal. The reactors were seeded with both aerobic toluene degraders and denitrifying bacteria. The initial DO at 8.2 mg/L did not inhibit the activity of denitrifying bacteria. When DO was not sufficient for a completely aerobic toluene removal, nitrate was the next available electron acceptor for denitrifying bacteria to utilize toluene. Nitrate concentration significantly decreased from 128 mg/L to 25 mg/L after the depletion of DO. During the aerobic stage, the DO consumption was 6.2 mg/L (from 8.2 to 2 mg/L). Theoretically, 6.2 mg/L DO consumption should result in 4 mg/L toluene removal. However, the results in Figure 1 show that toluene was completely removed. Therefore, the remaining 27 mg/L (31- 4= 27 mg/L) toluene should be removed by denitrifying bacteria. The results may imply that toluene could be effectively removed by denitrifying bacteria, as soon as the DO concentration was not high enough to form a highly oxidized environment to inhibit the activity of denitrifying bacteria. Table 1 summarizes the results of aerobic and anoxic studies.

**Table 1. Toluene and nitrate consumptions for aerobic and anoxic studies**

| Initial DO (mg/L) | Toluene Removed (mg/L) | Nitrate Consumed (mg/L) |
|---|---|---|
| 8.2 (without $H_2O_2$) | 31 | 103 (e-acceptor) |
| 21.9 (with $H_2O_2$: 31.3 mg/L) | 16.1 | 4.2 (N-source) |
| 35.2 (with $H_2O_2$: 62.5 mg/L) | 23.9 | 7.1 (N-source) |
| 54.9 (with $H_2O_2$: 101.5 mg/L) | 32.8 | 0 ($NH_4^+$:N-source) |

In the next study, hydrogen peroxide was used as an extra oxygen source to supply more but still limited DO. In order to avoid the formation of a highly oxidized environment, the concentrations of hydrogen peroxide were controlled at 10.2 and 24.3 mg/L. With the original DO, the initial DO concentrations were 12.3 and 18 mg/L for the addition of hydrogen peroxide at 10.2 and 24.3 mg/L, respectively. The results showed that the addition of hydrogen peroxide enhanced the removal of toluene in reactors containing both aerobic toluene degraders and denitrifying bacteria. The toluene removal can be divided into two stages, aerobic and anoxic toluene removals. For reactors with the presence of hydrogen peroxide at 10.2 mg/L, toluene concentration decreased from 52.9 to 40.4 mg/L during the aerobic stage with the decrease in DO from 12.3 to 1.5 mg/L. The 1.5 mg/L DO was not further used as the electron acceptor. Theoretically, the consumption of 10.8 mg/L DO (from 12.3 to 1.5 mg/L) should have resulted the toluene removal and nitrate consumption of 6.9 and 2 mg/L, respectively. However, the results showed that toluene removal was 13.4 mg/L, instead of 6.9 mg/L. Because the reactor was seeded with both aerobic toluene degraders and denitrifying bacteria, it is reasonable to assume that toluene could be removed by the mechanisms other than aerobic biodegradation. The results indicated that

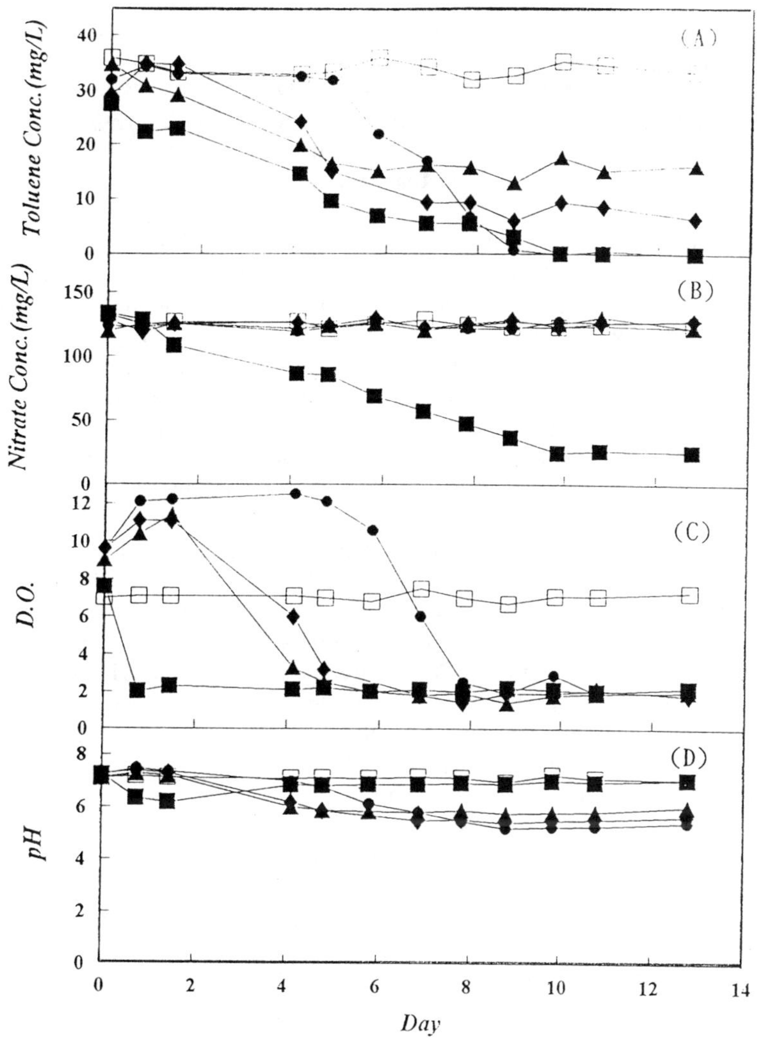


**FIGURE 1. Biodegradation of toluene in reactors seeded with both aerobic toluene degraders and denitrifying bacteria and contained both $NO_3^-$ and DO (with $H_2O_2$ conc.: 0 – 101.5 mg/L). (A): toluene, (B) $NO_3^-$, (C): DO, (D): pH ($H_2O_2$ conc.: ■, 0; ▲, 31.3; ◆ , 62.5; ●, 101.5 mg/L; □, Blank).**

nitrate consumption was 20 mg/L during the aerobic stage, but not 2 mg/L. The consumption of extra 18 mg/L of nitrate theoretically resulted in 5.5 mg/L toluene removal by denitrifying bacteria. Theoretically, total toluene removal through aerobic biodegradation and denitrification was 12.4 mg/L (5.5 + 6.9=12.4 mg/L), which was very close to the measured value, 13.4 mg/L. Again, the results showed that toluene was simultaneously removed by both aerobic toluene degraders and denitrifying bacteria with the presence of both nitrate and limited DO. After the depletion of DO, toluene was significantly removed with a significant decrease in nitrate concentration. During the anoxic stage, toluene removal and nitrate consumption were 39.5 and 83 mg/L, respectively. The nitrate utilization efficiency was 0.47 mg-toluene/mg-nitrate. Again, the nitrate utilization efficiency was higher than the calculated value, 0.26 mg-toluene /mg-nitrate.

When more hydrogen peroxide was added, aerobic biodegradation of toluene was more significant. The addition of 24.3 mg/L hydrogen peroxide resulted in the initial DO concentration of 18 mg/L. Toluene removal can be divided into two stages in reactors containing 24.3 mg/L hydrogen peroxide. In the aerobic stage, the presence of 18 mg/L DO resulted in 18.1 mg/L toluene removal (from 54.9 to 36.8 mg/L). Theoretically, the consumption of 16.5 mg/L DO (from 18 to 1.5 mg/L) should result in 10.5 mg/L toluene removal and 3.1 mg/L nitrate consumption. However, the results showed that toluene removal and nitrate consumption were 18.1 and 27.2 mg/L (from 214 to 186.8 mg/L), respectively. Again, it is reasonable to expect that toluene could be simultaneously removed by the mechanisms other than aerobic biodegradation. Theoretically, the consumption of 24.1 mg/L (27.2 – 3.1 =24.1 mg/L) nitrate through denitrification could result in 7.4 mg/L toluene removal. Therefore, total toluene removal through aerobic biodegradation and denitrification was 17.9 mg/L (7.4 + 10.5= 17.9 mg/L). The theoretical value (17.9 mg/L) was very close to the measured value (18.1 mg/L). Again, toluene was simultaneously removed by aerobic toluene degraders and denitrifying bacteria with the presence of both nitrate and limited oxygen. After the depletion of DO, toluene was further removed by denitrifying bacteria with a significant decrease in nitrate concentration. In the anoxic stage, toluene removal and nitrate consumption were 36.8 and 95 mg/L, respectively. The nitrate utilization efficiency was 0.39 mg-toluene/mg-nitrate, which is higher than the theoretical one (0.26 mg-toluene/mg-nitrate). The addition of 24.3 mg/L of hydrogen peroxide resulted in a more oxidized environment. Therefore, the nitrate utilization efficiency (0.39 mg-toluene/mg-nitrate) was relatively less than that with the presence DO at 8.6 mg/L and that with the addition of 10.2 mg/L hydrogen peroxide.

Table 2 summarizes the nitrate utilization efficiency in different operating conditions. Table 2 shows that the presence of both nitrate and limited DO could increase the nitrate utilization efficiency. When the initial DO concentrations increased from 4.6 to 12.3 mg/L, the overall nitrate utilization efficiency (contained both aerobic and anoxic stages) increased from 0.31 to 0.51 mg-toluene/mg-nitrate. The anoxic nitrate utilization efficiency also increased from 0.3 to 0.47 mg-toluene/mg-nitrate. However, a further increase in the initial DO concentrations decreased the nitrate utilization efficiency. When the initial DO concentration increased to 18 mg/L, the overall and anoxic nitrate utilization efficiencies were 0.45 and 0.39 mg-toluene/mg-nitrate, respectively.

**TABLE 2. Nitrate utilization efficiency in the different operating conditions**

| Operating condition | | Nitrate utilization efficiency (mg-toluene/mg-nitrate) |
|---|---|---|
| Theoretical value | | 0.26 |
| Anoxic (denitrifying bacteria) | | 0.22[(a)] |
| Aerobic cells + denitrifying bacteria | | |
| DO (4.6 mg/L) + Nitrate | | |
| | Overall | 0.31[(a)] |
| | After depletion of DO | 0.30[(a)] |
| DO (6.5 mg/L) + Nitrate | | |
| | Overall | 0.52[(b)] |
| | After depletion of DO | 0.46[(b)] |
| DO (12.3 mg/L) + Nitrate | | |
| | Overall | 0.51 |
| | After depletion of DO | 0.47 |
| DO (18 mg/L) + Nitrate | | |
| | Overall | 0.45 |
| | After depletion of DO | 0.39 |
| DO (21.2 mg/L) + Nitrate | | ---[(c)] |
| DO (35.8 mg/L) + Nitrate | | ---[(c)] |

(a): previous study (Lu *et al.*, 1999); (b): toluene was not completely utilized; (c): nitrate was not utilized as the electron acceptor

Table 2 shows that an increase in the initial DO concentration to 21.2 or 38.5 mg/L inhibited the activity of denitrifying bacteria. Therefore, nitrate was not further used as the electron acceptor by denitrifying bacteria.

## CONCLUSION

This study focused on the nitrate utilization efficiency in the different biodegradation processes. The experimental results showed that toluene could be simultaneously biodegraded by aerobic toluene degraders and denitrifying bacteria in the reactor containing both nitrate and limited dissolved oxygen. When extra dissolved oxygen was supplied to form a highly oxidized environment but insufficient to support completely aerobic biodegradation, toluene was partially removed. The results suggested the presence of both nitrate and limited dissolved oxygen increased the overall and anoxic nitrate utilization efficiency. When the initial DO concentrations increased from 4.6 to 12.3 mg/L, overall and anoxic nitrate utilization efficiencies increased from 0.31 and 0.3 mg-toluene/mg-nitrate to 0.51 and 0.47 mg-toluene/mg-nitrate, respectively. However, a further increase in the initial DO concentration did not increase nitrate utilization efficiency, because of the formation of a highly oxidized environment inhibiting the activity of denitrifying bacteria. The results suggested that an initial DO concentration in the range of 6.5 to 12.3 mg/L resulted the highest nitrate utilization efficiency, reducing the total amount of nitrate requirement.

## REFERENCES

Barbaro, J. R., J. F. Barker, and B. J. Butler. 1997. "In Situ Bioremediation of Gasoline Residuals under Mixed Electron-Acceptor Conditions." In B. C. Alleman and A. Leeson (Eds.), *In Situ and On Site Bioremediation*, , Vol. 5, pp. 21-26. Battelle Press, Columbus, OH.

Barenschee, E. R., P. Bochem, O. Helmling, and P. Weppen. 1991. "Effectiveness and Kinetics of Hydrogen Peroxide and Nitrate-Enhanced Biodegradation of Hydrocarbons." In R. E. Hinchee and R. F. Olfenbuttel (Eds.), *In Situ Bioreclamation: Application and Investigations for Hydrocarbon and Contaminated Site Remediation*, pp. 103-124. Butterworth-Heinemann, Stoneham, MA.

Hutchins, S. R. and J. T. Wilson. 1994. "Nitrate-Based Bioremediation of Petroleum-Contaminated Aquifer at Park City, Kansas: Site Characterization and Treatability Study." In R. E. Hinchee, B. C. Alleman, R. E. Hoeppel, and R. N. Miller (Eds.), *Hydrocarbon Bioremediation*, pp. 80-92. Lewis Publishers, Boca Raton, FL.

Lu, C. J., C. M. Lee, S. H. Liu, and Y. H. Du. 1999. "The Effect of Electron Acceptor on the Nitrate Utilization Efficiency in Groundwater." In B. C. Alleman and A. Leeson (Eds.), *In Situ Bioremediation Hydrocarbon and Other Organic Compounds*, pp. 469-474. Battelle Press, Columbus, OH.

Schreiber, M. E., J. M. Bahr, M. Zwolinski, Y. Shi, and W. J. Hickey. 1997. "Field and Laboratory Studies of BTEX Bioremediation under Denitrifying Conditions." In B. C. Alleman and A. Leeson (Eds.), *In Situ and On-Site Bioremediation,* Vol. 5,pp. 13-18. Battelle Press, Columbus, OH.

# SIMULTANEOUS AEROBIC AND ANAEROBIC DEGRADATION OF WASTEWATER IN A FIXED BED BIOREACTOR

B.A. Rivas-Lucero (UACH, Chihuahua, Mexico)
G.V. Nevárez-Moorillón (UACH, Chihuahua, Mexico)
R.A. Saucedo-Terán (CIMAV, Chihuahua, Mexico)
***R.G. Bautista-Margulis*** (CIMAV, Chihuahua, Mexico)

**ABSTRACT:** An experimental-scale fixed bed reactor, which combines an aerobic and anaerobic processes in a single unit, was developed to evaluate the removal efficiency of biodegradable material from industrial and domestic effluents. The morphology of the support material (silica sand) was also examined before and after treatment by using a scanning electron microscope (SEM). Experimental tests were carried out continuously for a period of 15 days each, varying the mean particle diameter ($d_s$), the static bed height ($H_s$) and the air superficial velocity ($U_g$). By measuring various physicochemical parameters, the experimental results showed that the highest removal efficiency of organic matter (94-95 %) was attained when the bioreactor operated at $d_s$= 1.87 mm, $H_s$= 0.65 m and $U_g$= 19.1 m/h. The SEM analysis confirmed that high carbon concentrations were adsorbed by the silica sand after the treatment process, while heavy metals were found to have no association with the sand matrix. The removal of organic matter in the proposed bioreactor was found to be influenced by the in-bed hydrodynamic and operating conditions in the following order: static bed height, air superficial velocity, and bed particle size.

## INTRODUCTION

The irrigation of agricultural crops with wastewater is a common practice in developing countries, especially for arid regions like the north of Mexico. However, it is well known that the application of both industrial and domestic wastewaters represents a serious hazard for human health due to undesirable microorganisms and organic and inorganic matter content (Fifield and Haines, 1995). This situation has prompted research to develop and study new biological reactors of either fixed or fluidized bed design, for the treatment of such wastewaters. Development of biofilms in these reactors has proven to be beneficial due to their high removal efficiencies, stability, and economy. The operation of these bioreactors has led to a lower hydraulic retention time, improved removal efficiency of organic matter, lower sludge production, and reduced space requirements. Nonetheless, the application of these reactors has been reported to have some constraints when organic loads, measured as chemical oxygen demand, exceed 300-350 mg/L (Tavares et al., 1995).

Biomass formation is a complex factor in bio-contact oxidation reactors since the overall density of the bioparticles is known to gradually decrease as biofilm grows. In this respect, microbial attachment may be affected by several factors, among them: characteristics of the support material, liquid phase hydrodynamics, nature and concentration of the wastewater, operating conditions, etc. Furthermore an increase in biofilm thickness, beyond a critical value, has been found to be useless for the treatment efficiency with a constant influent concentration. Several theoretical and experimental investigations have concluded that the optimal biofilm thickness may range between 100 and 300 μm for various environments (Zhang and Huck, 1996).

Microbiologically viewed, the combined activity of facultative bacteria and methanogens competing for the same substrate has been reported in the literature (Kato et al., 1994). The coexistence of anaerobic and aerobic bacteria was attributed to the intrinsic tolerance that some methanogenic species possess due to their ability to produce two enzymes, superoxide dismutase and catalase. A poor penetration of oxygen into the biofilms has also suggested that anaerobic microniches may be created inside the granules where the contact with methanogens is limited or non-existent.

**Objective.** The objective of this research work was to develop a new bioreactor system that incorporates both aerobic and anaerobic regions in a single unit. Likewise, aeration and clarification processes were combined, aimed at simplifying the washing operation by collecting most of the solids in the reactor. Furthermore, the influence of the hydrodynamics (particle size and bed height) on the bioreactor performance was examined to ascertain specific information for the design and performance of advanced fixed bed bioreactors.

## MATERIALS AND METHODS

The pilot bioreactor was made of a 1.80 m high polyacrylate material of 0.20 m internal diameter, and allowing for hydrodynamic behavior to be observed. The reactor was divided in four sections and had a 50-L working volume. The bed material used for this work was an irregularly shaped silica sand. A distributor plate of the "standpipe" type (vertical nozzle) was specifically designed to inject the incoming air and the wastewater through nine and four nozzles, respectively. A conical clarifier was also installed next to the reactor to improve the solid removal. Figure 1 illustrates the bioreactor system.

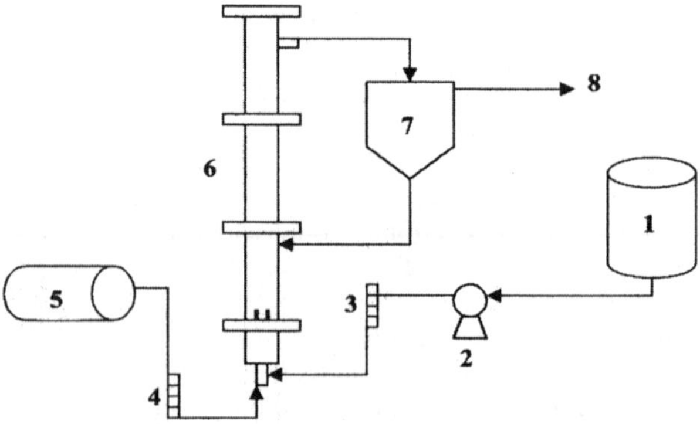


**FIGURE 1. Schematic diagram of the bioreactor system. (1) Wastewater reservoir. (2) Wastewater pump. (3) Wastewater rotameter. (4) Air rotameter. (5) Air compressor. (6) Bioreactor. (7) Clarifier. (8) Effluent.**

The wastewater used in this study was obtained from a discharge channel frequently employed for irrigating agricultural fields at Delicias City, Chihuahua

(North of Mexico). The composition of such wastewater is given in Table 1. The measurements of the influent stream showed that average concentrations of chemical oxygen demand (COD), biological oxygen demand ($BOD_5$), total suspended solids (TSS) and turbidity were 518 mg/L, 168 mg/L, 168 mg/L and 273 NTU, respectively.

| Constituents | Units | Mean ± Std. Dev. |
|---|---|---|
| pH | dimensionless | 7.68 ± 0.209 |
| Ammonia ($N\text{-}NH_3$) | mg/L | 35.12 ± 1.452 |
| Nitrates ($N\text{-}NO_3$) | mg/L | 1.487 ± 0.439 |
| Nitrites ($N\text{-} NO_2$) | mg/L | 0.137 ± 0.089 |
| Total nitrogen (TKN) | mg/L | 33.765 ± 2.52 |
| Biochemical oxygen demand (BOD) | mg/L | 168.5 ± 75.31 |
| Chemical oxygen demand (COD) | mg/L | 517.85 ± 248.22 |
| Total suspended solids (TSS) | mg/L | 167.85 ± 55.66 |
| Conductivity | mmhos/cm | 2.71 ± 0.543 |
| Turbidity | NTU | 273.9 ± 174.41 |
| Arsenic (As) | mg/L | 0.742 ± 1.05 |
| Boron (B) | mg/L | 1.019 ± 0.94 |
| Cadmium (Cd) | mg/L | 0.0148 ± 0.014 |
| Chromium (Cr) | mg/L | 0.019 ± 0.02 |
| Copper (Cu) | mg/L | 0.283 ± 0.141 |
| Mercury (Hg) | mg/L | 0.284 ± 0.264 |
| Lead (Pb) | mg/L | 1.29 ± 1.65 |
| Zinc (Zn) | mg/L | 0.666 ± 0.635 |

**TABLE 1. Influent wastewater composition.**

**Experimental Trials.** Prior to the injection in the reactor, the wastewater was homogenized in the storage tank for about 2 h and kept cool. At this point, physicochemical parameters were either measured "in situ" or samples were taken to be analyzed via either atomic absorption (AAS) or atomic emission (AES) spectrometry techniques. For a given test the air flow rate ($F_g$) was kept constant at 5 or 10 L/min, whilst the wastewater flow rate ($F_w$) was fixed at 1 L/min. Experiments were performed for three different static bed heights ($H_s$ = 0.40, 0.65 and 0.90 m) and mean particle sizes ($d_s$ = 0.92, 1.87 and 4.43 mm). Based on preliminary tests, continuous operation of the reactor was achieved for a 15 days period at room temperature for each experiment. The hydraulic retention time (HRT), with sand particles in the reactor, varied depending on bed particle size and bed height.

At the end of the experimental run three samples of the sand bed were taken for SEM analysis. Three different bed heights (top, middle and bottom) were chosen for the sampling of bioparticles. This microscopic analysis was intended to assess the presence of both organic and inorganic material in the morphological structure of the sand after the wastewater treatment. A previous SEM analysis of the silica sand before treatment was also made as a baseline.

**Bioreactor Performance**. To verify the reliability of the experimental results, three long-term tests were repeated for each experiment at given hydrodynamic and operating conditions. The performance of the reactor was evaluated from the difference in organic loads between the influent and effluent streams. All removal efficiencies were calculated using the following formula:

$$\% \text{ Removal} = [(C_{\text{influent}} - C_{\text{effluent}}) / C_{\text{influent}}] \times 100$$

where $C$ represents the concentration of the measured parameter. The total removal efficiency of the organic matter was then calculated in terms of the average value that resulted from the individual efficiencies with organic load. With a JEOL JMS-5800 SEM, scanning electron micrographs were also obtained in order to observe the morphological structure of the bioparticles.

## RESULTS AND DISCUSSION

**Removal Efficiency of Organic Matter.** Figure 2 shows the effect of $d_s$ on the COD removal efficiency, while other parameters such as $F_g$ (5 L/min) and $F_w$ (1 L/min) were kept constant. COD removal efficiencies up to 92 % were obtained for $d_s$ = 1.87, and 4.43 mm, having the $H_s$ constant at 0.65 m. When $d_s$ = 0.92 mm, however, a lower removal of organic matter (60-75 %) was observed in the bioreactor. This may be explained by a thicker biofilm formation attached to the smaller particles that, at the operating air superficial velocity ($U_g$ = 9.5 m/h), caused an overload of organic material in the sand bed. In fact, biomass was visually observed to leave the reactor and settle in the clarifier. Similar observations have been reported elsewhere (Carlson and Amy, 1998).

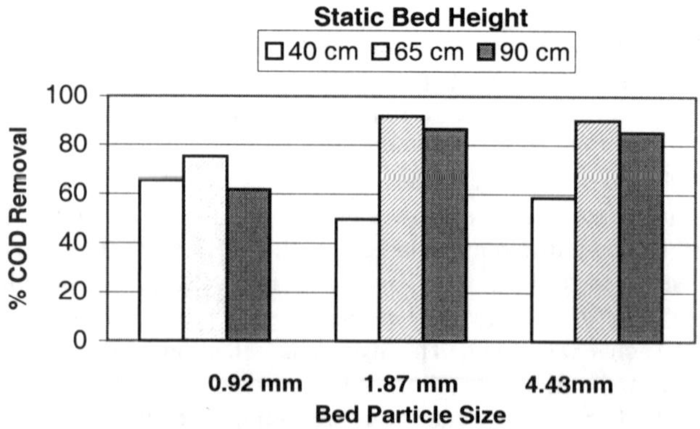


**FIGURE 2. COD removal efficiencies under various bed heights ($H_s$) and sand particle sizes ($d_s$).**

Figure 3 shows the effect of $H_s$ on the COD removal efficiency. In this case, two particular curves can be distinguished for each air superficial velocity operated in the reactor ($U_g$ = 9.5 and 19.1 m/h). The higher COD removal efficiencies (90 to 95 %) were obtained when $H_s$ = 0.65 m and $U_g$ = 19.1 m/h. The increase in $U_g$ may have caused an intensification of shear stresses and friction forces that acted on the bioparticle surfaces, giving rise to a more uniform formation of biomass in the sand particles. In agreement with this view, it has been reported that the creation of uniform layers of biomass resulted in the formation of dense and thin biofilms that enhanced bioreactor performance (Tavares et al., 1995). Alternatively, the higher $U_g$ value could have enhanced oxygen addition to the bioreactor, resulting in a higher oxygen gradient in the biofilm. From the results shown in Figure 3, it appeared that no significant enhancement of the removal process would take place for $H_s > 0.65$ m in this type of reactors. Additionally, a drop in COD removal efficiency was observed when $H_s < 0.65$ m. Also, at $H_s > 0.65$, differences in COD removal were negligible for the two $U_g$ values tested.

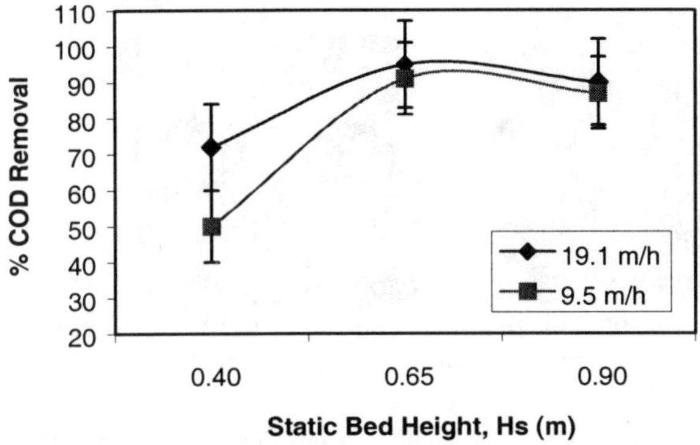


**FIGURE 3. Influence of the $H_s$ on the COD removal efficiency. Error bars are ± one standard deviation.**

**Coexistence of Anaerobic and Aerobic Activity**. Although anaerobes have been reported to be oxygen intolerant, the blackish color observed in the bioreactor may be an indication that an anaerobic metabolism also took place at the high dissolved oxygen values mentioned above. In fact, previous investigations have demonstrated that some methanogens such as *Methanobacterium* are insensitive to high concentrations of dissolved oxygen (Fetzer and Conrad, 1993). Also, during oxygen consumption by aerobic or facultative bacteria, anaerobic microniches could be created inside the particles in such a way that the oxygen could not reach strict anaerobes. Following this hypothesis, it is thought that the higher anaerobic activity took place in the current bioreactor with the larger sand particles since oxygen could not diffuse deeply into the biofilm. However, the extent of substrate competition between methanogens and facultative bacteria remains unclear and warrants further research.

**SEM Analysis of Bed Particles.** The SEM micrographs of sand particles were taken before and after the wastewater was treated in the reactor, as shown in Figure 4. Particles were classified into four groups according to their original color, in order to detect a possible physical or chemical difference related to the sand color, which could influence on the elemental composition of the particles. At this point, it is worth mentioning that each micrograph was taken with different magnification (i.e. 110X, 140X, etc..) in order to observe and quantify the presence of both organic and inorganic elements attached to the sand particles.

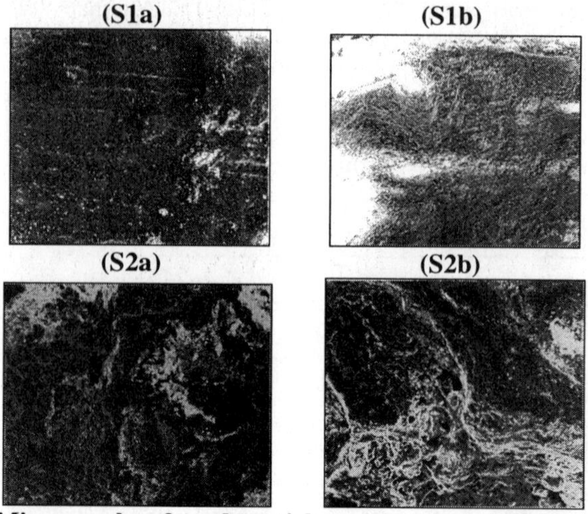


**FIGURE 4. Micrographs of sand particles analyzed via SEM before (S1) and after (S2) wastewater treatment. S1a & S2a: 140X; S1b & S2b: 110X.**

A semi-quantitative analysis of the elemental composition of these particles is shown in Table 2. Elemental carbon was found to be in the range of 36.80 to 49.75 % after wastewater treatment, while clean particles varied from 5.16 to 28.82 % carbon. These results produced clear evidence that a certain amount of the organic material was adsorbed at the particle surface. Organic nitrogen and phosphorous content observed in the particles after treatment also indicates the presence of biomass.

| Element | Before Treatment ( wt % ) | | | | After Treatment ( wt % ) | | | |
|---|---|---|---|---|---|---|---|---|
| | Brown | Gray | Orange | Red | Brown | Gray | Orange | Red |
| C | 12.71 | 28.82 | 5.16 | 14.50 | 49.75 | 40.21 | 42.24 | 36.80 |
| O | 51.68 | 48.99 | 52.24 | 48.79 | 35.81 | 48.75 | 42.25 | 44.82 |
| N | n.d. | n.d. | n.d. | n.d. | 11.79 | 11.45 | 12.58 | 9.80 |
| Na | 0.23 | 0.43 | 0.32 | 0.40 | 0.20 | 0.52 | 0.64 | 0.75 |
| Mg | 0.28 | 0.21 | 0.31 | n.d. | 0.10 | 0.18 | 0.15 | 0.33 |
| Al | 4.94 | 1.34 | 4.50 | 5.31 | 0.18 | 0.70 | 0.44 | 0.89 |
| Si | 26.89 | 18.66 | 33.38 | 25.53 | 0.91 | 7.28 | 1.25 | 4.45 |
| S | 0.20 | 0.60 | 0.20 | 0.51 | 0.48 | 0.91 | 0.23 | 0.83 |
| P | n.d. | n.d. | n.d. | n.d. | 0.45 | 0.53 | 0.22 | 0.50 |
| K | 0.31 | n.d. | 0.32 | 0.51 | n.d. | n.d. | n.d. | 0.11 |
| Ti | 0.49 | n.d. | 0.37 | n.d. | n.d. | n.d. | n.d. | n.d. |
| Fe | 2.26 | n.d. | 2.80 | 4.45 | n.d. | n.d. | n.d. | n.d. |
| Ca | n.d. | 0.62 | n.d. | n.d. | 0.32 | 0.79 | n.d. | 0.58 |

n.d. = not detected

**TABLE 2. Semi-quantitative analysis of sand particles via SEM before and after wastewater treatment.**

AAS/EAS analysis of four biomass samples also revealed a high concentration of inorganic matter (Table 3) in comparison with the influent concentrations (Table 1). From these results, it appears that some toxic metals contained in the influent wastewater (i.e. As, Cd, Cu, Hg, Pb, Zn) were absorbed into the biomass during the operation of the reactor. However, any association with heavy metal adsorption at the particle surface after wastewater treatment remains to be proved.

| Sample No. | As (ppm) | B (ppm) | Cd (ppm) | Cu (ppm) | Hg (ppm) | Pb (ppm) | Zn (ppm) |
|---|---|---|---|---|---|---|---|
| 1 | 5.21 | 105.3 | 0.80 | 297.2 | 1.70 | 32.26 | 1782.2 |
| 2 | 3.35 | 281.6 | 0.28 | 321.2 | 2.30 | 44.38 | 909.0 |
| 3 | 1.29 | 92.4 | 0.22 | 26.52 | n.d. | 3.34 | 120.3 |
| 4 | 4.55 | 1.74 | 0.22 | 212.2 | 0.101 | 20.12 | 1006.2 |

n.d. = not detected

**TABLE 3. Analysis of inorganic elements via AAS/EAS of four biomass samples obtained in the fixed bed bioreactor under the following conditions: $H_s = 0.65$ m, $d_s = 1.87$ mm, $U_g = 9.5$ m/h, $U_w = 1.9$ m/h**

## CONCLUSIONS

The proposed biological reactor has been proved to be effective for the treatment of wastewaters used in agricultural crops. The anaerobic regions visually observed in the reaction zone may suggest that certain anaerobes tolerate high levels of dissolved oxygen (4.45 mg/l) without affecting treatment performance.

For relatively low HRT (32 min), a high COD removal efficiency (92 %) was obtained when the reactor operated at $d_s$=1.87 mm, $H_s$= 0.65 m, $U_g$= 9.5 m/h and $U_w$= 1.9 m/h. At the expense of doubling the air superficial velocity ($U_g$= 19.1 m/h) under similar hydrodynamic conditions, a slight improvement of bioreactor performance may be achieved in terms of COD removal efficiency (2-3 %) for a fixed bed regime and distributor plates of the standpipe type.

**Acknowledgments**. The authors would like to thank the Mayor of Delicias City in Chihuahua-Mexico for financing the construction of the experimental installation. The senior author also wishes to acknowledge the technical assistance of M.C. Marcos Delgado, M.C. Hilda Esparza and Ing. Silvia Miranda.

**REFERENCES**

Carlson K.H., and Amy G.L. 1998. "BOM Removal During Biofiltration." *J. Am. Wat. Works Assoc. 90*(12): 42.

Fetzer S., and Conrad R. 1993. "Effect of Redox Potential on Methanogenesis by Methanosarcina Barkeri." *Arch. Microbiol. 160*: 108-113.

Fifield F.W., and Haines P.J. (Eds.). 1995. *Environmental Analytical Chemistry*. Blackie Academic and Professional. Chapman and Hall. Glasgow, U.K.

Kato M.T., Field J.A., Versteeg P., and Lettinga G. 1994. "Feasibility of Expanded Granular Sludge Bed Reactors for the Anaerobic Treatment of Low-Strength Soluble Wastewaters." *Biotechnol. and Bioengrg. 44*: 469-479.

Tavares C.R.G., Santanna Jr. G.L., and Capdeville B. 1995. "The Effect of Air Superficial Velocity on Biofilm Accumulation in a Three-Phase Fluidized-Bed Reactor." *Wat. Res. 20*(10): 2293-2298.

Zhang S, and Huck P.M. 1996. "Removal of AOC in Biological Water Treatment Processes." *Wat. Res. 30*: 1195-2004.

# BIOPROCESS OF WASTEWATER AROMATIC DYES: COST AND SIGNIFICANT VARIABLES

Roeb García Arrazola (CINVESTAV-IPN)
**Refugio Rodríguez Vázquez** (CINVESTAV-IPN)
Emilio Javier Manrique Ramírez (SYMCA, S.A. de C.V.)
Rubén Joel García Barajas (ESIQIE-IPN)
Ma. Dolores Díaz Cervantes and J. Manuel Fernández Sánchez (CINVESTAV-IPN)

**Abstract:** In order to contribute to the solution of the pollution of aromatic dyes in wastewaters, research was conducted for the development of an industrial bioreactor using the ligninolytic fungus *Trametes versicolor* (*T v.*), inoculated in pine sawdust, as the internal reactor biomass. The study was carried out at laboratory scale, accordingly to the Statistical Design of Experiments method to evaluate the interactions among seven proposed factors: Concentration of Inoculum (A), Presence/Absence of Glucose in the Medium (B), Days of Incubation (C), Concentration of Dyes (D), pH (E), Temperature (F) and Agitation (G) to find out the most important variables of the bioprocess. The kinetics of the *T v.* growth was examined and a two level fractional factorial design with resolution III ($2_{III}^{7-4}$) was applied. The statistical analysis (with $\alpha$=0.5) indicated: -A, -C, D, -E and –G as the significant variables and B and F were negligible. The cost of the proposed bioprocess was also estimated, coming from $0.20 to $0.80 (USD) per liter of wastewater treated at the first year of work, for capacities between 10 to 50 $m^3$ /day.

## INTRODUCTION

The removal of aromatic dyes from wastewater effluents has been the goal of many studies during last years. The reason is the carcinogenic and mutagenic effects in humankind (Cortinas and Vega, 1993). The complex structure of aromatic dyes and their phenolic nature make them extremely resistant to microbial degradation. Conventional treatment methods, such as aerated lagoons and activated sludge process, are ineffective in removing color. However, physical and chemical treatment methods, including ultrafiltration, ion-exchange, lime precipitation, are expensive and less efficient (Garg and Modi, 1999). Therefore it is necessary to develop and improve physical, chemical and biological treatment technologies.

Biological treatments represent an excellent option because the cost is generally four times less than the other two technologies, due to lower energy requirements (Levin and Gealt, 1997). Many microorganisms have been studied. The white-rot fungi are of interest because of their capability to degrade aromatic compounds into non-hazardous secondary compounds that are easily managed. Under anaerobic reducing condition, the azo dyes transform into colorless, deadend aromatic amines, which are generally more hazardous than the parent compounds (Swamy and Ramsay, 1999).

Previous studies have been limited to single decoloration of dyes, and confined to enriched cultures. Since an industrial effluent contains a range of dyes, successful decoloration does not adequately indicate the suitability of an organism for a decoloration treatment process (Swamy and Ramsay, 1999).

The present work evaluates the combined effect of several important variables in the biotreatment of a synthetic wastewater for the development of an industrial biorreactor, proposing a ligninolytic fungus inoculated in a natural support as the internal reactor's biomass. The goal was to identify the most important variables that affect the decoloration bioprocess in order to consolidate a conceptual engineering approach completed by an economic evaluation.

## EXPERIMENTAL DESIGN

This research was conducted accordingly to the Statistical Design of Experiments method. Synthetic wastewater, white-rot fungus and the natural support, were defined (see materials and methods). Once the system was defined, the next step was the choice of factors and levels to be evaluated in the decoloration bioprocess. The results are presented in Table 1.

A series of experiments performed in duplicate was planned in order to reach the proposed objective, resulting in three stages. In the first stage a verification of the capability of the system to decolorize a synthetic wastewater was carried out. There are reports the use of sugarcane bagasse pith in decoloration wastewater treatment. The wastewater increased its color because the loosening of the support's lignin (Díaz, 1999). In the second stage, the kinetics of the *T v.* growth in the pine sawdust was examined, to evaluate the performance of the fungus in the lignocellulose support. In the third stage, a two level fractional factorial design with resolution III ($2_{\text{III}}^{7-4}$) (Montgomery, 1991) was applied, to identify the most important variables of the bioprocess that affect the decoloration of the dyes and the levels of the variables that favored decoloration.

The preliminary estimation of the cost of biotreatment was calculated per $m^3$ of wastewater to be treated at the first year of work, for capacities between 10 to 50 $m^3$/day evaluated in this case.

**TABLE 1. Selected Factors and Levels to Evaluate the Bioprocess of Decoloration.**

| FACTORS | LEVELS | |
|---|---|---|
| | LOW(-) | HIGH (+) |
| A= Concentration of Inoculum (num. Of inoculum unit (I.U.)*) | 3 | 6 |
| B= Presence/Absence of Glucose in the Medium (g/L) | 0 | 10 |
| C= Days of Incubation | 4 | 8 |
| D= Concentration of Dyes (mg/L) | 100 | 500 |
| E= Initial pH | 5 | 6 |
| F= Aeration (rpm) | 100 | 200 |
| G= Temperature (°C) | 25 | 30 |

* equivalent to 1 g of inoculum fungus / 3 g of pine sawdust (approximately)

## MATERIALS AND METHODS

**Ligninolytic Fungus.** *Trametes versicolor* has displayed the greatest extent and rate of decoloration, and is sustained by repeated additions of several dyes and dye mixtures (Swamy and Ramsey, 1999). *Tv* was examined for its potential for use in treatment of dye-contaminated effluents

**Natural Support.** Pine sawdust is a type of softwood that contains a lignin with low reactivity to brown fungus degradation (Rodríguez, 1991). This probably could mean low ligninocellulose degradation and hence a long support lifetime. The pine sawdust is managed as a waste product, its use represents a low cost recycling option.

**Synthetic Wastewater.** A mixture made up of three dyes disolved in water. The three dyes were congo red (azoic), vat blue 4 (anthraquinonic), and malachite green (triarylmethane). This dyes were defined categories by the Environmental Defense Fund (EDF) and the Environmental Protection Agency (EPA) to determine specific constituent concentration levels in wastes at which the dyes will be hazardous; this includes industries such as textile, automotive, printing and plastics.

**Culture and Culture Maintenance.** The fungus *Tv* H-1051 CDBB-500, provided by CINVESTAV-IPN, was maintained on 2% malt agar plates at 4°C. Solid medium in Petri plates were prepared using EMA medium and incubated at 30°C. The culture in pine sawdust was prepared by adding 2 ml of modified Kirk's medium into 3g of inoculated support and incubated at 30°C.

**Kinetic of the Growing of *T. versicolor* Studies.** *Tv.* growth in the pine sawdust could be detected visually and was indirectly measured chromatographically by the $CO_2$ production in the system.

**Dyes and Monitoring of Decoloration.** Decoloration in liquid medium of the dye mixture could be determined spectrophotometrically at the maximum visible wavelength of absorbance for individual dyes: congo red (495nm), vat blue 4 (278nm), and malachite green (614nm).

**Statistical Analysis.** In order to identify the most important variables of the bioprocess, a statistical analysis was carried out by the software SAS 6.01.

**Economical Analysis.** In order to identify the conceptual engineering costs, an economic analysis was carried out using the software EVEPI 1.0 developed at ESIQIE-IPN (García et. al., 1997).

## RESULTS AND DISCUSSION

### 1st Stage: Verification of the Availability of the System.

In this stage there was no observed coloration in the synthetic wastewater by the effect of the natural support. This means no loosening of the support's lignin.

### 2nd Stage: Kinetic Growth of *Tv.* in Growth in the Pine Sawdust.

The growth of *Tv.* was measured indirectly by porcentage $CO_2$ diary. This data enable to analyze the effect of the factors on the develoment of *Tv.* and thus with the decoloration bioprocess. These factors are: concentration of inoculum, and presence/absence of glucose in the medium.

Figure 1 presents the results of the *Tv.* growth per day in pine sawdust without interaction with natural biota. The effect of glucose was not an important parameter for the same I.U. (see table 1), and the effect of I.U. was important up to seven days; after that, the difference was not significant.

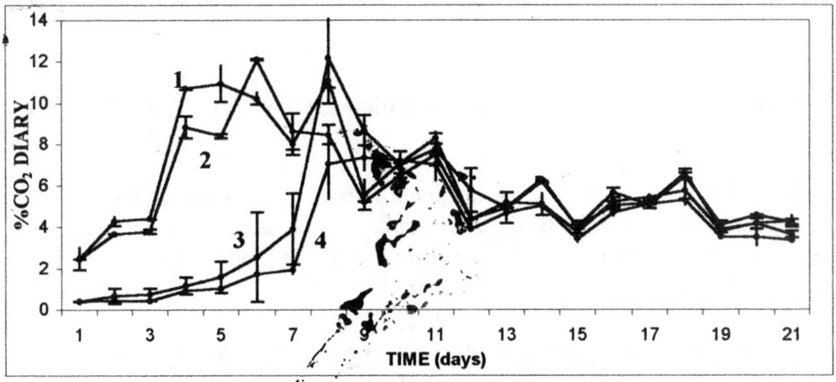


**1.** HIGH I.U./ GLUCOSE. **2.** HIGH I.U. / NO GLUCOSE
**3.** LOW I.U./ GLUCOSE **4.** LOW I.U./ NO GLUCOSE

**FIGURE 1. Kinetics Growth of *T. versicolor* in Pine Sawdust without interaction with the Natural Biota**

Figure 2 presents the results of *T v.* growth in pine sawdust with interaction with the natural biota. The effect of glucose and I.U. result not so important as can be observed from the similar trends of all the experiments.

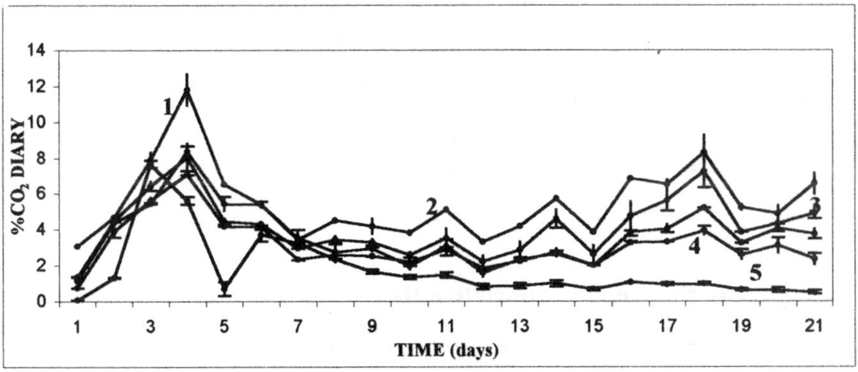


**1.** HIGH I.U./ GLUCOSE. **2.** HIGH I.U. / NO GLUCOSE
**3.** LOW I.U./ GLUCOSE **4.** LOW I.U./ NO GLUCOSE **5.** NATURAL BIOTA

**FIGURE 2. Kinetics Growth of *T. versicolor* in Pine Sawdust with interaction with Natural Biota**

In Figure 2 it can also be observed that the natural biota grew during the first four days of the experiment. This could mean a competence with the fungus for the nutrients from the Kirk's medium. From fourth day to the end of the experiment the trend of natural biota is to disappear. This could mean the decreasing of nutrients in the medium and the usage of the carbon source mainly from the support by the fungus.

**3rd Stage: Relevance and Levels Recommended for the Factors evaluated in the Bioprocess.**

The data of all the experiments originated from the selected statistical model design were analyzed by SAS 6.01. The results are presented for each dye and the dye mixture in Table 3 and Table 4. An empirical linear model was also obtained for each dye and the dye mixture (see Table 5).

**TABLE 3. Relevance of the Factors evaluated in the Bioprocess.**

| DYE | FACTORS | | | | | | |
|---|---|---|---|---|---|---|---|
| | A | B | C | D | E | F | G |
| Vat blue 4 | | | | | | | |
| Congo Red | * * | * | * * * | * * * | * * | * | * * * |
| Malaquite Green | * * * | * | * * * | * * * | * * * | * | * * * |
| Mixture | * * * | * | * | * * * | * * * | * | * * * |

* No significant ** significant *** highly significant.

**TABLE 4. Levels recommended of the factors evaluated in the Bioprocess.**

| DYE | FACTORS | | | | | | |
|---|---|---|---|---|---|---|---|
| | A | B | C | D | E | F | G |
| Vat blue 4 | | | | | | | |
| Congo Red | - | | + | + | - | | - |
| Malaquite Green | - | | - | + | - | | - |
| Mixture | - | | | + | - | | - |

- Low Level + High Level

**TABLE 5. Linear Equations obtained to estimate the Bioprocess' Decoloration.**

| DYE | MODEL |
|---|---|
| Vat Blue 4 | -------------- |
| Congo Red | $\%Decoloration^a = 51.63 - 2.47A + 3.14C + 10.01D - 2.40E - 3.33G$ |
| Malaquite Green | $\%Decoloration^b = 20.65 - 4.87A - 3.20C + 15.62D - 4.43E - 7.46G$ |
| Mixture | $\%Decoloration^c = 28.91 - 2.34A + 8.45D - 2.17E - 3.37G$ |

a) $R^2 = 0.97$ b) $R^2 = 0.98$ c) $R^2 = 0.99$

It is important to mention that the vat blue 4 could not be decolorized, neither by biodegradation nor by adsorption. This can be due to its structure: six aromatic rings make difficult the enzymatic degradation by white-rot fungus, additionally the ketonic groups lowered the solubility and the adsorption of the dye.

**Economic Evaluation.**

An engineering project, *per sé,* must consider the economic aspect. The software used in this research makes the economic evaluation in three phases: (1) estimation of the cost of the bioprocess equipment, (2) estimation of the total-capital investment and, (3) estimation of the cost of the biotreatment.

The estimation was set based on the proposed bioreactor (made up of concrete) and an air injector (see Figure 3).

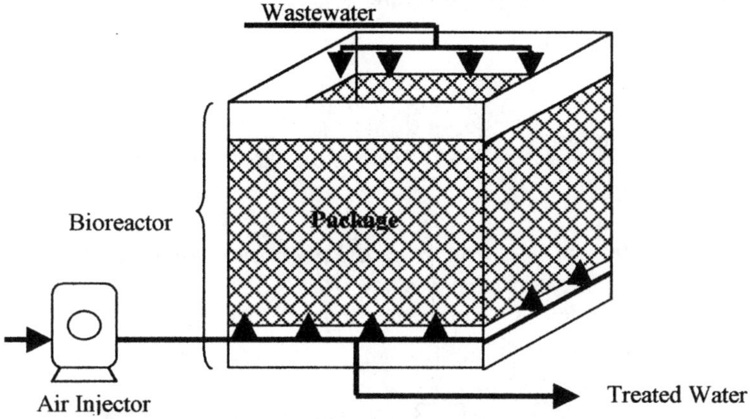


**FIGURE 3. General Scheme of the Bioprocess**

A synopsis of the total cost result for the biotreatment per m$^3$ of wastewater to be treated is presented in Table 6.

**TABLE 6. Costs in USD of the Bioprocess of Decoloration per m$^3$ of Wastewater to be Treated.**

| COMPONENT | CAPACITY | | | | |
|---|---|---|---|---|---|
| | 10 m$^3$/day | 20 m$^3$/day | 30 m$^3$/day | 40 m$^3$/day | 50 m$^3$/day |
| **Cost of Production** | 20,535 | 24,321 | 27,603.41 | 29,977 | 34,037 |
| **General Expenses** | 3,863 | 4,197 | 4,463.29 | 4,656 | 4,965 |
| **Contingency** | 4,879 | 5,703 | 6,413.34 | 6,927 | 7,800 |
| **Total Cost of the Biotreatment** | 29,278 | 34,222 | 38,480.04 | 41,560 | 46,803 |
| **Total Cost of the Biotreatment* (USD/ m$^3$ )** | **8.0** | **4.7** | **3.5** | **2.8** | **2.6** |

* The costs are in the first year of process operation (350 day per year, 24 hours per day).

## CONCLUSIONS

The bioreactor, proposing the ligninolytic fungus *Trametes versicolor* inoculated in pine sawdust, has shown high probabilities to be effective for the removal of wastewater dyes. The knowledge of the significant variables and a preliminary cost estimation of the bioprocess allows the recommendation for continuance of the research on a real industrial effluent and on a bench scale bioreactor.

It is important to develop a research program for the recalcitrant compound vat blue 4 and continue to evaluate new kinds of supports: natural, synthetic or an hybrid of both materials.

## REFERENCES

Cortinas N. C., Vega G., S., 1993., *"Hazardous Wastes in the World and Mexico"*, INE, Mexico.

Díaz C., M.D.,1999., "*Decoloration of a Paper Industry Wastewater Effluent by Trametes versicolor in Sugarcane Bagasse Pith"*, Laboratory of Xenobiotics of the Dept. of Biotechnology y Bioengineering of CINVESTAV-IPN, Personnel Communication, Mexico.

García B., R. J., Gómez F., J. F., Rodríguez A., M. S., Simancas A., A.., 1997., "*Development of EVEPI 1.0, a Software for the Economic Evaluation on Investment Projects of Chemical Plants"*, ESIQIE-IPN, Mexico.

Garg, S.K. y Modi, D.R., 1999., "Decolorization of Pulp-Paper Mill Effluentes by White-Rot Fungi", *Critical Reviews in Biotechnology*, 19 (2): 85-112.

Levin, M. A., Gealt, M. A., 1997., *"Biotreatment of Industrial and Hazardous Waste"*, Ed. Mc Graw Hill / Interamericana, España, 1997.

Montgomery, D. C., 1991., "*Design and Analysis of Experiments",* Third Ed., John Wiley & Sons Inc., Singapore.

Rodríguez V., R., 1991., "*Chemical Pretreatment of Biomass and Comparison with Brown Rot Decay Mechanism"*, PhD Thesis, Colorado State University, Fort Collins Co., USA.

Swamy, J., Ramsay, J.A., 1999., "The Evaluation of White Rot Fungi in the Decoloration of Textile Dyes", *Enz. and Microb. Technol.*, 24: (3-4).

# CELLULAR AUTOMATA MODELING OF ADSORPTION-DESORPTION WITH DIFFUSION OF POLLUTANT.

***Glicina Merino-Castro*** (ITESM,Toluca, México)
M. Sánchez (Universidad Complutense de Madrid, Madrid, Spain)
C. Garbi (Universidad Complutense de Madrid, Madrid, Spain)
R. Alonso (Universidad Politecnica de Madrid, Madrid, Spain)
M. Martín (Universidad Complutense de Madrid, Madrid, Spain)

**ABSTRACT**: The development of immobilized-cell systems, using isolated soil bacteria able to degrade different organic pollutants, could solve some of the problems showed by continuous fermentation processes. The use of adsorbed viable cells on solid supports presents advantages for the design of industrial bioreactors. Progress in the study of the factors involved in the performance of the proposed system and modeling will allow implementation in a variety of treatment and recycled water processes.
These processes involve adsorption-desorption and diffusion phenomena of particles suspended in a liquid that form a dynamic system. Cellular automata can model natural processes like adsorption-desorption and diffusion on many different levels of detail; in particular, they are useful in modeling dynamic systems to describe the global performance in terms of local interactions using different conditions.

## INTRODUCTION

Groundwater and soil pollution with herbicides have recently become a subject of increasing concern. The concept of bioremediation provides a potentially cheap alternative to traditional disposal techniques, in addition to representing a genuine removal of contaminants by microbial degradation rather than the relocation of contaminants. Our group has developed biological systems used in treatment and reuse of water contaminated by herbicides, such as Propachlor or Simazine (Martin et al. 2000 a, b). We have isolated different bacterial strains from soils contaminated by herbicides, the specific catabolic pathways have been characterized, and the intermediates of the degradation identified (Martin et al. 2000b). Cells have been immobilized in a ceramic support by using a reactor system, in order to get most effective use of bacteria in water bioremediation.

Cellular automata were first introduced by von Neumann (J. von Neumann 1963) as possible idealization of biological systems. Since then they have been applied to model a wide variety of systems. In particular, physical systems containing many discrete elements with local interactions are conveniently modeled as cellular automata.

Cellular automata are discrete dynamical systems whose behavior is completely specified in terms of local relation, as in the case of a large class of continuous dynamical systems defined by partial differential equations. A cellular automata can be understood as a stylized universe. Space is represented by a uniform grid, with each site or *cell* containing a few bits of data; time advances in discrete steps; and the laws of the universe are expressed by a single recipe- "a small look-up table"-through which at each step each cell computes its new state from that of its closest neighbors.

The phenomenon of adsorption-desorption is an example of a dynamic system with transition phase without balance. Dickman and Burschka (1988) presented a cellular automata model for a single species of particles with adsorption in an unoccupied cell and desorption in the occupied cells, under certain conditions of the neighborhood. The hypotheses of the model established that if the probability of adsorption is small, the covered surface is limited, but when the probability of the adsorption is high and the transition phase is saturated causing the surface to be covered completely by the particles, it is said that the system is in a state of adsorption. A way to diffuse particles on a surface that is modeled with cellular automata, is applying the rules first for the adsorption-desorption in the phase of upgrade of the automata followed by the diffusion rules in the lattice of the automata.

## OBJECTIVE

The objective of this report was to study the adsorption-desorption process in a bioreactor. In this report we describe a cellular automata model generated with experimental results obtained from bioreactors working with water polluted by Propachlor or Simazine.

## MATERIALS AND METHODS

Reactors containing 12.4 g of ceramic support were fed with 20 ml of water containing 105 ppm Propachlor or 3.25 ppm Simazine. Samples were taken during 60 min, and the herbicide concentration was analyzed by HPLC method (Martin et al. 2000a).

In this study we used previously published reports to implement the cellular automata called MODEL A, for the phenomena of adsorption-desorption. The conditions imposed to the cellular automata are:

a) Use of a square grid with periodic conditions of frontier (closed model) to simulate the process of adsorption-desorption.

b) The adsorbed particles are represented in the grid with the number 1. The empty cells represented with 0, symbolized the desorption state.

c) A Von Neumann neighborhood.

d) The automata's rules include in each cell the probabilistic local values of adsorption and we can manage different percentages of concentration of the pollutant.

For this model, if the probability of adsorption increases, a transition of second order happens. If Pc is the probability of adsorption and the following condition applies:

$$X_A > 1 \text{ for } P < P_c \wedge X_A = 1 \text{ for } P \geq P_c \text{ then } 1 - X_A = (P_c - P)^b$$

The above-mentioned equation allows the development of probabilistic cellular automata of synchronous operation, which simulates the process of adsorption-desorption for different concentrations of the pollutant in the fluid. The developed model provide the following type of results:

a) Numeric. Percentage of adsorption in each one of the evolution states in the time of the automaton.

b) Graphic. Graph each one of the evolution states in the time of the cellular automata.

For the diffusion process a model is used in two dimensions, synchronous, of periodic conditions of frontier (closed model) with a Margolous neighborhood.

The particles in the neighborhood move according to symmetry operations that mantain the number of particles in each block. The application of the operation is carried by randomness.

The iterative application of the process allows for the random propagation of the particles in the system. Even more, since the particles in each block are strongly correlated, the method we used permits that this correlation disappears quickly. The obtained automaton allows the graphing of each one of the states of the cellular automaton.

The previous models are integrated in an only cellular automata with different local rules that produce the wanted global actions. For the appropriate operation of the model it is considered that the frontiers of the layer of the adsorbent basis interact with the frontiers of the liquid layers. It is only in the area of the adsorbent basis and in the bordering layers of the liquid area where the phenomenon is considered adsorption-desorption. In the rest of the automaton diffusion phenomena are simulated. The scheme of the simulation is shown in the figure 1.

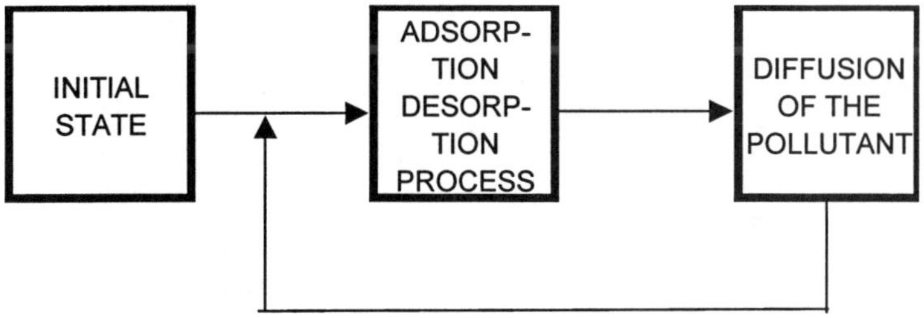


**FIGURE 1.** Scheme of the simulation.

## RESULTS AND DISCUSSION

Simulations were made using the proposed cellular automata creating a matrix of 100x100 cells. The tests were carried out for values different from the concentration of the pollutant. We observe that:

a) In low concentrations of the pollutant, the behavior of the adsorption-desorption-diffusion phenomenon is very oscillating. Figure 2.
b) The process tends to have a less unstable behavior at concentrations of at least 5.5 ppm. Figure 3.
c) The obtained pattern provides results that sufficiently approach the data obtained experimentally.

Comparative results are shown between the experimental data for Simazine and those of the cellular automata in Figure 4.

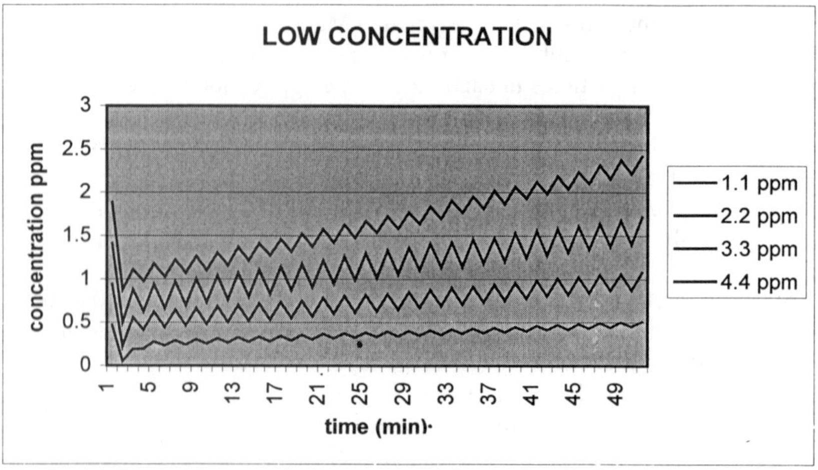


**FIGURE 2.** Simulation for Simazine at low concentration.

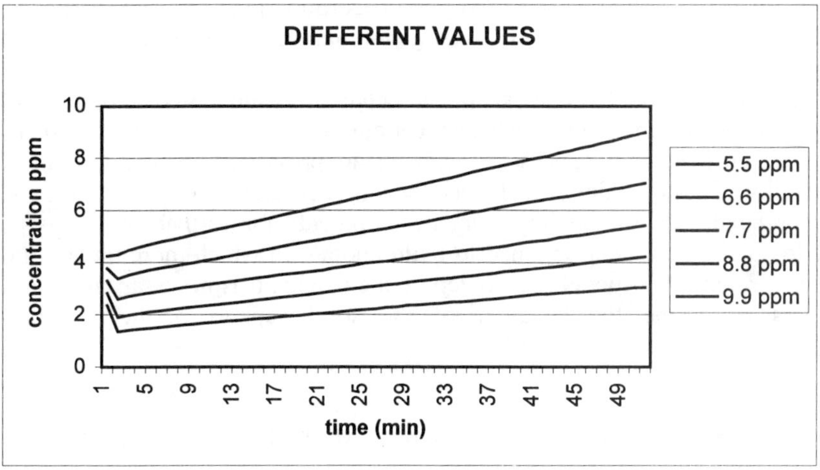


**FIGURE 3.** Behavior in the simulation of the adsorption-desorption-diffusion phenomena for concentration >5 ppm

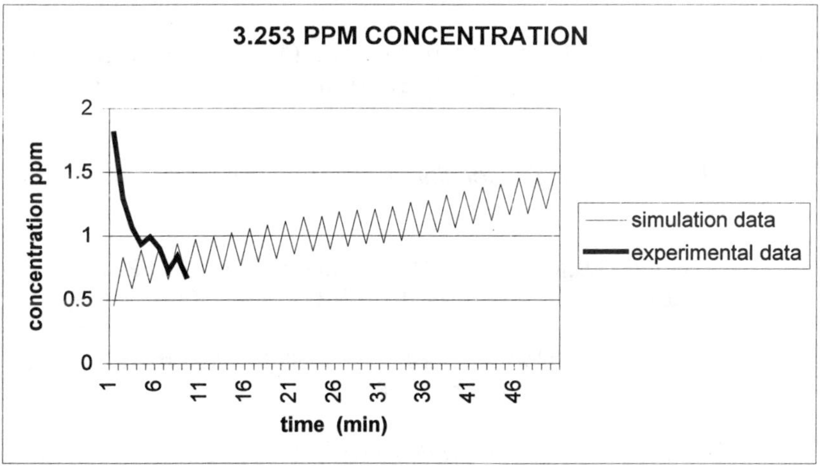


**FIGURE 4.** Comparation data for Simazine.

## CONCLUSIONS

The obtained cellular automata model generates results that approach experimental results and observations, including the area of oscillatory behavior. In the tests performed in the laboratory it has been observed that at low concentrations below 5 ppm the model does not stabilize the adsorption-desorption process. This part of the simulation is one of the lines of future investigation that may allow to understand the circumstances of the phenomenon.

The obtained results make the analysis of the behavior of the system possible over time and to predict the moment in which a certain percentage of adsorption of the pollutant is achieved, since the necessary elements are provided to predict the time in which a quantity of polluted liquid will reach the target cleanup level.

On the other hand, one of the important developments for further studies is the possibility to adjust the model obtained to build another in which more variables of the phenomenon than it is modeling. To depure the results, it is necessary to include characteristics to define the behavior of the solid support beside biological components.

## ACKNOWLEDGMENTS

This work was partially supported by grant CAM07M/0030/1999 (Comunidad Autónoma de Madrid, Spain).

## REFERENCES

Benney,P.,Droz,M. and Frachebourg,L. 1990. "On the Critical Behaviour of Cellular Automata Models of Non-equilibrium Phase Transitions." *Phys. A: Math. Gen.* 23: 3353-3359.

Bagnoli,F.,Chopard.B.,Droz.M.,and Frachebourg.L. (1992): Critical Behaviour of a diffusive Model with one Adsorbing State. J. Phys. A: Math. Gen. 25: 1085-1091.

Chopard,B. and Droz,M. 1998. *Cellular Automata Modeling of Physical Systems.* Cambridge University Press.

Dickmaman.R. and Burschka,M.A. 1988. "Nonequilibrium Critical Poisoning in a single- species Model." *Physics Letters A.* 127(3): 132-137.

Di Gregorio,S.,Serra,R. and Villani,M. 1997. "A Cellular Automata Model of Soil Bioremediation." *Complex Systems.* 11: 81-54.

Gaylor,R.J. and Nishidate,K. 1966 *Modelling Nature with Cellular Automata.* Telos/Springer-Verlag.

Martín,M.,Mengs, G.,Martín,D., Sánchez,M., Garbi,C., Plaza,E., Ferrer,E., Fortún, A., Allende,J.L., Alonso, R., Martínez-Iñigo, M.J. and Lobo,C. 2000a. "Bioremediation of Soils and Waters contaminated by Herbicides." In:

Remediation of Chlorinated and Recalcitrant Compounds. Vol. C2-7, p 237. Battelle Press. Columbus.

Martín,M., G.Mengs, E.Plaza, C.Garbi, M.Sánchez, A.Gibello, F.Gutierrez, and E.Ferrer. 2000b "Propachlor Removal by *Pseudomonas* Strain GCH1 in an Immobilized-Cell System." *Appl. Environ. Microbiol.* 66(3): 1190-1194.

Merino,G. and Bustillos,O. 1998. Simulation of Pollutant Difusion between two media by means of cellular automata. VII International congress of Ecology. Florencia, Italia.

Merino,G. and Bustillos,O. 1996. Autómata Celular con Diferentes Reglas Locales. XXIX Congreso de la Sociedad Matemática Mexicana.

Serra,R.,Villani,M.,Oricchio,D.,Di and Gregorio,S. 1998. "Recent Advances in Dynamical Models of Biodegradation." In Proceeding of the Third Conference on Cellular Automata for Research and Industry. S. Bandini, R. Serra, F. Suggi. (eds.). Springer.

Toffoli,T. and Margolous,N. 1991. *Cellular Automata Machines.* MIT press.

# 2001 AUTHOR INDEX

This index contains names, affiliations, and volume/page citations for all authors who contributed to the ten-volume proceedings of the Sixth International In Situ and On-Site Bioremediation Symposium (San Diego, California, June 4-7, 2001). Ordering information is provided on the back cover of this book.

The citations reference the ten volumes as follows:

**6(1):** Magar, V.S., J.T. Gibbs, K.T. O'Reilly, M.R. Hyman, and A. Leeson (Eds.), *Bioremediation of MTBE, Alcohols, and Ethers*. Battelle Press, Columbus, OH, 2001. 249 pp.

**6(2):** Leeson, A., M.E. Kelley, H.S. Rifai, and V.S. Magar (Eds.), *Natural Attenuation of Environmental Contaminants*. Battelle Press, Columbus, OH, 2001. 307 pp.

**6(3):** Magar, V.S., G. Johnson, S.K. Ong, and A. Leeson (Eds.), *Bioremediation of Energetics, Phenolics, and Polycyclic Aromatic Hydrocarbons*. Battelle Press, Columbus, OH, 2001. 313 pp.

**6(4):** Magar, V.S., T.M. Vogel, C.M. Aelion, and A. Leeson (Eds.), *Innovative Methods in Support of Bioremediation*. Battelle Press, Columbus, OH, 2001. 197 pp.

**6(5):** Leeson, A., E.A. Foote, M.K. Banks, and V.S. Magar (Eds.), *Phytoremediation, Wetlands, and Sediments*. Battelle Press, Columbus, OH, 2001. 383 pp.

**6(6):** Magar, V.S., F.M. von Fahnestock, and A. Leeson (Eds.), *Ex Situ Biological Treatment Technologies*. Battelle Press, Columbus, OH, 2001. 423 pp.

**6(7):** Magar, V.S., D.E. Fennell, J.J. Morse, B.C. Alleman, and A. Leeson (Eds.), *Anaerobic Degradation of Chlorinated Solvents*. Battelle Press, Columbus, OH, 2001. 387 pp.

**6(8):** Leeson, A., B.C. Alleman, P.J. Alvarez, and V.S. Magar (Eds.), *Bioaugmentation, Biobarriers, and Biogeochemistry*. Battelle Press, Columbus, OH, 2001. 255 pp.

**6(9):** Leeson, A., B.M. Peyton, J.L. Means, and V.S. Magar (Eds.), *Bioremediation of Inorganic Compounds*. Battelle Press, Columbus, OH, 2001. 377 pp.

**6(10):** Leeson, A., P.C. Johnson, R.E. Hinchee, L. Semprini, and V.S. Magar (Eds.), *In Situ Aeration and Aerobic Remediation*. Battelle Press, Columbus, OH, 2001. 391 pp.

# KEYWORD INDEX

This index contains keyword terms assigned to the articles in the ten-volume proceedings of the Sixth International In Situ and On-Site Bioremediation Symposium (San Diego, California, June 4-7, 2001). Ordering information is provided on the back cover of this book.

In assigning the terms that appear in this index, no attempt was made to reference all subjects addressed. Instead, terms were assigned to each article to reflect the primary topics covered by that article. Authors' suggestions were taken into consideration and expanded or revised as necessary. The citations reference the ten volumes as follows:

**6(1):** Magar, V.S., J.T. Gibbs, K.T. O'Reilly, M.R. Hyman, and A. Leeson (Eds.), *Bioremediation of MTBE, Alcohols, and Ethers*. Battelle Press, Columbus, OH, 2001. 249 pp.

**6(2):** Leeson, A., M.E. Kelley, H.S. Rifai, and V.S. Magar (Eds.), *Natural Attenuation of Environmental Contaminants*. Battelle Press, Columbus, OH, 2001. 307 pp.

**6(3):** Magar, V.S., G. Johnson, S.K. Ong, and A. Leeson (Eds.), *Bioremediation of Energetics, Phenolics, and Polycyclic Aromatic Hydrocarbons*. Battelle Press, Columbus, OH, 2001. 313 pp.

**6(4):** Magar, V.S., T.M. Vogel, C.M. Aelion, and A. Leeson (Eds.), *Innovative Methods in Support of Bioremediation*. Battelle Press, Columbus, OH, 2001. 197 pp.

**6(5):** Leeson, A., E.A. Foote, M.K. Banks, and V.S. Magar (Eds.), *Phytoremediation, Wetlands, and Sediments*. Battelle Press, Columbus, OH, 2001. 383 pp.

**6(6):** Magar, V.S., F.M. von Fahnestock, and A. Leeson (Eds.), *Ex Situ Biological Treatment Technologies*. Battelle Press, Columbus, OH, 2001. 423 pp.

**6(7):** Magar, V.S., D.E. Fennell, J.J. Morse, B.C. Alleman, and A. Leeson (Eds.), *Anaerobic Degradation of Chlorinated Solvents*. Battelle Press, Columbus, OH, 2001. 387 pp.

**6(8):** Leeson, A., B.C. Alleman, P.J. Alvarez, and V.S. Magar (Eds.), *Bioaugmentation, Biobarriers, and Biogeochemistry*. Battelle Press, Columbus, OH, 2001 255 pp.

**6(9):** Leeson, A., B.M. Peyton, J.L. Means, and V.S. Magar (Eds.), *Bioremediation of Inorganic Compounds*. Battelle Press, Columbus, OH, 2001. 377 pp.

**6(10):** Leeson, A., P.C. Johnson, R.E. Hinchee, L. Semprini, and V.S. Magar (Eds.), *In Situ Aeration and Aerobic Remediation*. Battelle Press, Columbus, OH, 2001. 391 pp.

## A

## B

# E

# F



# G



# H



# I

## J



## K



## L



## M

# N



# O

# P

## R



## S



## T

## U



## V



## W



## X

## Y



## Z